PHYSICS RESEARCH AND TECHNOLOGY

OLD PROBLEMS AND NEW HORIZONS IN WORLD PHYSICS

PHYSICS RESEARCH AND TECHNOLOGY

Additional books and e-books in this series can be found
on Nova's website under the Series tab.

PHYSICS RESEARCH AND TECHNOLOGY

OLD PROBLEMS AND NEW HORIZONS IN WORLD PHYSICS

VOLODYMYR KRASNOHOLOVETS
VICTOR CHRISTIANTO
AND
FLORENTIN SMARANDACHE
EDITORS

nova
science publishers
New York

Library of Congress Cataloging-in-Publication Data

ISBN: 978-1-53615-430-6

Published by Nova Science Publishers, Inc. † New York

CONTENTS

PREFACE

"This is your last chance. After this, there is no turning back. You take the blue pill—the story ends, you wake up in your bed and believe whatever you want to believe. You take the red pill—you stay in Wonderland and I show you how deep the rabbit-hole goes."

– Morpheus in *The Matrix.*[1]

The above phrase is quoted from a legendary movie, *The Matrix*. While this book is surely not intended to discuss philosophical ramifications of that movie, the same phrase surprisingly echoes what happens in modern physics, especially in physicists' effort to describe the dynamics of microworld.

Illustrious figures of Quantum Physics, including Bohr and Heisenberg, famously declared that it is not physicists' job to speak about the reality of Nature, but only what they can tell about experiments. Such a viewpoint is called "*operationalism*" as a strand of philosophy of science.[2]

Furthermore, the so-called Copenhagen interpretation of QM includes a number of bizarre views, such as: (a) probabilistic cloud model of atoms and electrons, (b) assumption about collapsing wave function (without clear mechanism), (c) observer dependent reality, (d) it predicts quantum jumps without trajectories[3], and so on.

Some physicists and philosophers alike have argued against such bizarre notions, by proposing *Quantum Mechanics without observer* (notably Karl Popper), but it sounds more like voices in the wilderness. In a similar tone, Sir Roger Penrose - an eminent mathematical physicist – wrote in his new book:

> "…the wave function involves a functional freedom that is far higher than that which manifests itself in reality, or at least in that aspect of reality that is revealed as the result of a quantum measurement. Are we to regard ψ as actually representing physical reality? Or

---

[1] https://www.thoughtco.com/morpheus-wisdom-quotes-from-the-matrix-2832834

[2] https://www.britannica.com/topic/operationalism

[3] …but Wilson chamber experiments prove quite the opposite, that particles leave trajectories.

is it to be viewed as being merely a calculational tool for working out probabilities of the results of experiments that might be performed, the results of these being "real", but not the wave function itself? As mentioned in §2.4, it was part of the Copenhagen interpretation of quantum mechanics to take this latter viewpoint, and, according to various other schools of thought also, ψ *is to be regarded as a calculational convenience with no ontological* status other than to be part of the state of mind of the experimenter or theoretician, so that the actual results of observation can be probabilistically assessed."[4]

The present book is a collection of papers written by researchers who have been working for defining new concepts in science for years. Among new approaches there are views developed on the basis of new mathematical principles, possible relationships between physical processes and also ideas resting on a firsthand experience in penetrating elusive realms, which finally result in explanatory theoretic models. In the frame of new proposed theories, matter and energy may be different characteristics of a physical system and "equivalence" between matter and energy becomes not so obvious. The entire Quantum Mechanics was developed based on the assumption that electron mass is constant. A variable mass of electron automatically rule out the entire quantum mechanics. An electron mass can change during chemical and biological processes, then other characteristics modify correspondingly. It is accepted that Special Theory of Relativity (STR) does not contradict quantum mechanics, but in reality, the opposite is true. Even for a non-rocket scientist this contradiction becomes evident at a simple analysis of mass energy equivalence formula. In simple words the formula assumes that: If energy is quantized, mass have to be quantized too. How do atomic particles know how much mass to convert into energy and keep the same proportion of conversion? Maybe one proton or one neutron converts more mass than his neighbors! If protons and neutrons can be fragmented and chopped using energies of order of MeV, why do we need CERN or other large nuclear facilities?

***

In order to address the above deep questions, in this uncompromising book we will discuss not about various interpretations of QM, which are often confusing. Instead, we will argue towards a more realistic description of microworld based on classical wave equation (Kreidik & Shpenkov). Then the second chapter argues in favor of *cellular automaton microphysics* or even a deeper view – a mathematical lattice of primary topological balls, the tessellatice (Christianto, Krasnoholovets & Smarandache).

Moreover, we also present some novel arguments: consciousness research (inner workings of reality, Boyd & Klein). In the next two papers, Christianto described an

---

[4] Sir Roger Penrose. *Fashion, Faith and Fantasy in the New Physics of the Universe*. Princeton: Princeton University Press, 2016, p. 198.

extension of Smarandache's neutropsychology and he also gave 12 simple advices to younger scientists. Then the seventh and 8th paper discuss ghosts in theoretical physics and also the role of empiricism in modern physics (Kanda et al.). Neil Boyd also gives a response to Prof. Kanda et al.'s papers. He also gives an exposition of philosophy of physics (Neil Boyd).

Then Arnold Gorgels presents his hypothetical model of Cantor's continuum. Meanwhile, Ying-Qiu Gu presents a chapter devoted to spiral structure and collective motion of stars. Though Gu's chapter is written in conventional terms, it clearly shows the necessity of an additional potential Φ that must be presented in the system of interacting stars, though it still eludes the researchers. Moreover, the potential is able to generate a wave behavior similar to that described in electrodynamics, i.e., alike the Maxwell equations. What is the nature of such a potential? Does gravity has a dynamic nature? But in this case the ordinary physical space must be considered as a substrate with its own structure (maybe the *tessellattice* as it is examined in a monograph of V. Krasnoholovets published in 2017).

In the next paper, Christianto & Smarandache wrote a paper dedicated to reality concerning innovation waves in sciences.

Finally, the book includes a short chapter written by V. Krasnoholovets, a proposal to launch a new branch of astronomy/astrophysics, namely, *Inerton Astronomy.*

We hope that this book will inspire more theoretical and experimental investigations on the hidden aspects of Nature both in microscopic and macroscopic realms. A careful reader may find out that this book follows a progression just like what the readers may listen from J.S. Bach or Beethoven's *Symphony no.9*, i.e., from the wave mechanics, to neutropsychology and consciousness, to philosophy of physics, to continuum hypothesis and also astrophysics and cosmology modeling etc.

***

While generally this book is in some ways quite in tune with Penrose's *Fashion, Faith and Fantasy in the New Physics of Universe*, our proposed solution for QM's problems is different from Penrose's Bohmian mechanics. In this book, we argue on how we can go beyond the standard QM. To summarize, the message of this book is quite simple: "The whole *Quantum Mechanics is inadequate, leave it to historians.*"[5]

This book will be suitable for a wide variety of readers, including: scholars, mathematical-physics professors and college students, theologians, cosmologists, theoretical physicists, mathematicians, and perhaps psychology, engineering and social science college students too.

---

[5] Thanks to Dr. George Shpenkov who wrote this catchy phrase.

We would like to extend our deepest gratitude to all contributors to this volume:

- George Shpenkov
- Leonid Kreidik
- Akira Kanda
- Mihai Prunescu
- Renata Wong
- Robert Neil Boyd
- Adrian Klein
- Arnold Gorgels
- Ying-Qiu Gu
- and also to *Nova Science Publishers*.

Allow us to end this preface with another quote from Morpheus:

Morpheus: *Welcome...To the real world.*[6]

Ver.1.0: Medio June 2018
Ver.1.1: 1st July 2018
Ver.1.2: 18th July 2018
Ver.1.3: 28th July 2018
Ver 1.4: 4th Aug. 2018
Ver 1.5: 9th Aug. 2018
Ver 1.6: 21st Sept. 2018
Ver 1.7: 1st October 2018
Ver 1.8: 4th October 2018
Ver 1.9: 2nd Nov. 2018
Ver 1.10: 5th Feb. 2019

VC, VK, FS

[6] https://www.rottentomatoes.com/m/matrix/quotes/

In: Old Problems and New Horizons in World Physics ISBN: 978-1-53615-430-6
Editors: Volodymyr Krasnoholovets et al.

Chapter 1

LET'S START JAMMING

Victor Christianto *
Satyabhakti Advanced School of Theology – Jakarta Chapter,
Jakarta, Indonesia

ABSTRACT

In this short chapter, this author explained the process of preparation of this book. At times, heated discussions were inevitable, but we kept *"dialogic spirit",* because we realized the necessity to learn from each other's different fields of expertise, different experiences etc. In essence, we viewed the process of this book as somewhat like "jamming" in jazz music.

This group of contributors spent many days in debates and arguing on each other's standpoints, but in the end we agreed to publish rather than perish.

Our last email conversation with Prof. Akira Kanda went as follows:

I understood that as a mathematician you want full perfection in all fields that you touch...

Well that is part of a mathematician's attitude, just reminding me of an old joke:

> "So a mathematician, an engineer, and a physicist are out hunting together. They spy a deer in the woods. The physicist calculates the velocity of the deer* and the effect of gravity on the bullet, aims his rifle and fires. Alas, he misses; the bullet passes three feet behind the deer. The deer bolts some yards, but comes to a halt, still within sight of the trio.

* E-mail: victorchristianto@gmail.com.

'Shame you missed', comments the engineer, 'but of course with an ordinary gun, one would expect that'. He then levels his special deer-hunting gun, which he rigged together from an ordinary rifle, a sextant, a compass, a barometer, and a bunch of flashing lights which don't do anything but impress onlookers, and fires. Alas, his bullet passes three feet in front of the deer, who by this time wises up and vanishes for good.

'Well,' says the physicist, 'your contraption didn't get it either'.

'What do you mean?' pipes up the mathematician. 'Between the two of you, that was a perfect shot!'

*How they knew it was a deer:

The physicist observed that it behaved in a deer-like manner, so it must be a deer.

The mathematician asked the physicist what it was, thereby reducing it to a previously solved problem.

The engineer was in the woods to hunt deer, therefore it was a deer"[1].

What I mean, is that this book has the title: "old problems and new horizons in world physics". That way, we mean that we will do our best to expose the greatest unsolved problems in physics, and offer a solution, if at all possible.

What you discuss is mostly the mathematical foundations of modern physics, such as definition of point, continuum etc. They are indeed interesting, that is why we invite you and your team to participate.

So, I guess we should try our best to keep in mind that we have different attitudes on things, but we shall keep the dialogue tone. In response, he replies:

You are too kind to mathematics.

1. An engineer discovered that 2 is even, 4 is even, and so is 8. He concluded that all numbers are even.
2. A physicist discovered that 2 is prime, 3 is prime, 5 is prime and 7 is prime. He concluded that all numbers are prime.
3. When a mathematician was asked to catch a lion in the field, he made a fence around him and said, by definition I caught a lion.
   -Akira

That's it: we believe in *spirit of dialogue*, just like the title of Galileo Galilei's classic book[2].

In essence, the process of this book is more or less like jamming in jazz music, where contradictions exist but can lead to new ideas:

---

[1] Jokes - University of Maryland Observatory <https://www.astro.umd.edu/~avondale/extra/Humor/ScienceHumor/PhysicistJokes.html>.

[2] https://en.wikipedia.org/wiki/Dialogue_Concerning_the_Two_Chief_World_Systems.

It is the stretched soul that makes music, and souls are stretched by the pull of opposites—opposite bents, tastes, yearnings, and loyalties. Where there is no polarity—where energies flow smoothly in one direction— there will be much doing but no music. — Eric Hoffer[3]

JAMMING

Bob Marley, Bob Marley and the Wailers

We jammin' (Bob Marley, Bob Marley)
I wanna jam it with you
We jammin' (Bob Marley, Bob Marley)
I hope you like jammin' too
In rows or around, we can jam it anyhow
I'n'I, I will see you through
Everyday we pay the price, with a livin' sacrifice
Jammin' 'til the jam is through
We jammin', to think that jammin' was a thing of the past
We jammin and I hope the jam is gonna last
We jammin' on the for the daughters and sons
The struggle is not over until the battle is won
Hip-Hop is life but it ain't all to me
Now where would we be if they take our mics from we.
Or better yet strip on might itself
We done it before so I sit by the door
Those that claim but don't really know the game
Bob Marley, learn the man behind the name
Oh Holy Mount Zion
Oh Holy Mount Zion
Oh Holy Mount Zion
Oh Holy Mount Zion
We see it in Mount Zion
And Rules all creation
Were jammin' [Repeats]
Hope you like jammin' to [Repeats]
J-A-Double M-I-N
We be jammin' from Brooklyn down to Kingston

---

[3] Paul Rinzler. *Contradictions of jazz*. The scarecrow Inc.

We be jammin' in the name of the Lord sweet Allah
We no the blessings from above I and are be no rap star
Exodus when i bust,
See the struggle i must,
Realizing the just but whats next on the cusp
All soldiers of war steadily ignore
Still we ain't sure what the battle is for
Years ago when we toured
Ready to explore
Entered through back doors
Treated like stank manure
Spoke up for rights, we were cracked in the jaw
Elvis made a bundle while we remained poor
We remained poor
but we be jammin'
I'n'I will see you through [Repeat: x2]
Come along way but we're still not there yet
I'n'I will see you through
Jah knows the mighty cry, the truth it cannot hide
To keep you satisfied
Love that does exist, is the love I can't resist
So jam by my side
We jammin'
and we're jammin' in the name of the lord
we're jammin'
we're jammin right straight from York
holy mount zion
I'n'I will see you through
[Repeats]

Songwriters: Mary C. Brockert

In: Old Problems and New Horizons in World Physics
Editors: Volodymyr Krasnoholovets et al.
ISBN: 978-1-53615-430-6

Chapter 2

AN ANALYSIS OF THE BASIC CONCEPTS OF QUANTUM MECHANICS AND NEW (DIALECTICAL) SOLUTIONS FOR THE FIELD OF A STRING AND H-ATOM*

L. Kreidik[1] and G. Shpenkov[2,*]
[1]Minsk, Belarus
[2]Institute of Mathematics and Physics, UTA, Poland

ABSTRACT

In this chapter, the authors submit wholeheartedly that the procedures that Schrodinger took while deriving his equation (known as "*wave mechanics*") are deeply mistaken. It has resulted in so many paradoxes and confusions throughout the last 90 years, including the physical meaning of quantum wave function. Actually, the root cause of such endless debates and confusion is the unphysical nature of Schrödinger's wave function. Moreover, the authors show convincingly that it is possible to derive a periodic table of elements based on the spherical classical wave equation, which is impossible to obtain from the wave mechanics.

* Corresponding Author's Email: g.shpenkov@gmail.com.

1. SCHRÖDINGER'S EQUATION AND THE MYTHS OF QUANTUM MECHANICS

Schrödinger's equation appeared in the years of a wild bloom of formalism, which was represented, first of all, by positivism, machism, pragmatism, and other trends of philosophical thought, denying the objective world. Arbitrary mathematical constructions, on the basis of a free game of notions, were the characteristic result of these philosophical currents for physics. By virtue of this, a reasonable logic in such constructions (theories-myths) was almost absent or insignificant. Schrödinger, a representative of those years, has designed his equation, exactly following the spirit of aforementioned ideological trends in physics.

Nevertheless, we should give Schrödinger his due because the positivistic style did not satisfy his. He had a propensity for actual realism and restrainedly regarded to new fashion trends – the physical mathematical advance-guard. But he was under the influence of that time.

In the 1920's, Schrödinger's equation began to be regarded as the major achievement of scientific thought. It became the basis for lectures on atomic physics in universities. The common opinion is that Schrödinger's equation proved its validity by the conformity of its solutions with the vast amount of experimental data and the co-ordination with general physical notions. However, all this is only emotional judgements.

Now, at the beginning of the 21st century, let us look at the events of those years from the point of view of dialectics, and give an objective estimation of the past. Unfortunately, the mathematical model, put forward by Schrödinger, is currently represented in the form of generalized and extended equations, including the relativistic invariant of them.

In the initial variant, Schrödinger's equation had the following form

$$\Delta\Psi + \frac{2m}{\hbar^2}\left(W + \frac{e^2}{4\pi\varepsilon_0 r}\right)\Psi = 0 . \qquad (1.1)$$

Its structure had a quite artificial character and rested upon the operator and variational methods.

The wave Ψ–function, satisfying the wave equation (1.1), is represented as

$$\Psi = R(r)\Theta(\theta)\Phi(\varphi)T(t) = \psi(r,\theta,\varphi)T(t), \qquad (1.1a)$$

where $\psi(r,\theta,\varphi) = R(r)\Theta(\theta)\Phi(\varphi)$ is the complex amplitude of the wave function, because $\Phi_m(\varphi) = C_m e^{\pm im\varphi}$.

The multiplicative form of the amplitude ψ–function allows dividing of Schrödinger's equation (1.1) into the equations of the radial $R(r)$, polar $\Theta_{lm}(\theta)$, and azimuth $\Phi_m(\varphi)$ functions:

$$\frac{d^2R}{dr^2}+\frac{2}{r}\frac{dR}{dr}+\left(\frac{2m}{\hbar^2}\left(W+\frac{Ze^2}{4\pi\varepsilon_0 r}\right)-\frac{l(l+1)}{r^2}\right)R=0, \tag{1.2}$$

$$\frac{d^2\Theta_{l,m}}{d\theta^2}+ctg\theta\frac{d\Theta_{l,m}}{d\theta}+\left(l(l+1)-\frac{m^2}{\sin^2\theta}\right)\Theta_{l,m}=0, \tag{1.2a}$$

$$\frac{d^2\Phi_m}{d\varphi^2}=-m^2\Phi_m. \tag{1.2b}$$

An equation for the time component of Ψ is

$$\frac{d^2T}{dt^2}=-\omega^2T. \tag{1.2c}$$

The simplest solution of Eq. (1.2c) defines the function of time of the form $T=e^{\pm i\omega t}$. The last three equations were known in a theory of wave fields. Hence, these equations presented nothing new. Only the radial equation (1.2) was a new one. Its solution turned out to be the divergent one. However, Schrödinger together with H. Weyl (1885-1955, German mathematician), contrary to the logic and all experience of theoretical physics, artificially cut off the divergent power series of the radial function $R(r)$ at a κ-member. This allowed them to obtain the radial solutions, which, as an effect of the cut off operation, actually are the fictitious solutions. For hydrogen-like atoms, the radial function has the following form

$$R_{\kappa l}=N_{\kappa l}e^{-\rho/(\kappa+l+1)}\left(\frac{2\rho}{\kappa+l+1}\right)^l L_\kappa^{2l+1}\left(\frac{2\rho}{\kappa+l+1}\right),\ \rho=\frac{r}{a}, \tag{1.3}$$

where $L_\kappa^{2l+1}\left(\frac{2\rho}{\kappa+1+l}\right)$ is Laguerre polynomial of the power κ. At that, the power κ is simultaneously the parameter of cutting off of the divergent series, $a=\frac{a_0}{Z}$ is the large half-axis of the orbit, a_0 is the Bohr radius, and

$$N_{\kappa l}=\frac{2}{(\kappa+1+l)^2}\sqrt{\frac{\kappa!}{(\kappa+1+2l)!}}\cdot\left(\frac{1}{a}\right)^{3/2} \qquad (1.4)$$

is the normalizing multiplier.

Since the radial equation (1.2) contains the energy of interaction of the electron with the nucleus $-\frac{Ze^2}{4\pi\varepsilon_0 r}$, therefore, it was natural to wait that Eq. (1.2) will "give out" this energy as a result of the solution under the definite conditions. Indeed, the formal cutting off leads to the discrete series of values of the total electron energy

$$W=-\frac{Ze^2}{8\pi\varepsilon_0 a_0(\kappa+l+1)^2}=-\frac{Ze^2}{8\pi\varepsilon_0 an^2}, \qquad (1.5)$$

where the sum $n=\kappa+l+1$ has the name of the main quantum number and is equal to 1, 2, 3, ... The formula (1.5) gives birth to an illusion of the solution to the problem. Actually, in a strictly scientific sense, here we deal with the plain mathematical adjustment to Bohr postulates. The power two (as in the case of n^2) is a usual result for the many second order equations.

In the case of H-atom ($Z=1$), the expression for energy (1.5) is simplified

$$W=-\frac{e^2}{8\pi\varepsilon_0 a_0(\kappa+l+1)^2}=-\frac{e^2}{8\pi\varepsilon_0 a_0 n^2}. \qquad (1.5a)$$

The radial solution (1.3) for the hydrogen-like atom takes the form

$$R_{nl}=N_{nl}e^{-\rho/n}\left(\frac{2\rho}{n}\right)^l L_{n-l-1}^{2l+1}\left(\frac{2\rho}{n}\right), \qquad (1.6)$$

where

$$N_{nl}=\frac{2}{n^2}\sqrt{\frac{(n-l-1)!}{(n+l)!}}\cdot\left(\frac{1}{a}\right)^{3/2}, \qquad (1.7)$$

and

$$L_{n-l-1}^{2l+1}\left(\frac{2\rho}{n}\right)=\frac{1}{(n-l-1)!}e^{2\rho/n}\left(\frac{2\rho}{n}\right)^{-(2l+1)}\frac{d^{n-l-1}}{d\left(\frac{2\rho}{n}\right)^{n-l-1}}\left\{e^{-2\rho/n}\left(\frac{2\rho}{n}\right)^{l+n}\right\} \tag{1.8}$$

is the standard Laguerre polynomial.

With use of the accepted designations, Schrödinger's wave function (1.1a) is represented as

$$\Psi=R_{nl}(r)\Theta_{lm}(\theta)\Phi_m(\varphi)T(t)=\psi_{nl}(r,\theta,\varphi)T(t), \tag{1.9}$$

where

$$\psi_{nl}(r,\theta,\varphi)=R_{nl}(r)\Theta_{lm}(\theta)\Phi_m(\varphi). \tag{1.9a}$$

Schrödinger's radial equation (1.2) contains only the wave number l. The number κ has an auxiliary character. Accordingly, the radial function should be presented in the form $R_l(\rho,\kappa)$. Taking into consideration these remarks, Schrödinger's elementary Ψ-function (1.9) can be rewritten as

$$\Psi_{lm,\kappa}=\psi_{lm,\kappa}(\rho,\theta,\varphi;\kappa)T(t)=R_l(\rho;\kappa)\Theta_{lm}(\theta)\Phi_m(\varphi)T(t), \tag{1.9b}$$

where

$$\psi_{lm,\kappa}(\rho,\theta,\varphi;\kappa)=R_l(\rho;\kappa)\Theta_{lm}(\theta)\Phi_m(\varphi) \tag{1.9c}$$

is the spatial $\psi_{lm,\kappa}$-function.

According to the quantum mechanics conception, extremes of radial function R_{nl} define the radii of shells of the most probable states: $r=a\rho_{nl,\max i}$, where i is the number of the root of the extremum. However, for the overwhelming number of cases, *these roots are not equal to the integers squared*, i.e., $\rho_{nl,\max i}\neq m^2$ (where m=1, 2, 3, …), and hence, *they deny the cutting off condition* (1.5). Such roots define the energetic levels, non-existent in nature:

$$W=-\frac{e^2}{8\pi\varepsilon_0 a\rho_{nl,\max i}}. \tag{1.10}$$

Thus, according to the condition (1.5), the energetic levels (states) (1.10) must not exist. If we suppose that they exist, then these *levels must formally transform the radial function into a divergent functional series* because Eq. (1.10), where $\rho \neq m^2$ (m = 1, 2, 3,...), does not satisfy the cutting off condition (1.5), where $\rho = n^2$ (n = 1, 2, 3,...). Such an absurdity appeared because of the cutting off, artificial and invalid, of the power series.

The wave numbers of Schrödinger's equation are usually compared with the quantum numbers in Bohr-Sommerfeld's generalized theory of a hydrogen atom. Between 1913 and 1926, the Bohr-Sommerfeld's theory took roots in minds; as a result, the superficial resemblance of its quantum numbers was groundlessly used by founders of quantum mechanics.

As an analog of the azimuth number m, the magnetic number m of Bohr-Sommerfeld's theory was accepted. The number l plays the role of the azimuth number n_φ, which defines (along with the main quantum number n) the smaller half-axis of the elliptical electron orbit $b = a_0 n n_\varphi$. The larger half-axis of the orbit a, defining the electrons total energy on the orbit, in Bohr's-Sommerfeld's theory, depends only on the main quantum number n: $a = a_0 n^2$. Such a formal juxtaposition must mean that the wave function in Schrödinger's equation (1.1) contains the elliptical orbits in the form of "electron clouds". All these definitions are a fruit of fantasy. In fact, Schrödinger's equation describes only the *circular* electron orbits, but not mystic clouds-orbitals. We will make sure of it below.

Creation of any theory requires, at the initial stage, very important preliminary research work. This work, as a rule, is never published and remains forever unknown. Moreover, it can contain errors, which are hidden in the final variant of the theory. In publications, only the results of the work are presented, which, too often, are far from the initial sketches of the theory, both in the form and contents.

Let us remember the birth of Maxwell's electromagnetic theory. His contemporaries noted that Maxwell constructed his theory not following strict logic. Maxwell can plainly turn down a member, replace an unsuitable sign by the opposite one in an expression, substitute a meaning of a letter, etc. H. Poincaré noted, in connection with this, that since Maxwell's arguments and calculations contain difficult to correct errors, all six of Maxwell's equations should be accepted as an initial hypothesis, as postulates upon which we must rest in electromagnetic theory.

Schrödinger acted in approximately the same way. And we can assume that, at the initial stage of his work, of course, he cannot do without the use of the ordinary wave equation (for an arbitrary periodic process, running in space and time):

$$\Delta\Psi = \frac{1}{\upsilon_0^2}\frac{\partial^2\Psi}{\partial t^2}. \tag{1.11}$$

Presenting the Ψ-function in the form $\Psi = \psi(x,y,z)e^{i\omega t}$, where $\psi(x,y,z)$ is its amplitude (a complex magnitude, in a general case), we obtain

$$\Delta\Psi = \frac{1}{\upsilon_0^2}\frac{\partial^2\Psi}{\partial t^2} = -\frac{\omega^2}{\upsilon_0^2}\Psi = -k^2\Psi .$$

Hence, the wave equation (1.11) can be presented as

$$\Delta\Psi + k^2\Psi = 0, \tag{1.12}$$

where $k = \frac{\omega}{\upsilon_0} = \frac{2\pi}{\lambda} = \frac{1}{\lambda\!\!\!^{-}}$ is the wave number of the field.

Comparing equations (1.12) and (1.1), we find that the wave number *k* in Schrödinger's equation is:

$$k = \pm\sqrt{\frac{2m}{\hbar^2}\left(W + \frac{e^2}{4\pi\varepsilon_0 r}\right)} . \tag{1.13}$$

It means that the *wave number* (entered in Schrödinger's radial equation) is a *continuously variable, in the radial direction, quantity*. Is it possible to image a field, where the wave number and, accordingly, the frequency change from one point to another in the space of the field? Of course, it is not possible. *Such wave objects do not exist in nature*!

The wave number *k* is a constant parameter of wave objects. It can take a definite series of discrete values only in dependence on the boundary conditions.

According to the cutting off condition (1.5), the wave number (1.13) is defined by the following formula

$$k = e\sqrt{\frac{Zm}{2\pi\varepsilon_0\hbar^2}\left(\frac{1}{r} - \frac{1}{2an^2}\right)} . \tag{1.13a}$$

From this it follows that the wave number *k* is a real number only under the condition $r < 2an^2$. Therefore, one should mention the limiting sphere of wave processes in the atom. The radius of the sphere is equal to the doubled radius of *n*-th Bohr orbit (orbital):

$$r_{\max} = 2r_n = 2an^2 . \tag{1.14}$$

In a case when the wave number k also takes the imaginary values, the field will not be a wave field and hydrogen-like atoms will be surrounded, beyond their spheres of the radius r_{max}, with the field of the aperiodic structure. However, this is completely contradictory to reality. Thus, the limiting sphere bounds around the Schrödinger atom. Beyond the sphere, it is impossible to speak about the structure and wave properties of the atom.

The *normalizing factors of radial functions have a conditional character*, because they are determined by the integrals with an upper limit of integration, equal to infinity but not to the limiting radius (1.14). These remarks are valid also for formulae of mean values. In particular, a mean value of inverse distances, defined by the integral

$$\left\langle \frac{1}{r} \right\rangle = \int_0^\infty R_{nl}^2 r dr = \left(\frac{Z}{a_0} \right) \frac{1}{n^2}, \tag{1.15}$$

has a conditional character. Moreover, *this expression is incorrect*. As a matter of fact, the radial functions R_{nl} define the shells of the most probable values of atomic distances in accordance with the quantum mechanical interpretation of the wave function. These distances form a *discrete series*, which *cannot be averaged*, as it is impossible to average an inverse series of distances. Indeed, suppose we need to know the mean wavelength of a hydrogen atom spectrum, for example, Balmer series. Of course, we can calculate it, but it is a senseless operation, because such a mean wave does not exist in nature.

In spite of all fittings, the mean radius of an electron orbit,

$$\langle r \rangle = \int_0^\infty R_{n,l}^2 r^3 dr = a \frac{3n^2 - l(l+1)}{2}, \tag{1.15a}$$

is not proportional to n^2. Moreover, the radial spheres define the orbits of the most probable states; therefore, the radii of stationary electron orbits are constant within the corresponding spheres. Thus, the averaging (1.15a) has no sense. Further, even if we assume that the electron's motion can be the elliptic one then such orbits must pierce the shells of the stationary states. Accordingly, when an electron recedes from H-atom, moving along a stationary elliptic orbit, it must absorb energy, at the transition from one shell to another, and, at the approaching to H-atom, it must emit energy of the same value. The energetic transitions within the orbit will be determined by irrational numbers that is not observed in reality. Apart from this, such strange orbits cannot be regarded as stationary.

Let us turn now to Schrödinger's initial report, where the equation of quantum mechanics (1.1) was first presented on the basis of the operator and variational methods. We will consider this in an elementary form.

Any material object is characterized by the kinetic and potential energies which define its total energy

$$E = \frac{p^2}{2m} + U . \tag{1.16}$$

We introduce the Ψ-function of zero dimensionality, complex in a general case; its field gradient is the momentum of a microparticle, defined by the equation

$$\mathbf{p} = i\hbar\left(\frac{\partial\Psi}{\partial x}\mathbf{e}_x + \frac{\partial\Psi}{\partial y}\mathbf{e}_y + \frac{\partial\Psi}{\partial z}\mathbf{e}_z\right) = i\hbar\nabla\Psi , \tag{1.17}$$

where $\hbar$ is some elementary action, needed for realization of the law of equality of dimensionalities of the left and right parts in Eq. (1.17); i is the imaginary unit. Inasmuch as, in a general case, the Ψ-function is a complex one, components of the momentum are also, in a general case, complex ones. However, their real parts (by definition) represent the ordinary projections of the momentum along the coordinate axes. Thus, the real part of the complex momentum defines the momentum of a microparticle:

$$\mathbf{p} = \mathrm{Re}\left(i\hbar\nabla\Psi\right). \tag{1.18}$$

Relying on the expression (1.17), we can represent the energy (1.16) in the following way

$$E = -\frac{\hbar^2}{2m}\left(\left(\frac{\partial}{\partial x}\right)^2 + \left(\frac{\partial}{\partial y}\right)^2 + \left(\frac{\partial}{\partial z}\right)^2\right)\Psi + U ,$$

or

$$E = -\frac{\hbar^2}{2m}\Delta\Psi + U . \tag{1.19}$$

Let us now introduce operators of the total and potential energies, $\hat{H}$ and $\hat{U}$, according to the following expressions:

$$E = \hat{H}\Psi , \; U = \hat{U}\Psi . \tag{1.20}$$

Substituting E and U, in Eq. (1.19), with these operator expressions, we will have

$$\hat{H}\Psi = -\frac{\hbar^2}{2m}\Delta\Psi + \hat{U}\Psi\,. \tag{1.21}$$

or

$$\Delta\Psi + \frac{2m}{\hbar^2}(\hat{H} - \hat{U})\Psi = 0\,. \tag{1.21a}$$

In a case with hydrogen-like atoms, we seek the field of such a Ψ-function for which the following equalities must exist:

$$\hat{H} = W\,,\ \hat{U} = -\frac{Ze^2}{4\pi\varepsilon_0 r}\,. \tag{1.22}$$

As a result, if we accept $Z = 1$, we arrive (as Schrödinger assumed) at the wave equation for the electron in the H-atom

$$\Delta\Psi + \frac{2m}{\hbar^2}\left(W + \frac{Ze^2}{4\pi\varepsilon_0 r}\right)\Psi = 0\,. \tag{1.23}$$

Taking into consideration the expression (1.13), we deduce Eq. (1.23) to the standard form:

$$\Delta\Psi + k^2\Psi = 0\,. \tag{1.23a}$$

The following question arises: why did Schrödinger not use the standard wave equation? There are several answers to this question. They reflect the different aspects of a situation, fully developed in physics in the 1920's.

Before the appearance of Schrödinger's works, the quantum ideology has already exerted its influence on the minds of physicists. Quanta, as the elements of discreteness, as the atoms of properties, had been considered separately from the wave nature of matter-space-time. And this situation still remains, in physics, up to this day.

In fact, nature is the symmetrical structure of contradictions inseparable from each other. This also concerns such a contradictory pair as discontinuity-continuity. Wave fields represent continuity; periods-measures of properties of wave objects and fields represent its opposite side – discontinuity. These periods-measures are the physical manifestation of what is accepted as the discrete nature of matter.

In wave processes, transitions from one dynamically stable state to another usually occur through jumps. These jumps by themselves represent transient processes, occurring during a very short time, which physics, all too often, cannot detect. A specific complicated frequency structure is inherent for a jump.

For example, if we would attempt to hear the constant sound of a cross-cut saw (or a plane metal plate), beginning from a low frequency and progressing to higher frequencies, while fluently bending the saw, we will find that frequency will not have changed continuously. Frequency can suddenly increase, by jump, up to a relatively high value. We can also find such an unstable state when an insignificant change of the pressure on a fiddlestick changes the note[1].

Quantum formalism has interpreted quantum jumps (in accordance with Einstein's views) as transitions from one energy level to another without intermediate states, i.e., as processes occurring out of space and time. That is, they are equivalent to transitions with infinite speed. Einstein suggested such a primitive model. His energy quantum arises instantly with the wave speed of motion. And H. Weil, in the spirit of the age, "helped" Schrödinger (by means of the formal cutting off of the infinite solutions) to obtain a theoretical justification for such a myth. However, Schrödinger had understood all the conditionality of this solution, and he had rightly called the quantum jumps, out of space and time, the "cursed" jumps.

As known, *any wave equation is the equation of mass processes*. It describes the result of the interaction of particles and subparticles in space, from which the waves arise. Wave mass processes represent the *kinematic level of motion*, or the *level of superstructure*, below which is the *level of interaction*, or the *level of basis*.

Because of this, Schrödinger's equation is unable to describe the motion of the only electron. In spite of this, at that time, physicists groundlessly ascribed to Schrödinger's equation a non existent aptitude (unnatural for wave equations in principle). They assumed it must describe the motion of the single electron in the hydrogen atom. This was a gross blunder.

An introduction, in kinematic wave equations, of potentials or potential energies of an interaction means a lack of understanding of discriminate differences between the *dynamic basis of wave, i.e., the level of mass coordinated interaction, and the level of superstructure of wave, i.e., the ordered kinematic motion*.

The divergence of the power series of the radial function in Schrödinger's equation is the effect of mixing the kinematic and dynamic levels of motion, which were formally (incorrectly in essence) joined together in Schrödinger's equation.

---

[1] Gentill K., Acoustics, 4, 58, 1957.

Such an approach exerted significant influence on the development of microworld theories. It gave birth to the phenomenological constructions and, correspondingly, definite scientific schools. The latter are represented through numerous scientific journals, which take away science from the real picture of the microworld. Things reached a crisis point; therefore, it is necessary to analyze in detail the errors of quantum mechanics, and, most of all, the logically and physically erroneous interpretation of complex Ψ-functions.

For this purpose, we again return to operator formalism, which was created by "lazy" theorists, since operator methods are simpler than functional calculations.

Let us present Eq. (1.23a) in the form of the product of the Ψ-function by the operator binomial as

$$(\Delta + k^2)\Psi = 0 . \qquad (1.24)$$

Because the Ψ-function is unequal to zero, we obtain the operator equation for the simplest value of the operator Δ (regarding Δ as a variable operator magnitude): $\Delta + k^2 = 0$. Solving this quadratic equation, we have

$$\Delta = -k^2 \text{, and } \boldsymbol{\nabla} = -i\mathbf{k} , \qquad (1.25)$$

since $\Delta = \boldsymbol{\nabla}^2$. In the wave Ψ-field, the momentum (1.17) of an arbitrary particle takes the form

$$\mathbf{p} = i\hbar\boldsymbol{\nabla}\Psi = \mathbf{k}\hbar\Psi , \qquad (1.26)$$

or, in the scalar form,

$$p = i\hbar\nabla\Psi = k\hbar\Psi . \qquad (1.27)$$

What does this equality represent by itself?

Any physical parameter P (of an arbitrary physical wave field) has its own fundamental wave measure, or a period-quantum P_q. Using this quantum, the value of a parameter P can be presented by the quantitative relative Ψ-measure:

$$\Psi = \frac{P}{P_q} . \qquad (1.28)$$

In a general case, the parameter P is the complex quantity

$$P = p_k + ip_p \,. \tag{1.29}$$

Let us agree to call the "real" part of Eq. (1.29) the "kinetic" component and the "imaginary" part, the "potential" component of *P*-parameter. (The usefulness of this terminology will be demonstrated later on)

By virtue of this, Ψ-measure of the zero physical dimensionality will be a complex wave function with the argument,

$$\arg = i(\omega t - \mathbf{kr}) = i(\omega t - k_x x - k_y y - k_z z)\,, \tag{1.30}$$

which indicates that the quantitative measure of *P*-parameter is changed in space and time. The presence, in (1.30), of the imaginary unit *i* is not casual. It simplifies calculations and has a deep philosophical sense, which will be revealed in the following chapters.

The argument (1.30) meets general physical principles. Thus, the wave structure of any physical parameter *P* is presented by the following scalar measure:

$$P = P_q \Psi(i(\omega t - k_x x - k_y y - k_z z))\,. \tag{1.31}$$

If *P* is the momentum, then Eq. (1.27) can be written as

$$p = k\hbar \Psi(i(\omega t - k_x x - k_y y - k_z z))\,. \tag{1.32}$$

The *relative elementary harmonic measure*

$$\Psi = \Psi_m \exp(i(\omega t - k_x x - k_y y - k_z z)) \tag{1.32a}$$

of any parameter *P* satisfies the differential equations with: (1) the spatial partial derivatives of the second order

$$\frac{\partial^2 \Psi}{\partial x^2} = -k_x^2 \Psi\,,\ \frac{\partial^2 \Psi}{\partial y^2} = -k_y^2 \Psi\,,\ \frac{\partial^2 \Psi}{\partial z^2} = -k_z^2 \Psi\,, \tag{1.33}$$

or

$$\Delta \Psi = -(k_x^2 + k_y^2 + k_z^2)\Psi = -k^2 \Psi\,, \tag{1.33a}$$

and (2) the partial time derivative of the second order

$$\frac{\partial^2\Psi}{\partial t^2}=-\omega^2\Psi\,. \tag{1.33b}$$

Equations (1.33a) and (1.33b) form the *wave equation of the harmonic Ψ-function*:

$$\Delta\Psi=\frac{1}{\upsilon_0^2}\frac{\partial^2\Psi}{\partial t^2}\,. \tag{1.34}$$

Obviously, the sum of elementary measures constitutes the measure of the general character; therefore, we assume that Eq. (1.34) also defines the wave field of the measure of an arbitrary parameter. Because in any point under the steady-state wave motion the product of its spatial (amplitude) $\psi(\mathbf{kr})$ and time $\hat{T}(\omega t)$ components represents Ψ-function, the wave equation (1.34) therefore falls into the amplitude and time equations:

$$\Delta\psi+k^2\psi=0\,,\ \frac{\partial^2\hat{T}}{\partial t^2}=-\omega^2\hat{T}\,. \tag{1.34a}$$

The constant parameters, k and ω, are determined on the basis of boundary conditions. But if these parameters are difficult to determine, then we have to remember that nature itself solves this problem. We need only find, empirically, a series of the discrete values of k and ω and try to understand it.

Since equations (1.34) and (1.34a) describe Ψ-measures of arbitrary physical parameters, the difference of their wave structure comes down to the difference of kinematic types of the corresponding wave fields. The basic wave fields are the plane, cylindrical, spherical, and complicated (spherical-cylindrical) fields. Therefore, these fields, to an equal degree, successfully describe not only the atomic structure, but also the structure of megaobjects.

2. Ψ-Functions of the Wave Field of a String

First, for the sake of simplicity and clarity, we will consider the logic of errors by quantum mechanics, analyzing oscillations of a homogeneous string. We are interested in small transversal oscillations of the string near the equilibrium (Figure 3.1).

Every point of the string is defined by the coordinate z, the state of its motion in time t – by the displacement from the equilibrium, $x(z,t)$, and the motion itself – by the complex Ψ-function

$$\Psi(z,t) = x(z,t) + iy(z,t)\,, \tag{2.1}$$

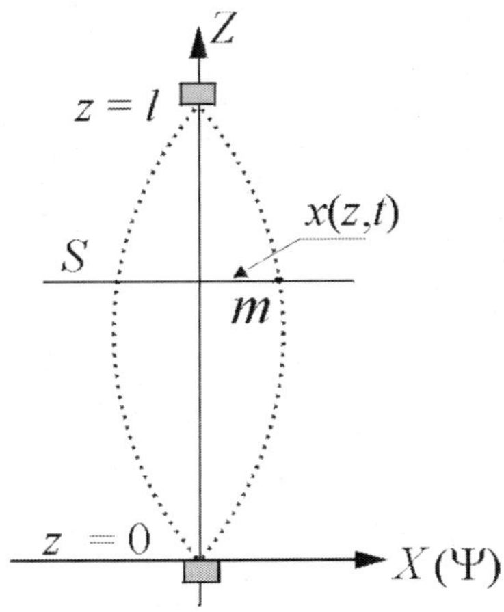

Figure 3.1. On the problem of oscillations of a string with the length *l*, fixed at both ends; *S* is a cross-section of oscillations, *m* is the mass of an arbitrary part of the string in the cross-section *S*.

Every point of the string is defined by the coordinate *z*, the state of its motion in time *t* – by the displacement from the equilibrium, $x(z,t)$, and the motion itself – by the complex Ψ-function

$$\Psi(z,t) = x(z,t) + iy(z,t)\,, \tag{2.1}$$

Displacements of points of the string from the equilibrium are equal to the real part of Ψ-function:

$$x(z,t) = \mathrm{Re}\,\Psi(z,t)\,. \tag{2.2}$$

For the description of oscillations of the string, we use Euler's formula $e^{i\varphi} = \cos\varphi + i\sin\varphi$ with its famous *imaginary unit i*. In order to get rid of this unit, creators of quantum mechanics proposed the well-known interpretation of the wave Ψ-function for *H*-atom. On the occasion of this interpretation, an epigram devoted to Schrödinger was spread among physicists. The epigram, composed in English and German languages, ascribed to E. Hückel (1896-1980, German physicist-theorist), is as follows:

Erwin with his psi can do
Calculations quite a few
But one thing has not been seen
Just what does psi really mean?

Gar Machens rechnet Erwin schon
Mit seiner Wellenfunktion
Nur wissen möcht'man gerne ohl
Was man sich dabei vorstell'n soll?

This epigram rightly comments on the real situation concerning the interpretation of the "psi" function. And what is more, a careful analysis brings to light the fact that Schrödinger's calculations are not in conformity with experimental data, which these functions (as is usually accepted) must describe. And we will demonstrate this below.

In the case of the wave field of the string, Ψ-function satisfies the wave equation for the string

$$\Delta\Psi = \frac{1}{v_0^2}\frac{\partial^2\Psi}{\partial t^2} . \tag{2.3}$$

Here, $\Delta = \frac{\partial^2}{\partial z^2}$ is the one-dimensional Laplacian operator; v_0 is the wave speed in the string defined by the following expressions

$$v_0 = \sqrt{\frac{T_S\, l}{M}} , \; T_S = E\frac{\Delta l}{l}S , \tag{2.4}$$

where T_S is the tension, l is the length, M is the mass, E is Young modulus, Δl is the lengthening, and S is the area of the cross-section of the string.

Let us agree to call the "real" component of the complex displacement $\mathrm{Re}\,\Psi(z,t) = x(z,t)$ the *potential displacement* and the "imaginary" component $\mathrm{Im}\,\Psi(z,t) = y(z,t)$, the *kinetic displacement.*

The conjugated displacements make it possible to more completely describe the wave field of the string, as the potential-kinetic wave field. An introduction of these notions entirely complies with the demands of the laws of dialectics, i.e., with the philosophical and logical system treating the Universe as a symmetrical system of oppositions.

We will seek an elementary solution of the wave equation in the form of the product of the string's space displacements, represented by the function $\psi(kz)$, and the space of time of wave events, expressed by the function $\hat{T}(\omega t)$:

$$\Psi(z,t) = \psi(kz)\hat{T}(\omega t) , \tag{2.5}$$

where $k = \frac{\omega}{v_0} = \frac{2\pi}{\lambda}$ is the wave number, ω is the circular frequency of oscillations, and the sign ^ over the symbol of the time function $\hat{T}(\omega t)$ does not allow confusion with the period T.

The function (2.5) is the mathematical expression of the indissoluble bond of material and time spaces, or (rather) the fields of material space and physical time, which are the major oppositions in the Universe.

The function $\hat{T}(\omega t)$ expresses the alternating physical time field by means of the argument t, which represents the ideal mathematical time of the imaginary absolute uniform motion, which is non-existent in nature. Timers of different types realize this imaginary time with an approximation. Such a preliminary explanation of $\hat{T}$ -function, as an image of the alternating physical field of time, is quite abstract; therefore, further, this notion will be supplemented with concrete content.

The existence of two spaces-fields (2.5) allows representation of the *wave equation of the superstructure* (2.3) through *equations of the space and time of the superstructure*:

$$\Delta\psi + k^2\psi = 0, \tag{2.6}$$

$$\frac{\partial^2 \hat{T}}{\partial t^2} = -\omega^2 \hat{T}. \tag{2.6a}$$

These equations define two elementary plane-polarized transversal waves, travelling towards each other:

$$\Psi^+ = Ae^{i(\varphi_0 - kz)}e^{i\omega t} = Ae^{i(\omega t - kz + \varphi_0)} = A(\cos(\omega t - kz + \varphi_0) + i\sin(\omega t - kz + \varphi_0)),$$

$$\Psi^- = Ae^{-i(\varphi_0 - kz)}e^{i\omega t} = Ae^{i(\omega t + kz - \varphi_0)} = A(\cos(\omega t + kz - \varphi_0) + i\sin(\omega t + kz - \varphi_0)).$$

(2.7), (2.7a)

Upper indexes of Ψ-functions indicate the direction of waves motion: the sign "+" designates motion in the positive direction and the sign "-", in the negative direction along the Z-axis with the speed v_0; A is the amplitude of waves, which, generally speaking, can take any complex value. In a particular case, under the name the amplitude, we will mean any real number.

Traveling waves Ψ^+ and Ψ^- define the potential and kinetic displacements in the following forms:

$$x^+ = \operatorname{Re}\Psi^+ = A\cos(\omega t - kz + \varphi_0), \ y^+ = \operatorname{Im}\Psi^+ = A\sin(\omega t - kz + \varphi_0). \tag{2.8}$$

$$x^- = \operatorname{Re}\Psi^- = A\cos(\omega t + kz - \varphi_0), \ y^- = \operatorname{Im}\Psi^- = A\sin(\omega t + kz - \varphi_0). \tag{2.9}$$

The addition of the two traveling waves results in the standing wave:

$$\Psi = \Psi^+ + \Psi^- = a\cos(kz - \varphi_0)e^{i\omega t}. \tag{2.10}$$

It defines the potential and kinetic displacements:

$$x = x^+ + x^- = \operatorname{Re}\Psi = a\cos(kz - \varphi_0)\cos\omega t, \tag{2.10a}$$

$$y = y^+ + y^- = \operatorname{Im}\Psi = a\cos(kz - \varphi_0)\sin\omega t, \tag{2.10b}$$

where $a = 2A$ is the amplitude of the standing wave.

The boundary condition for the string, fixed on the ends, is

$$\operatorname{Re}\Psi(kz,t) = a\cos(kz - \varphi_0)\big|_{z=0} = a\cos(kz - \varphi_0)_{z=l} = 0. \tag{2.11}$$

This condition is realized if

$\varphi_0 = \pi/2$ and $k = \dfrac{\omega}{v_0} = \dfrac{2\pi}{\lambda} = \dfrac{n\pi}{l}$, or

$$\omega = \frac{2\pi v_0}{\lambda} = \frac{n\pi}{l}v_0, \text{ or } l = \frac{n}{2}\lambda = \nu\lambda, \tag{2.12}$$

where $n = 1, 2, 3,\ldots$ is the number of half-waves and $\nu = \dfrac{n}{2}$ is the number of waves placed on the string.

The condition $n = 1$ defines the parameters of the fundamental tone of the string:

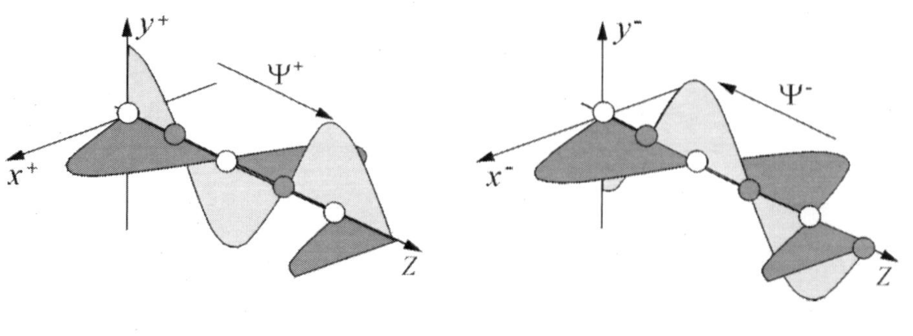

Figure 3.2. Traveling (in space and time) potential-kinetic waves Ψ^+ and Ψ^- at the instant $t = 0$; white circles define points of extremes of the kinetic displacements and dark circles – points of extremes of the potential displacements.

$$k_1 = \frac{\omega_1}{v_0} = \frac{2\pi}{\lambda_1} = \frac{\pi}{l}, \ \omega_1 = \frac{2\pi v_0}{\lambda_1} = \frac{\pi}{l} v_0 \text{ and } l = \frac{1}{2}\lambda_1 . \quad (2.13)$$

Under the realization of the conditions presented above, Ψ-function describes the following elementary traveling waves of the space-time:

$$\Psi^+ = Ae^{i(\omega t - kz + \pi/2)} = A(-\sin(\omega t - kz) + i\cos(\omega t - kz)), \quad (2.14)$$

$$\Psi^- = Ae^{i(\omega t + kz - \pi/2)} = A(\sin(\omega t + kz) - i\cos(\omega t + kz)), \quad (2.14a)$$

with the potential and kinetic displacements

$$x^+ = \operatorname{Re}\Psi^+ = -A\sin(\omega t - kz), \ y^+ = \operatorname{Im}\Psi^+ = A\cos(\omega t - kz), \quad (2.15)$$

$$x^- = \operatorname{Re}\Psi^- = A\sin(\omega t + kz), \ y^- = \operatorname{Im}\Psi^- = -A\cos(\omega t + kz). \quad (2.15a)$$

Following the solutions obtained, the potential and kinetic displacements of traveling (in space and time) waves have the phase lag by a quarter of the period, 90°. Let us agree on the graphs of the potential and kinetic displacements to present with the two mutually perpendicular axes, x^+ (x^-) and y^+ (y^-) (Figure 3.2).

The traveling waves form the standing wave (Figure 3.3):

$$\Psi = \Psi^+ + \Psi^- = a\sin kz e^{i\omega t} = x + iy . \quad (2.16)$$

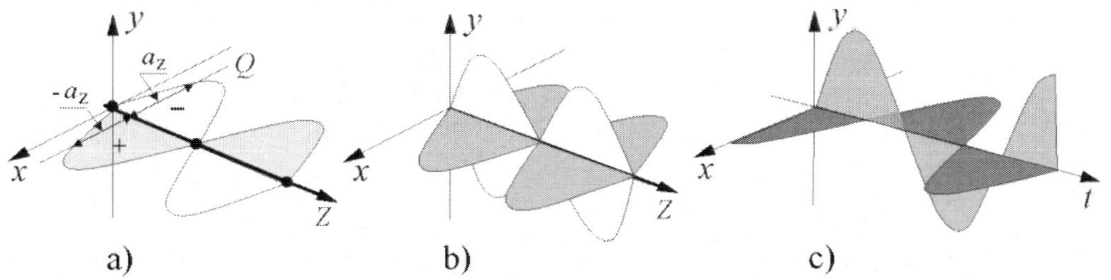

Figure 3.3. (a) A graph of the potential-kinetic standing Ψ-wave of the field of the string at the instant $t = 0$ (signs "+" and "-" show the domains of opposite phases of oscillations, $\pm a_z = \pm a\sin kz$ which are amplitudes of oscillations of opposite phases in a cross-section Q); (b) a graph of the potential-kinetic wave (in space and time) with the potential and kinetic displacements, x and y, at the instant $t \neq n\pi$, where n is an integer (displacements x and y, in space, coincide in phase and are shifted, in time, by a quarter of the period, 90°); (c) a graph of the potential-kinetic oscillations in time in the standing wave, in an arbitrary point z, satisfying the condition $kz \neq n\pi$.

In the Eq. (2.16),

$$x = \operatorname{Re}\Psi = a\sin kz\cos\omega t\,,\;\; y = \operatorname{Im}\Psi = a\sin kz\sin\omega t\,, \tag{2.16a}$$

or, briefly,

$$x = \operatorname{Re}\Psi = a_z\cos\omega t\,,\;\; y = \operatorname{Im}\Psi = a_z\sin\omega t\,, \tag{2.16b}$$

where $a_z = a\sin kz$ is the amplitude of oscillations of the standing wave in the point z. It defines the points of the string, in which potential displacements are maximal. We call such points the *potential points of the wave field of the string*.

On Z-axis, points of the string are characterized by maximal kinetic displacements; therefore, we call Z-axis the axis of *kinetic points of the wave field of the string*.

On the basis of (2.16a), we find the *kinetic speed* equal to the first derivative of the *potential displacement* with respect to time:

$$v_k = \frac{dx}{dt} = \frac{\operatorname{Re}\Psi}{dt} = -\omega a\sin kz\sin\omega t = -\omega y\,, \tag{2.17}$$

which is proportional to the *kinetic displacement y*.

By analogy to the definition of the kinetic speed, we call the derivative of the *kinetic displacement* with respect to time the *potential speed*:

$$v_p = \frac{dy}{dt} = \frac{d\operatorname{Im}\Psi}{dt} = \omega a\sin kz\cos\omega t = \omega x\,, \tag{2.17a}$$

which is proportional to the *potential displacement x*.

Kinetic and potential speeds form the kinetic-potential complex speed

$$v = \frac{d\Psi}{dt} = v_k + iv_p = -\omega y + i\omega x\,, \tag{2.18}$$

which defines the *kinetic-potential momentum* of the material object:

$$p = mv = m\frac{d\Psi}{dt} = mv_k + imv_p = -m\omega y + im\omega x\,, \tag{2.19}$$

where

$$p_k = mv_k = -m\omega y\ , \tag{2.19a}$$

$$p_p = mv_p = m\omega x \tag{2.19b}$$

are the kinetic and potential momenta, correspondingly.

As we see, the kinetic displacement defines the kinetic momentum. In turn, the potential displacement defines the potential momentum. The kinetic displacement and speed, y and v_k, correspond to the conventional classical notions of the displacement and speed.

Let us clarify the meaning of the potential displacement and speed.

Any atom of the string, as a physical point of its space of the mass m_A (Figure 3.1), in the wave field of the standing wave is characterized by the additional kinetic ε_k and potential ε_p energies, which are determined over the, correspondingly, kinetic and potential parameters:

$$\varepsilon_k = \frac{m_A v_k^2}{2} = \frac{p_k^2}{2m_A} = \frac{\mu y^2}{2} = \frac{m_A v_m^2}{2} \sin^2 kz \cdot \sin^2 \omega t\ , \tag{2.20}$$

$$\varepsilon_p = \frac{m_A v_p^2}{2} = \frac{p_p^2}{2m_A} = \frac{\mu x^2}{2} = \frac{m_A v_m^2}{2} \sin^2 kz \cdot \cos^2 \omega t\ , \tag{2.20a}$$

where $\mu = m_A \omega^2$ is the elasticity coefficient and $v_m = \omega a$ is the limiting speed of oscillations. These expressions represent the standing waves of energy. The symmetric representation of energies points out the definite symmetry of rest-motion.

The definition of the potential energy (2.20a) is the classical definition; it has arithmetic character. Whilst, strictly speaking, it should be determined on the basis of the potential speed with the measure iv_p. In such a case, the potential energy will be a negative value

$$\varepsilon_p = \frac{m_A (iv_p)^2}{2} = -\frac{\mu x^2}{2} = -\frac{p_p^2}{2m_A} = -\frac{m_A v_m^2}{2} \sin^2 kz \cdot \cos^2 \omega t\ . \tag{2.20b}$$

It is an algebraic definition of energy: kinetic and potential energies must have different signs, because these express the opposite states of the field – motion and rest. This is why, as soon as we turn to motion in a central field, we must operate with the negative potential energy.

The sum of potential and kinetic energies will be the variable value:

$$\varepsilon=\varepsilon_k+\varepsilon_p=-\frac{m_A v_m^2}{2}\sin^2 kz\cos 2\omega t\,. \tag{2.21}$$

The total energy, in the wave field of a standing wave of the string, performs harmonic oscillations with double frequency. It confirms the fact that rest and motion are shifted with respect to each other, by phase, to a quarter of the period. On the other hand, the difference of the kinetic and potential energies is the constant quantity equal the amplitude of oscillations of the energies:

$$\varepsilon_m=\varepsilon_k-\varepsilon_p=\frac{m_A v_m^2}{2}\sin^2 kz\,. \tag{2.22}$$

Every atom is also characterized by the kinetic and potential actions, defined by measures

$$\hbar_k=p_k y \text{ and } \hbar_p=p_p x\,, \tag{2.23}$$

and by the amplitude action

$$\hbar=m_A v_z a_z\,, \tag{2.23a}$$

where

$$a_z=a\sin kz\,,\ v_z=a_z\omega=a\omega\sin kz\,. \tag{2.23b}$$

The difference of symmetrical energies defines the amplitude of oscillations of energy of an atom in a cross-section Q of the string with the coordinate z:

$$\varepsilon_m=\varepsilon_k-\varepsilon_p=\frac{\mu a^2\sin^2 kz}{2}=\frac{\mu a_z^2}{2}=\frac{\hbar\omega}{2}\,. \tag{2.24}$$

The potential speed and energy are maximal, in value, in potential points, while the kinetic ones are maximal on the Z-axis.

As the potential displacement increases in value, the potential momentum and potential energy increase; the kinetic displacement, kinetic momentum, and kinetic energy decrease.

The potential displacement and speed are measures of the potential wave field of the string. And the kinetic displacement and speed are measures of the kinetic wave field of the string. The complex displacement, speed, and momentum are potential-kinetic parameters, accurately describing the potential-kinetic wave field of the string.

It is possible to say that the *kinetic speed* is the measure of the *intensity of motion*, i.e., the *kinetic field*, and the *potential speed* is the measure of the *intensity of rest*, i.e., the *potential field*.

Potential and kinetic speeds and, corresponding to them, momenta, like potential and kinetic displacements, are also shifted in phase with respect to each other by 90°. This phase lag explains the unbalanced state of kinetic and potential fields as well as their mutual transformation one into another, which, under the ideal conditions of absence of energy loss, lasts eternally.

On the basis of the conjugated kinetic and potential parameters, the formulae of energies obtain a symmetrical form, reflecting the fundamental feature of all physical fields, as potential-kinetic fields (for the complete description of which the conjugated parameters are necessary).

We perceive some parameters of kinetic and potential fields visually, other ones – physically. For example, a potential displacement is perceived visually, whereas a potential momentum and a potential speed are perceived physically only over the influence upon an object of a compressed elastic system.

The potential-kinetic exchange of motion-rest in the wave kinetic-potential field is described by the potential-kinetic rate of exchange of the motion-rest state, or briefly, by the *kinema* ("force") F, i.e., by the rate of change of potential-kinetic momentum. The rate of exchange of momentum is

$$F = m_A \frac{d^2\Psi_v}{dt^2} = \frac{dp}{dt} = F_p + F_k = m_A \frac{dv_k}{dt} + i m_A \frac{dv_p}{dt} = -m_A \omega^2 (x + iy), \qquad (2.25)$$

where

$$F_p = \frac{dp_k}{dt} = m_A \frac{dv_k}{dt} = -m_A \omega^2 x = -\mu x \qquad (2.25a)$$

is the potential kinema, or the potential power, in a wide sense of this word (we will consider these notions in detail further, in Chapter 4), or the "potential force" of the exchange of motion, and

$$F_k = \frac{dip_p}{dt} = m_A \frac{div_p}{dt} = -m_A \omega^2 iy = -\mu iy \qquad (2.25b)$$

is the kinetic kinema, or the kinetic power of exchange of rest, or the "kinetic force".

An integral action of the potential kinema (2.25a) defines the potential energy

$$\varepsilon_p = \int_0^x F_p dx = -\int_0^x \mu x dx = -\frac{\mu x^2}{2} = \frac{m_A (i v_p)^2}{2}, \tag{2.26}$$

and an integral action of the kinetic kinema (2.25b) defines the kinetic energy

$$\varepsilon_k = \int_0^x F_k diy = -\int_0^x \mu iy diy = \frac{\mu y^2}{2} = \frac{m_A v_k^2}{2}. \tag{2.27}$$

Thus, the kinetic and potential energies have different signs that is in conformity with the above-accepted definitions. In some cases, when it does not matter, we will omit the minus sign of the potential energy.

It should be noted that *the sign of any quantity, including kinetic and potential energies, depends on the choice of a zero-point (e.g., a zero level of energy) and the direction of reading. This concerns all physical quantities. Accordingly, none absolutism of signs can be: it is only important that the opposite measures will have the different signs and nothing more.*

For example, we describe the wave field of the string by Ψ-function. However, *i*Ψ-function (i.e., Ψ-function multiplied by the "imaginary" unit) also is the solution of the wave equation. But now, the kinetic speed will be the "imaginary" quantity and the potential speed – the "real" one. As a result, the kinetic energy will be the negative quantity and the potential energy will be the positive one. And this is rightfully.

When the points of the string pass through kinetic points, the "elastic force", i.e., the potential kinema, vanishes; while the "motion force", i.e., the kinetic kinema, reaches maximal value and it is perceived, visually, as the inertia of motion.

We see, thus, that the kinema, as the physical parameter of exchange of momentum, is the broader notion than the "force" related, in a definite degree, to physiological physics. This is why, the integrals, (2.26) and (2.27), were called the integral actions of kinemas, but not the integrals of "work".

As follows from the formulae of definition of kinemas, the extremal rate of change of kinetic momentum, i.e., the potential kinema of exchange of motion F_p, takes place in potential points of the field of the string. Analogously, the extremal rate of change of potential momentum, i.e., the kinetic kinema of exchange of rest F_k, takes place on the line of kinetic points.

The above-considered *potential and kinetic energies are measures of the transversal potential and kinetic wave field of the string, which relates to the superstructure of the field of the string (to the oscillatory level). These are transversal energies in the plane-polarized wave.*

Now, we will supplement the transversal energies with the *conjugated longitudinal energies* of the *basis of the wave field. The basis is related with the dynamics of transmission of the wave signal.*

Generally, it is necessary for all the parameters of the superstructure of the wave to be supplemented with the conjugated parameters of the basis of the wave. A general structure of the potential-kinetic wave can be expressed by the graph shown in Figure 3.4).

The potential energy of the field of basis of the string W_p has the form

$$W_p = \frac{\mu(\Delta l)^2}{2} = \frac{ES(\Delta l)^2}{2l}. \tag{2.28}$$

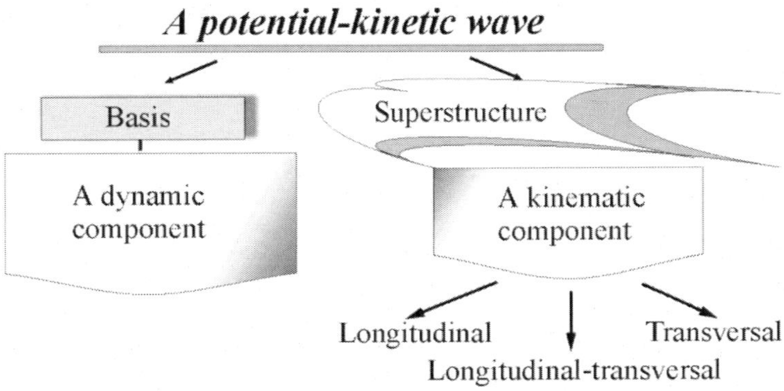

Figure 3.4. A graph of the potential-kinetic wave, representing its generalized model.

Taking into consideration Eq. (2.4) and assuming $M = \rho Sl$, after elementary transformations, we find the longitudinal potential energy per one atom of the string ε_p:

$$\varepsilon_p = \frac{W_p}{M/m_A} = \frac{m_A E(\Delta l)^2}{2\rho l^2} = \frac{m_A \varepsilon_T v_0^2}{2}, \tag{2.28a}$$

where $\varepsilon_T = \dfrac{\Delta l}{l}$ is the relative lengthening of the string. As a result we have

$$\varepsilon_p = \frac{m_A \varepsilon_T v_0^2}{2} = \frac{m_A v_{0p}^2}{2}, \tag{2.29}$$

where $v_{0p} = v_0\sqrt{\varepsilon_T}$ is the potential speed.

Since every atom takes part also in the transmission of the wave signal along the string, hence, it has the longitudinal kinetic energy of the basis of the wave, which is defined by the speed of the wave field:

$$\varepsilon_k = \frac{m_A v_0^2}{2}. \tag{2.30}$$

It makes sense also to speak about the wave action of an atom during the transmission of a signal along the whole length of the string, which is defined by the equality: $h = m_A v_0 l$. It is the action at the level of the wave basis. Taking into account that

$l = \frac{n}{2}\lambda$, we obtain

$$h = m_A v_0 l = \frac{n}{2} h_\lambda, \tag{2.31}$$

where

$$h_\lambda = m_A v_0 \lambda \tag{2.31a}$$

is the wave action. If $n = 1$, we have

$$h = m_A v_0 l = \frac{1}{2} h_1, \tag{2.32}$$

where

$$h_1 = m_A v_0 \lambda_1 \tag{2.32a}$$

is the wave action of the fundamental tone. From these equalities, it follows that

$$h_\lambda = \frac{h_1}{n} \text{ or } \hbar_\lambda = \frac{\hbar_1}{n}, \tag{2.33}$$

where

$$\hbar_\lambda = \frac{h_\lambda}{2\pi} \text{ and } \hbar_1 = \frac{h_1}{2\pi}. \tag{2.33a}$$

On the basis of formulae (2.33) and (2.33a), the kinetic energy of any atom of the string, at the level of the wave basis of the field, can be presented in the following forms:

$$\varepsilon_k = \frac{1}{2} h_1 v_1 = \hbar\omega_1 = \frac{1}{2}\hbar_1\omega_1 = \frac{m_A v_0^2}{2}, \tag{2.34}$$

where

$$\hbar = \frac{m_A v_0 l}{2\pi} = \frac{1}{2}\hbar_1 = \frac{1}{2} m_A v_0 \lambda_1 . \tag{2.34a}$$

It is usual to call the measure (2.34) the *energy quantum*. The definite wave portions of any measures of potential-kinetic fields relate to quanta. In general, all quanta are effects of the wave nature of matter-space-time. Hence, strictly speaking, they should be called *wave quanta*, or *quanta-periods*. Such fundamental quanta of wave spaces, as quanta of the spatial extension of waves (represented by the wavelengths λ, the quanta of time length, the named periods of physical time waves T, etc.), relate to the wave quanta. Unfortunately, M. Born did not understand this and introduced the term "quantum mechanics", which began to press the wave theory of matter. It was a serious logical error.

The measure of energy (2.34) repeats the well-known Planck's energy for "quanta of energy – photons", which (as will be shown in the following chapters) do not exist in nature, being, at the most, mathematical fictions.

On the basis of all above stated, we can present the wave number squared k^2, in the following form

$$k^2 = \frac{\omega^2}{v_0^2} = \frac{2m_A\varepsilon_k}{\hbar_\lambda^2}. \tag{2.35}$$

Then, the amplitude equation for the wave motion of the string (2.6) takes the form of Schrödinger's equation

$$\Delta\psi + \frac{2m_A\varepsilon_k}{\hbar_\lambda^2}\psi = 0 . \tag{2.36}$$

Because the quanta of mass at the atomic level, as W. Prout noted, are masses of H-atoms (neutrons, protons), hence, the equation. (2.36) can be presented as

$$\Delta\psi + \frac{2m_0\varepsilon_k}{\hbar_\lambda^2}\psi = 0\,, \tag{2.36a}$$

where m_0 is the atomic mass unit, substituted in the above presented parameters for the mass of an atom of the string m_A.

Since the total potential-kinetic energy of basis in the wave field of the string, for every H-atom, is equal to $\varepsilon = \varepsilon_k + \varepsilon_p = \frac{m_0 v_0^2}{2} + \frac{m_0 v_0^2}{2}\varepsilon_T$, we can again rewrite the equation (2.36a) as

$$\Delta\psi + \frac{2m_0(\varepsilon - \varepsilon_p)}{\hbar_\lambda^2}\psi = 0\,. \tag{2.36b}$$

In this form, the wave equation (2.36b) looks spectacularly. But actually, this equation is senseless. Here, $\frac{2m_0(\varepsilon - \varepsilon_p)}{\hbar_\lambda^2}$ is the quantity, representing by itself a discrete series of values, which depend on the boundary conditions. Therefore, one should operate by the wave equation of superstructure in the form

$$\Delta\psi + k^2\psi = 0\,.$$

We can also present the wave equation of the harmonic Ψ-function (1.34) in the complex form:

$$\frac{1}{v_0^2}\frac{\partial^2\Psi}{\partial t^2} = \frac{i\omega}{v_0^2}\frac{\partial\Psi}{\partial t} \quad \text{and} \quad \frac{\partial^2\Psi}{\partial z^2} = \frac{i\omega}{v_0^2}\frac{\partial\Psi}{\partial t}\,.$$

Then, multiplying the equality (1.34) by the wave action $\hbar_\lambda$, we have

$$\frac{v_0^2\hbar_\lambda}{\omega}\Delta\Psi = \frac{v_0\hbar_\lambda}{k}\Delta\Psi = i\hbar_\lambda\frac{\partial\Psi}{\partial t}\,.$$

If we introduce the operator of the string

$$\hat{H} = \frac{v_0\hbar_\lambda}{k}\Delta = \frac{\hbar_\lambda^2}{m_0}\Delta\,, \tag{2.37}$$

then the equation of the string will have the form

$$\hat{H}\Psi = i\hbar_{\lambda} \frac{\partial \Psi}{\partial t}. \tag{2.38}$$

According to the fully developed opinion, in quantum mechanics, Schrödinger's wave equation in the form (2.38) is regarded as a particular equation, since it contains the first derivative of the wave function with respect to time and the imaginary unit. And because of this, its solution can be presented only by the complex wave Ψ-function. Of course, it is a fallacy. As has been shown above, the wave equation with the second derivative of the wave function with respect to time can be also presented with the first derivative of the same function with respect to time.

Previously, we devoted major attention, first to the spatial $\psi(kz)$-function. Now, the time comes when we must turn to the *detail analysis* of the function $\hat{T}(\omega t)$ *of the time field-space* of the string.

The material field-space $\psi(kz)$ is characterized by the *wave number of the space*, *k*, connected with the *wave quantum of the space* λ (the space wavelength) through the following ratio

$$\lambda = \frac{2\pi}{k}. \tag{2.39}$$

In the time field-space $\hat{T}(\omega t)$, the wave number is the circular frequency ω, which should be called the *wave number of the time field-space*. The time wave number ω is inseparable to the wave quantum of the time field – the period *T*. The *period of the time wave is the time wavelength T*, which is analogous to the space wavelength λ. Between the time wavelength *T* and the time wave number ω takes place the ratio, analogous to (2.39):

$$T = \frac{2\pi}{\omega}. \tag{2.40}$$

$\hat{T}$-Function, in the explicit form, is presented as

$$\hat{T} = e^{i\omega t} = \cos\omega t + i\sin\omega t.$$

In conformity with the potential and kinetic displacements

$x = a\sin kz\cos\omega t$ and $y = a\sin kz\sin\omega t$,

we will call the real part of the time $\hat{T}$-function, $\mathrm{Re}\,\hat{T}$, the *potential time displacement* t_p, or simply, the *potential time with the relative unit amplitude*; the imaginary part, $\mathrm{Im}\,\hat{T}$, – the kinetic *time displacement* t_k, or simply, the *kinetic time*. In such a case, we have

$$\hat{T} = e^{i\omega t} = t_p + it_k = \cos\omega t + i\sin\omega t, \tag{2.41}$$

where

$$t_p = \mathrm{Re}\,\hat{T}(\omega t) = \cos\omega t\,,\ t_k = \mathrm{Im}\,\hat{T}(\omega t) = \sin\omega t\,. \tag{2.41a}$$

Thus, the *physical time field* is the *potential-kinetic time wave*; its graph is presented in Figure 3.5.

The physical time wave field of the string is a particular case of the complicated ideal time wave field-space of the Universe with its infinite series of levels.

The following relation connects the spatial standing wave Ψ (see Eq. (2.16) and Figure 3.3) with the time wave:

$$\Psi = a\sin kz\cdot\hat{T} = a_z\hat{T}\,; \tag{2.42}$$

Therefore, they are structurally similar.

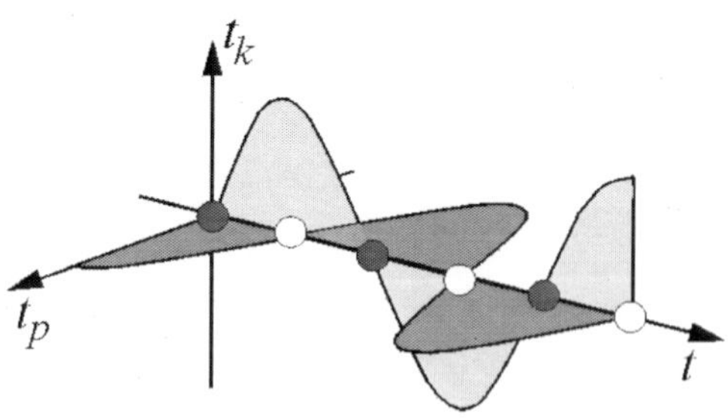

Figure 3.5. A graph of the time potential-kinetic wave: $t_p = \cos\omega t$ is the potential component and $t_k = \sin\omega t$ is the kinetic component of the wave of time.

The rate of change of time wave is determined by the potential-kinetic time speed γ:

$$\gamma = \frac{d\hat{T}}{dt} = i\omega\hat{T} = \gamma_k + i\gamma_p = -\omega\sin\omega t + i\omega\cos\omega t, \tag{2.43}$$

where

$$\gamma_k = \frac{dt_p}{dt} = -\omega t_k = -\omega \sin \omega t \tag{2.43a}$$

is the *kinetic rate of change of time* and

$$\gamma_p = \frac{dt_k}{dt} = \omega t_p = \omega \cos \omega t \tag{2.43b}$$

is the *potential rate of change of time.*

Amplitude of the time speed is equal to the time wave number, referred to as the circular frequency (or, under the circular motion, the angular speed).

The following relation takes place between the spatial and time speeds of a standing wave:

$$\mathsf{v} = \frac{d\Psi}{dt} = a \sin kz \frac{d\hat{T}}{dt} = a_z \cdot \gamma \ . \tag{2.44}$$

At that,

$$\mathsf{v}_k = a_z \cdot \gamma_k \, , \; i\mathsf{v}_p = a_z \cdot i\gamma_p \, . \tag{2.45}$$

According to the equations (2.29) and (2.30), the standing wave of amplitude energy of superstructure, at the level of *H*-masses, has the form

$$\varepsilon_m = \varepsilon_k - \varepsilon_p = \frac{m_0 a_z^2}{2}(\gamma_k^2 + \gamma_p^2) = \frac{m_0 a_z^2}{2}\omega^2 \, . \tag{2.46}$$

The wave field of a string can be described also in the space of motion and time. As the measure of the space of motion, it is possible to take the wave field-space of speed $\mathsf{v}(kz)$. If the time field-space is represented by the aforementioned function $\hat{T}(\omega t)$, then new Ψ_v-function takes the form

$$\Psi_\upsilon(z,t) = \mathsf{v}(kz)\hat{T}(\omega t) \, . \tag{2.47}$$

This function satisfies the wave equation of a string for the wave field-space of speed and time:

$$\Delta\Psi_v = \frac{1}{v_0^2}\frac{\partial^2 \Psi_v}{\partial t^2}. \tag{2.48}$$

Elementary solutions of the wave equation (2.48) define two traveling waves of speed, Ψ_v^+ and Ψ_v^-, correspondingly, in positive and negative directions along the Z-axis.

Taking into account the boundary conditions, Ψ_v^+-function takes the form

$$\Psi_v^+ = i\omega\Psi^+ = i\omega A e^{i(\omega t - kz + \pi/2)} = -\omega A e^{i(\omega t - kz)} = v_k^+ + i v_p^+, \tag{2.49}$$

where

$$v_k^+ = -\omega A\cos(\omega t - kz), \tag{2.49a}$$

$$v_p^+ = -\omega A\sin(\omega t - kz) \tag{2.49b}$$

are the traveling waves of the kinetic and potential speeds, correspondingly.

The traveling wave of Ψ_v^--speed, conjugated to Ψ_v^+, is

$$\Psi_v^- = i\omega\Psi^- = i\omega A e^{i(\omega t + kz - \pi/2)} = \omega A e^{i(\omega t + kz)} = v_k^- + i v_p^-, \tag{2.50}$$

where

$$v_k^- = \omega A\cos(\omega t + kz),\ v_p^- = \omega A\sin(\omega t + kz). \tag{2.50a}$$

On the basis of the equation (2.49), we obtain the expressions for the kinetic, potential, and amplitude energies of traveling Ψ_v^+-waves:

$$\varepsilon_k^+ = \frac{m_0(v_k^+)^2}{2} = \frac{m_0\omega^2 A^2}{2}\cos^2(\omega t - kz) = \frac{m_0 v_m^2}{2}\cos^2(\omega t - kz), \tag{2.51}$$

$$\varepsilon_p^+ = \frac{m_0(i v_p^+)^2}{2} = -\frac{m_0\omega^2 A^2}{2}\sin^2(\omega t - kz) = -\frac{m_0 v_m^2}{2}\sin^2(\omega t - kz), \tag{2.52}$$

$$\varepsilon_m^+ = \varepsilon_k^+ - \varepsilon_p^+ = \frac{m_0 v_m^2}{2}, \tag{2.53}$$

where $v_m = \omega A$ is the amplitude value of speed.

Potential and kinetic energies in traveling waves are shifted in phase by 90° (Figure 3.6a).

Analogous wave energies defines Ψ_v^--function:

$$\varepsilon_k^- = \frac{m_0 (v_k^-)^2}{2} = \frac{m_0 \omega^2 A^2}{2} \cos^2(\omega t + kz) = \frac{m_0 v_m^2}{2} \cos^2(\omega t + kz), \qquad (2.54)$$

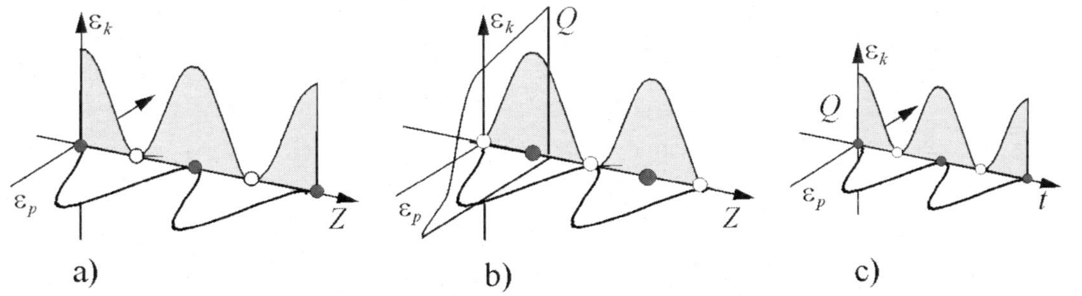

Figure 3.6. A graph of the traveling (a) and standing (b) potential-kinetic waves of energy; Q is the plane of cross-section of the standing wave, where potential-kinetic oscillations of the wave of time (c), shifted by 90°, occur.

$$\varepsilon_p^- = \frac{m_0 (i v_p^-)^2}{2} = -\frac{m_0 \omega^2 A^2}{2} \sin^2(\omega t + kz) = -\frac{m_0 v_m^2}{2} \sin^2(\omega t + kz), \qquad (2.55)$$

$$\varepsilon_m^- = \varepsilon_k^- - \varepsilon_p^- = \frac{m_0 v_m^2}{2}. \qquad (2.56)$$

Obviously, the sum of the traveling kinetic-potential waves, Ψ_v^+ and Ψ_v^-, generates the standing wave of speed

$$\Psi_v = i\omega a \sin kz e^{i\omega t} = v_k + i v_p, \qquad (2.57)$$

where

$$v_k = v_k^+ + v_k^- = -\omega a \sin kz \sin \omega t = -\omega a_z \sin \omega t \qquad (2.57a)$$

and

$$v_p = v_p^+ + v_p^- = \omega a \sin kz \cos \omega t = \omega a_z \cos \omega t. \qquad (2.57b)$$

Thus, the expressions for energies in the standing wave take the following form:

$$\varepsilon_k = \frac{m_A v_k^2}{2} = \frac{\hbar_k \omega}{2} = \frac{m_A(\omega a)^2}{2}\sin^2 kz \cdot \sin^2 \omega t\,, \tag{2.58}$$

$$\varepsilon_p = \frac{m_A(iv_p)^2}{2} = -\frac{\hbar_p \omega}{2} = -\frac{m_A(\omega a)^2}{2}\sin^2 kz \cdot \cos^2 \omega t\,, \tag{2.58a}$$

$$\varepsilon_m = \varepsilon_k - \varepsilon_p = \frac{\hbar\omega}{2} = \frac{m_A v_m^2}{2}\sin^2 kz\,, \tag{2.59}$$

where $v_m = \omega a$ is the amplitude value of speed.

The structure of standing waves of potential-kinetic energy is shown in Figure 3.6b,c. These waves of energy were earlier presented by the formulae (2.20) and (2.20a).

If in a string, the two mutually perpendicular plane-polarized waves, traveling in opposite directions, are excited then the circular-polarized wave arises. The energy of every particle in this wave has now the constant value, independent of time:

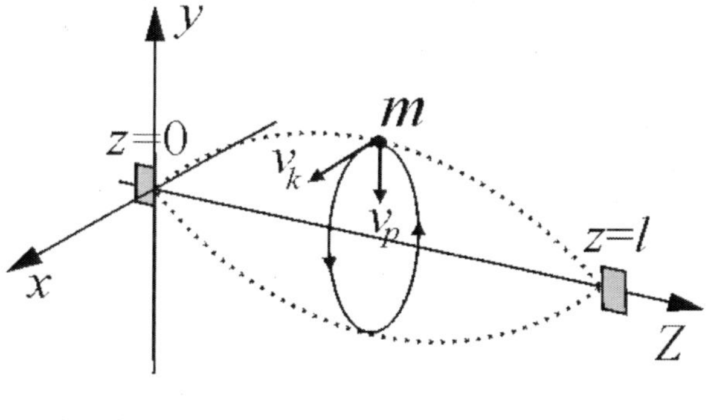

Figure 3.7. A circular potential-kinetic wave of the string.

$$\varepsilon_k = \frac{m_A v_k^2}{2} = \frac{m_A v_m^2 \sin^2 kz}{2}\,, \tag{2.60}$$

$$\varepsilon_p = \frac{m_A(iv_p)^2}{2} = -\frac{m_A v_m^2 \sin^2 kz}{2}\,. \tag{2.60a}$$

From this we find the amplitude energy

$$\varepsilon_m = \varepsilon_k - \varepsilon_p = m_A v_z^2 = \hbar_z \omega = m_A v_m^2 \sin^2 kz\,. \tag{2.61}$$

At that the total energy is equal to zero:

$$\varepsilon = \varepsilon_k + \varepsilon_p = 0 \,. \tag{2.62}$$

Thus, every atom of a string, in such a wave, performs the circular motion with the total energy equal to zero. At that, the kinetic speed is directed tangentially to the circular trajectory, whereas the potential speed, perpendicular to the kinetic one, is directed to the Z-axis. In this wave, the kinetic and potential fields are mutually perpendicular (Figure 3.7) and negate each other. Thus, it is clear why in nature the circular motion prevails.

Obviously, the description of the wave field can be realized also by the complex Ψ-function, which represents other physical parameters, such as the potential-kinetic momentum, potential-kinetic kinema, etc. It means that the wave Ψ-function of string can describe any potential-kinetic fields of the string: fields of displacements, speeds-time, momenta-time, etc. All these wave fields represent by themselves the different measures of a many-sided potential-kinetic field of a string.

In virtue of the above stated, in dialectics, the question about the interpretation of the wave Ψ-function makes no logical sense, because everything depends on which facets of the potential-kinetic field we mean to regard. If we would write the equation of a wave process, or a wave system, only formally and, further, be engaged in scientific guessing ("interpretation"), then we could come to the creation of myths. And myths, as is well known, take deep roots in the mass scientific consciousness, making it blind.

In this connection, let us once more turn to the generalized equation of the string (2.36), introducing in it some potential energy of interaction of atoms $\varphi(z)$ (that, unconditionally, is inadmissible; we have written about it already). Then, designating the total energy of oscillations of an atom of the string in a cross-section S by the letter W, we can write:

$$\varepsilon_k = W - \varphi(z) \,. \tag{2.63}$$

On the basis of such a "generalization", the wave number

$$k = \frac{\omega}{v_0} = \sqrt{\frac{2m\varepsilon_k}{\hbar_0^2}} = \sqrt{\frac{2m}{\hbar_0^2}\left(W - \varphi(z)\right)} \tag{2.64}$$

and, corresponding to it, frequency of the wave field of the string

$$\omega = v_0\sqrt{\frac{2m}{\hbar_0^2}\left(W - \varphi(z)\right)} \tag{2.65}$$

will be the functions of the coordinate z of points of the string, $k = k(z)$ and $\omega = \omega(z)$. Actually, the wave number k is the constant quantity, defining some frequency of the wave field ω, which bonds the wave system in a single whole.

According to such a "generalization", all points of the string must oscillate with different frequencies. This is an absurdity, possible only due to a free game of notions, unrestricted within the reasonable framework of logic.

This cardinal error has transformed quantum mechanics into a great caricature about the world of real wave processes. And the extensive publicity created an illusion as if mankind deals with a great theory. In fact, a phenomenological theory, with the definite fitting of it to the experiment, was built as a result. This theory (quantum mechanics) as much significantly distorted the real picture of the microworld that it became the world of theoretical monsters and quantum chaos, but not the world of real images.

The absurdity of the above-described formal "deducing" of the relation (2.64) (for the field of the string) is clear and no sane physicist will agree with the "generalized" wave equation of the string in the form as follows:

$$\Delta\psi + \frac{2m}{\hbar_0^2}\left(W - \varphi(z)\right)\psi = 0 . \qquad (2.66)$$

The falsity of such a formal generalization-derivation is obvious. But in the 1930's, the similar work for the hydrogen atom was (and still) declared as a greatest penetration in the mystery of the atomic world. Many physicists have objected to such a theory, but they had (and yet have) no apparatus to be able to "see" atoms; therefore, it was very difficult to object to the aggressive physical-mathematical decadence. And the decadence overcame reason. Since then, all right notices against such a formalistic arbitrariness were/are declared as the "classical backwardness".

3. Parameters of the Wave Field of a String and Balmer's Formula

The wave motion of a string with the frequency of the fundamental tone ν_1 and wavelength λ_1 generates an acoustic wave in the surrounding air of the same frequency:

$$\nu_1 = \frac{1}{T_1} = \frac{v_0}{\lambda_1} = \frac{c}{\lambda_c}, \qquad (3.1)$$

where λ_c is the acoustic wavelength, c is the speed of sound. From the equations (2.13) and (3.1), the following relation for the fundamental tone takes place:

$$\frac{1}{\lambda_c} = \frac{v_0 / c}{2l}. \tag{3.2}$$

Let us imagine that the ends of the string are joined together, i.e., we deal with a string circle with one node. If the radius of the string circle is equal to r_1 then the equation (3.2) will take the form

$$\frac{1}{\lambda_c} = \frac{v_0 / c}{4\pi r_1} = \frac{\nu_1}{c}, \tag{3.3}$$

where

$$\nu_1 = \frac{v_0}{4\pi r_1} \tag{3.4}$$

is the frequency of the fundamental tone of the string.

If we have a series of string circles of the length l_n, whose radii r_n are multiple to the radius of the first circle r_1, $r_n = nr_1$, and the wave speeds are inversely proportional to numbers of this series, $v_n = v_1 / n$, then such strings will have the following proper frequencies:

$$\nu_n = \frac{v_0}{4\pi r_1 n^2}. \tag{3.5}$$

These frequencies will define the spectrum of acoustic waves of the fundamental tone

$$\frac{1}{\lambda_n} = \frac{v_0 / c}{4\pi r_1} \frac{1}{n^2}. \tag{3.6}$$

The difference of the two inverse acoustic wavelengths (two arbitrary string levels of (3.6)) is presented by the following equality:

$$\Delta\left(\frac{1}{\lambda}\right) = \frac{v_0 / c}{4\pi r_1}\left(\frac{1}{m^2} - \frac{1}{n^2}\right) = R\left(\frac{1}{m^2} - \frac{1}{n^2}\right), \tag{3.7}$$

where

$$R = \frac{v_0 / c}{4\pi r_1} \tag{3.8}$$

is the wave constant of the string of the fundamental tone.

Let us remember now Balmer's wave formula for H-atom spectrum

$$\frac{1}{\lambda} = R\left(\frac{1}{m^2} - \frac{1}{n^2}\right) = \frac{v_0 / c}{4\pi r_0}\left(\frac{1}{m^2} - \frac{1}{n^2}\right), \tag{3.9}$$

where v_0 and r_0 are the Bohr speed and radius, correspondingly. Obviously, for the limiting wave of radiation, of the fundamental tone of H-atom ($n = \infty$ and $m = 1$), the following relation, analogous to the relation for the string (3.8), takes place:

$$\frac{1}{\lambda} = R = \frac{v_0 / c}{4\pi r_0} = \frac{\nu_1}{c}, \tag{3.10}$$

where

$$\nu_1 = \frac{1}{T_1} = \frac{v_0}{4\pi r_0} \tag{3.11}$$

is the frequency of the fundamental tone of the first electron orbit.

Conformity of the parameters, for the electron orbit and string circle, is obvious. And it is natural, because an electron on its orbit is the node of the electron circular cylindrical wave. A physical mystery of Rydberg constant is revealed here. The relation (3.10) has the fundamental wave character. It points out the existence of a series of Rydberg constants for many wave systems. In a particular case of a string, the Rydberg acoustic constant is

$$R = \frac{v_0 / c}{2l}. \tag{3.12}$$

Thus, the formula of the Rydberg constant uniquely rejects the quantum mechanical myth that electrons in atoms have no orbits (trajectories) and their probabilistic chaotic motion forms so-called "orbitals" ("charged clouds" of probability of electron's localization). All this is formalistic fantasy, yet we will analyze it in detail in the following chapters.

Earth's orbit is analogous to the electron orbit of the fundamental tone. Therefore, the formula (3.10) is also valid for the wave gravitational field of the solar system. Since the period of Earth's rotation is equal to one-half period of the fundamental tone of Earth's orbit (with its only gravitational *node*-Earth), therefore, it should be assumed the existence in our Galaxy of waves in two light years and half-waves in one light year. And actually, a half-wave of Earth's orbit shows its worth in the distances between stars.

The sand grains in Chladni's famous experiments[2] with oscillating plates are localized not in points of maxima of Ψ-function, but only in points of its zero values, i.e., in the *nodal* points of the field. Just so, we should assume that stars, in the galactic wave field, are situated in *nodes* of cosmic space. The astronomical data confirms this supposition. Bessel, who has measured the parallax of stars, carried out (in 1838) the first reliable determination of the distance to the star 61, Swan. It proven to be 11 light years (the calculated value, according to the parallax in $0.296''$, is 11.0247). Bernard's star has the parallax in $0.543''$ that corresponds to the distance in 6 light years (calculations give 6.0016). The star Procaine: the parallax is $0.297''$, the distance - 11 light years (10.9921), etc. Thus, within accuracy of measurements of parallaxes, the distances to the above-mentioned stars are multiple, precisely enough, to a half-wave of Earth's orbit.

A string, fixed from both ends, can be called (in a definite sense) the closed string. If one end of the string is fixed and another one is free (open), it will be then a half-open string. In such a string, an elementary standing wave also arises,

$$\Psi = \Psi^- + \Psi^+ = a\cos(kz - \varphi_0)e^{i\omega t}. \tag{3.13}$$

The boundary conditions for the fixed and free ends are

$$\mathrm{Re}\,\Psi(kz,t) = a\cos(kz - \varphi_0)\big|_{z=0} = 0, \tag{3.14}$$

$$\mathrm{Re}\,\Psi(kz,t) = a\cos(kz - \varphi_0)_{z=l} = d, \tag{3.15}$$

From the condition (3.14), we find the initial phase, $\varphi_0 = \pi/2$. The second boundary condition (3.15) gives the possible values of the wave number and frequency of the field of string:

$$k = \frac{1}{l}\arcsin\left(\frac{d}{a}\right). \tag{3.16}$$

[2] Waller M.D., *Cladni Plates*, London, Staples Press, 1960.

Let us assume that displacements of the free end of the string take the following series of values

$$\arcsin\left(\frac{d}{a}\right)=\pi\left(\frac{1}{m^2}-\frac{1}{n^2}\right), \tag{3.17}$$

where m and n are integers. This condition always can be realized. The discrete series of values of the wave number and circular frequency corresponds to the above condition:

$$k=\frac{\pi}{l}\left(\frac{1}{m^2}-\frac{1}{n^2}\right),\ \omega=\frac{\pi v_0}{l}\left(\frac{1}{m^2}-\frac{1}{n^2}\right). \tag{3.18}$$

As follows from (3.18), standing waves of the undertones must arise in the half-open string:

$$\Psi=a\sin\left(\frac{\pi}{l}\left(\frac{1}{m^2}-\frac{1}{n^2}\right)z\right)\exp\left(i\frac{\pi v_0}{l}\left(\frac{1}{m^2}-\frac{1}{n^2}\right)t\right). \tag{3.19}$$

Oscillations of the open string excite acoustic waves in the air of the same frequency

$$\nu=\frac{v_0}{\lambda}=\frac{c}{\lambda_c}=\frac{v_0}{2l}\left(\frac{1}{m^2}-\frac{1}{n^2}\right); \tag{3.20}$$

they form the acoustic spectrum of waves, analogous to the H-atom spectrum:

$$\frac{1}{\lambda_c}=\frac{v_0}{2cl}\left(\frac{1}{m^2}-\frac{1}{n^2}\right). \tag{3.21}$$

4. THE MATHEMATICAL BASIS OF THE HYDROGEN ATOM USING THE WAVE EQUATION OF MATTER-SPACE-TIME

Quantum mechanics, with its Schrödinger's Ψ-function, has actually no relation to the concrete presentation of the microworld. The philosophy of abstractionism lies in its base. Abstractionism exhibited itself first in art. As usual to regard, the first work of abstractionism was Candinsky's picture of 1910 (in water-color), painted in Munich. Ch. Bru, a French critic, has written, "For the first time in history of painting, it is impossible to find or recognize anything in this picture."[3] Bru noted also that Candinsky's abstractive idea arose when he, by chance, had leaned one of his pictures, looking at it under the angle: "the image on this picture has become incomprehensible, as much enigmatically paints have sparkled at once." According to another version (by A.A. Sidorov): one day, Candinsky, drying the brushes, paid attention to the combination of accidental stains on a rag and thought, for a moment, why not regard this rag as the work of abstract art, without any objective content.

According to theorists of the abstract view on the world, abstract painting is painting "...when we can know nothing from it about the objective reality, being the normal sphere of our life... From this it follows that abstract painting is a reproduction brought to such an extreme, when it is impossible to find there an initial natural object; it is the reproduction, which, at first sight on it, nothing reminds us of the original. Only this work will have the right to be called the abstraction."[4]

The physical-mathematical abstractionists have acted in the same spirit. Their philosophy sweeps aside, without a trace, any possibility of the concrete presentation of the microworld, in full agreement with *Mach's* and *Ostwald's* (F.W. Ostwald, 1853-1932, German chemist) ideas. Ostwald has written, in 1902, that the time, when all these "atoms and molecules will disappear in a dust of archives", is not so far. Mach has called the electron theory as a "respectable witches' Sabbath."

Therefore, it is no wonder that, in quantum mechanics, the extraordinary primitive abstract model, based on the formal mathematical group of waves, represents the correlation of waves and particles. The superposition of these waves forms an envelope, spreading in space with the speed, different from the speed of constituent waves. This envelope, called the *wave packet*, is regarded as a microparticle; and waves, related to the packet, as waves of probability. Thus, the foundation of nature is represented as probabilistic "vague" chaos.

We will present now, in a general outline, the abstract-concrete dialectical description of the microworld, which reflects their authors' works.[5]

The wave equation of matter-space-time

$$\Delta\Psi = \frac{1}{v^2}\frac{\partial^2\Psi}{\partial t^2}. \tag{4.1}$$

lies at the base of the dialectical description of the world. Ψ-Function represents the mathematical image of the potential-kinetic field of space-time of the atomic level (the

[3] Ch. Bru, *Esthétique de L'abstraction*, Paris, 1959, p.3.
[4] Dictionnaire de la abstraite, Paris, 1957, pp. 2-3.
[5] L.G. Kreidik and G.P. Shpenkov, *Alternative Picture of the World*, V.2, Bydgoszcz, 1996.

level of superstructure over the nearest subatomic level – the level of basis). The wave equation (4.1) falls, like Schrödinger's equation, into the equation of spatial ψ-function,

$$\Delta\psi + k^2\psi = 0\,, \tag{4.2}$$

and the equation of time $\hat{T}$ -function,

$$\frac{d^2\hat{T}}{dt^2} = -\omega^2\hat{T}\,. \tag{4.2a}$$

$\hat{T}$ -Function has the form

$$\hat{T} = e^{i\omega t}\,. \tag{4.2b}$$

4.1. The Spherical Wave Field of H-Atom

If a wave object and its wave field-space have the *spherical structure*, that is characteristic for all microparticles, then the elementary structure of the field of space-time of microobjects is expressed by the functions

$$\Psi_{l,m} = \psi_{l,m}(\rho,\theta,\varphi)\hat{T}(\omega t) = R_l(\rho)\Theta_{l,m}(\theta)\Phi_m(\varphi)\hat{T}(\omega t)\,, \tag{4.3}$$

where

$$\psi_{l,m}(\rho,\theta,\varphi) = R_l(\rho)\Theta_{l,m}(\theta)\Phi_m(\varphi) \tag{4.3a}$$

is the spatial $\psi_{l,m}$ -function, $\rho = kr$ is the relative radius, *k* is the wave number of atomic space, and $\hat{T}(\omega t)$ is the time function.

The structure of Ψ-functions (4.3) allows decomposing the spatial equation (4.2) into three equations:

$$\frac{d^2R}{d\rho^2} + \frac{2}{\rho}\frac{dR}{d\rho} + \left(1 - \frac{l(l+1)}{\rho^2}\right)R = 0\,, \tag{4.4}$$

$$\frac{d^2\Theta_{l,m}}{d\theta^2}+ctg\theta\frac{d\Theta_{l,m}}{d\theta}+\left(l(l+1)-\frac{m^2}{\sin^2\theta}\right)\Theta_{l,m}=0\,, \tag{4.4a}$$

$$\frac{d^2\Phi_m}{d\varphi^2}=-m^2\Phi_m\,, \tag{4.4b}$$

where $\rho = kr$ is the radial argument and $k=\frac{2\pi}{\lambda}$ is the wave radial number of the atomic space. The azimuth function has the form

$$\Phi_m(\varphi)=e^{\pm i(m\varphi+\varphi_0)}\,. \tag{4.4c}$$

Here, φ_0 is the initial phase, which is defined on the basis of the azimuth structure of a microobject.

Table 3.1. Polar-azimuth functions

L	m	$\Theta_{l,m}(\theta)\Phi_m(\varphi)$
0	0	1
1	0	$\cos\theta$
	±1	$\sin\theta \exp(\pm i\varphi)$
2	0	$\cos^2\theta$ - 1/3
	±1	$\sin\theta \cos\theta \exp(\pm i\varphi)$
	±2	$\sin^2\theta \exp(\pm 2i\varphi)$
3	0	$\cos\theta (\cos^2\theta$ - 3/5)
	±1	$\sin\theta (\cos^2\theta$ - 1/5) $\exp(\pm i\varphi)$
	±2	$\sin^2\theta \cos\theta \exp(\pm 2i\varphi)$
	±3	$\sin^3\theta \exp(\pm 3i\varphi)$
4	0	$\cos^4\theta$ - 6/7 $\cos^2\theta$ + 3/35
	±1	$\sin\theta \cos\theta (\cos^2\theta$ - 3/7) $\exp(\pm i\varphi)$
	±2	$\sin^2\theta (\cos^2\theta$ - 1/7) $\exp(\pm 2i\varphi)$
	±3	$\sin^3\theta \cos\theta \exp(\pm 3i\varphi)$
	±4	$\sin^4\theta \exp(\pm 4i\varphi)$
5	0	$\cos\theta (\cos^4\theta$ - 10/9 $\cos^2\theta$ + 5/21)
	±1	$\sin\theta (\cos^4\theta$ - 2/3 $\cos^2\theta$ + 1/21) $\exp(\pm i\varphi)$
	±2	$\sin^2\theta \cos\theta (\cos^2\theta$ - 1/3) $\exp(\pm 2i\varphi)$
	±3	$\sin^3\theta (\cos^2\theta$ - 1/9) $\exp(\pm 3i\varphi)$
	±4	$\sin^4\theta \cos\theta \exp(\pm 4i\varphi)$
	±5	$\sin^5\theta \exp(\pm 5i\varphi)$

The simplest polar-azimuth functions, within the constant factor, are presented in Table 3.1. The initial phase φ_0, for simplicity, is omitted.

Solutions of the wave equation (4.2) differ from the solutions of Schrödinger's equation because of their radial components, which are different, in principle. Elementary radial solutions of the atomic space do not diverge. They are presented by Bessel spherical functions, well known in the theory of wave fields. Only two wave numbers, l and m, characterize elementary functions of wave atomic space.

In the simplest case of the spherical field, elementary radial solutions are described by the radial functions of half-integer order:

$$\hat{R}_l(\rho) = \frac{A\hat{e}_l(\rho)}{\rho}, \tag{4.5}$$

where

$$\hat{e}_l(\rho) = \sqrt{\frac{\pi\rho}{2}} H^{\pm}_{l+\frac{1}{2}}(\rho) = \sqrt{\frac{\pi\rho}{2}} \left(J_{l+\frac{1}{2}}(\rho) \pm iN_{l+\frac{1}{2}}(\rho)\right) \tag{4.5a}$$

is the spherical "exponent".

Considering the solutions of the spherical field (4.5), one should keep in mind that the *argument* ρ *can take values within the interval from* $\rho_1 = kr_0$ *to* $\rho = \infty$. *The radial parameter* r_0 *is the radius of the sphere-shell, separating the proper space of H-atom (with its "atmosphere") from the surrounding field-space of matter. Thus, H-atom is regarded as the atom of the field of matter-space-time. Its shell of the radius* r_0 *is the boundary shell of the wave atomic space, from below, while the upper boundary shell is boundless.*

At $l = 0$, we obtain a simplest solution

$$\psi_{0,0}(\rho) = \frac{A\hat{e}_0(\rho)}{\rho} = \frac{A\sqrt{\frac{\pi\rho}{2}} H^{+}_{\frac{1}{2}}(\rho)}{\rho} = \frac{A(\sin\rho + i\cos\rho)}{\rho}, \tag{4.6}$$

where

$$\operatorname{Re}\psi_{0,0}(\rho) = \frac{A\sin\rho}{\rho} \tag{4.6a}$$

is the *potential displacement* and

$$\mathrm{Im}\,\psi_{0,0}(\rho) = \frac{A\cos\rho}{\rho} \tag{4.6b}$$

is the *kinetic displacement*. The condition

$$\mathrm{Re}\,\psi_{0,0}(kr) = \frac{A\sin kr}{kr} = 0 \tag{4.7}$$

defines the radii of potential spheres (shells), situated from each other at the distance of a radial half-wave,

$$kr = n\pi \text{ or } r = n\frac{\lambda_r}{2}, \tag{4.8}$$

and the radii of potential spheres, multiple to one wave,

$$kr = 2\pi n \text{ or } r = n\lambda_r. \tag{4.8a}$$

On the boundary shell, the condition (4.8) takes the form

$$r_0 = \frac{\lambda_r}{2} \text{ or } \lambda_r = 2r_0. \tag{4.8b}$$

Hence, the radii of stationary shells turn out to be multiple to the radius of the boundary shell,

$$r = n\frac{\lambda_r}{2} = r_0 n. \tag{4.9}$$

A radial wave of the boundary shell defines the azimuth wave of the shell,

$$\lambda_0 = 2\pi\lambda_r = 4\pi r_0. \tag{4.10}$$

Thus, the elementary wave λ_0 is the wave of the fundamental tone. Similar to the case of the wave field of a string, only a half-wave of the fundamental tone is placed on the electron orbit, and the electron is in the node of the wave.

The condition (4.9) defines the spherical wave field with the uniformly disposed potential spheres.

It should be noted, in this connection, that other solutions of the equation (4.5), unrelated to *H*-atom, with wave numbers $l \neq 0$ (we will discuss it in other chapters), define the discrete nodal structure of the others atoms. From these solutions it follows that the spherical space of the atoms is nonuniform. Roots of the Bessel functions determine radii of their stationary nodal spheres:

$$\mathrm{Re}\,\Psi = \sqrt{\frac{\pi\rho}{2}} J_{l+\frac{1}{2}}(\rho) = 0. \quad (4.11)$$

If we will denote the roots of the order $\nu = l + \frac{1}{2}$ by $z_{\nu,s}$, where *s* is the number of the root, then we have

$$kr = z_{\nu,s} \text{ or } r = \frac{z_{\nu,s}}{\pi}\frac{\lambda_r}{2}. \quad (4.12)$$

Under $kr >> 1$, the condition (4.11) is substituted for the approximate equality

$$\mathrm{Re}\,\Psi = \sqrt{\frac{\pi\rho}{2}} \cdot \sqrt{\frac{2}{\pi\rho}} \cos\left(\rho - \frac{\nu\pi}{2} - \frac{\pi}{4}\right) = 0 \quad (4.13)$$

and

$$kr - \frac{\nu\pi}{2} - \frac{\pi}{4} = \frac{\pi}{2}(2s-1) \text{ or } r = \left(\frac{l}{2} + s\right)\frac{\lambda_r}{2}. \quad (4.14)$$

For $l << s$, we have

$$r = s\frac{\lambda_r}{2}. \quad (4.15)$$

Thus, for $l \neq 0$, the spherical wave field-space approaches to the uniform one only at large distances.

Let us return again to the solutions for *H*-atom ($l = 0$). According to Eq. (4.6), Ψ-function of the spherical field of *H*-atom has the form

$$\Psi^{+} = \psi_{0,0}(\rho)e^{i\omega t} = \frac{iAe^{i(\omega t - \rho)}}{\rho} = \frac{iAe^{i(\omega t - kr)}}{kr}, \quad (4.16)$$

where the superscript "+" of the wave function shows that the traveling spherical wave moves in the positive direction, in which it spreads in all directions from H-atom. The corresponding convergent wave has the structure

$$\Psi^- = \psi_{0,0}(\rho)e^{i\omega t} = \frac{iAe^{i(\omega t+\rho)}}{\rho} = \frac{iAe^{i(\omega t+kr)}}{kr}. \tag{4.16a}$$

The sum of waves, (4.16) and (4.16a), defines the standing wave, in space, with respect to time:

$$\Psi = \Psi^+ + \Psi^- = \frac{2iA\cos\rho \cdot e^{i\omega t}}{\rho} = \frac{2iA\cos kr \cdot e^{i\omega t}}{kr} = \frac{ia\cos kr \cdot e^{i\omega t}}{kr}, \tag{4.17}$$

where $a = 2A$ is the amplitude of the standing wave.

At the potential shells ($kr = n\pi$), the kinetic and potential amplitudes are equal, correspondingly, to

$$a_m = \frac{ia}{kr}, \; a_m = -\frac{a}{kr}. \tag{4.18}$$

Within the signs of directions of displacements, ±, and the imaginary unit i (the indicator of belonging of displacements to the kinetic field), the values (modules) of potential and kinetic amplitudes are equal to the same numerical measure

$$a_m = \frac{a}{kr}. \tag{4.18a}$$

Sometimes, when only these will interest us, we will operate only with modules of amplitudes. For the intratomic field, $k = \omega / \upsilon_0$ and the amplitude of displacement in the spherical field of Ψ-function has the form

$$a_m = \frac{\mathrm{v}_0 a}{\omega r}. \tag{4.19}$$

The amplitude of displacement defines the amplitude kinetic and potential speeds

$$\mathrm{v}_m = \frac{i\omega i a}{\rho} = -\frac{\mathrm{v}_0 a}{r}, \; \mathrm{v}_m = \frac{i\omega a}{\rho} = \frac{i\mathrm{v}_0 a}{r}. \tag{4.20}$$

The sign "–" indicates the direction of motion at the instant, when the phase of the wave is equal to zero, $\omega t - kr = 0$; therefore, it does not matter and can be omitted for the sake of simplicity.

In the spherical wave field, the amplitude elementary momentum of a particle of mass m is

$$p_m = m\mathrm{v}_m = \frac{m\mathrm{v}_0 a}{r}. \tag{4.21}$$

From this we find the elementary wave action

$$\hbar = p_m r = m\mathrm{v}_m r = m\mathrm{v}_0 a. \tag{4.22}$$

If $a = r_0$ and υ_0 is the speed on the first Bohr orbit, *then this action is Planck's action.* The radial action (4.22) defines the azimuth action

$$h = 2\pi\hbar = p_m 2\pi r = m\mathrm{v}_m 2\pi r = 2\pi m\mathrm{v}_0 a. \tag{4.22a}$$

As we see, the *elementary actions in the spherical field are the constant parameters of the field.*

According to the equations (4.10) and (4.22a), the wave action of the wave of the fundamental tone h_λ will be presented as

$$h_\lambda = 4\pi\hbar = 4\pi m\mathrm{v}_0 a. \tag{4.23}$$

Obviously, azimuth and wave actions of the wave of fundamental tone are related as

$$h = \frac{1}{2}h_\lambda = 2\pi m\mathrm{v}_0 r_0. \tag{4.23a}$$

The elementary momentum (4.21) allows determining the potential energy of an arbitrary particle of this field with the mass m:

$$E = \frac{p_m^2}{2m} = \frac{m\mathrm{v}_m^2}{2} = -\frac{ma^2\mathrm{v}_0^2}{2r^2}. \tag{4.24}$$

According to the condition (4.8), energies of particles on the potential spheres are

$$E = \frac{p_m^2}{2m} = -\frac{2ma^2 v_0^2}{\lambda_r^2}\frac{1}{n^2}, \tag{4.24a}$$

where $a = 2A$ is the amplitude in the standing wave. The factor 2, in the denominator, shows that under other equal conditions the energy is half as much the energy of circular motion, the motion with two degrees of freedom resulted in the superposition of two mutually orthogonal potential-kinetic waves.

In the traveling wave, a distance between two sequential potential shells, in a simplest case, is equal to the amplitude of oscillations *a*. On the other hand, this distance is equal to a half-wave, hence,

$$a = 2A = \lambda_r / 2. \tag{4.25}$$

In particular, on the boundary shell, we have

$$r_0 = a = 2A = \lambda_r / 2. \tag{4.25a}$$

Taking into account the equation (4.25), the expression (4.24a) can be rewritten as

$$E = \frac{p_m^2}{2m} = \frac{mv_m^2}{2} = -\frac{mv_0^2}{2}\frac{1}{n^2}. \tag{4.26}$$

4.2. The Cylindrical Wave Field of H-Atom

An electron, moving along the wave orbit, represents by itself the *cylindrical wave field*. Therefore, its motion should be described by the wave equation (4.1) and for the cylindrical wave field-space.

We will suppose that the plane of the orbiting electron is perpendicular to *Z*-axis, then Ψ-function can be presented as

$$\Psi_l = \psi_l(\rho, \varphi, z)\hat{T}(\omega t) = R_l(\rho)\Phi_l(\varphi)\hat{Z}(k_z z)\hat{T}(\omega t). \tag{4.27}$$

In the cylindrical field, the wave equation (4.2) falls into the following equations:

$$\frac{d^2 R}{d\rho^2} + \frac{1}{\rho}\frac{dR}{d\rho} + \left(1 - \frac{l^2}{\rho^2}\right)R = 0, \tag{4.28}$$

$$\frac{d^2\Phi}{d\varphi^2}=-m^2\Phi\,,\ \frac{d^2\hat{Z}}{dz^2}=-k_z^2\hat{Z}\,,\ \frac{d^2\hat{T}}{dt^2}=-\omega^2\hat{T}\,,\tag{4.29}$$

where k_z is the wave number of the wave, continuing along the Z-axis.

Thus, the elementary space component can be presented by the following space-wave

$$\psi_l(\rho,\varphi,z)=R_l(\rho)e^{-il\varphi}e^{-ik_z z}\,.\tag{4.30}$$

Simplest solutions of the radial equation (4.28) are

$$R_l(\rho)=A\sqrt{\frac{\pi}{2}}(J_l(\rho)\pm iN_l(\rho))\,.\tag{4.31}$$

As a result, the cylindrical wave function takes the form

$$\Psi_l=A\sqrt{\frac{\pi}{2}}(J_l(\rho)\pm iN_l(\rho))e^{-i(l\varphi+\varphi_0)}e^{-ik_z z}e^{i\omega t}\,,\tag{4.32}$$

where φ_0 is the initial phase of the azimuth wave.

In the cylindrical field, the order of the radial function l defines the number of waves, which are placed on the orbit. In a simplest case, one electron, as the node of the wave, shows that only half-wave is placed on its orbit. So that such an orbit will be described by the function of the order $l=\frac{1}{2}$. As a solution, we choose the following function

$$\Psi_{\frac{1}{2}}=A\sqrt{\frac{\pi}{2}}(J_{\frac{1}{2}}(\rho)+iN_{\frac{1}{2}}(\rho))e^{-i(\frac{1}{2}\varphi+\varphi_0)}e^{-ik_z z}e^{i\omega t}\tag{4.33}$$

or

$$\Psi^+_{\frac{1}{2}}=Ai\frac{e^{i(\omega t-kr)}}{\sqrt{kr}}e^{-i(\frac{1}{2}\varphi+\varphi_0)}e^{-ik_z z}\,,\tag{4.34}$$

where φ_0 is the initial phase of the azimuth component of the radial divergent wave, defined on the basis of the boundary conditions. Naturally, the "radial divergent wave" is not the full name of the wave, because it represents the wave structure of radial, azimuth, and axial waves-spaces. The axial wave, represented by the function (4.34), propagates along *Z*-axis in the positive direction. The convergent radial wave $\Psi^-_{1/2}$ corresponds to the divergent one,

$$\Psi^-_{1/2} = Ai\frac{e^{i(\omega t+kr)}}{\sqrt{kr}}e^{-i(1/2\varphi+\varphi_0)}e^{-ik_z z} \ . \tag{4.35}$$

Both waves form *the dynamic stationary wave field in the radial direction*, expressed mathematically by the standing radial wave:

$$\Psi_{1/2} = \Psi^+_{1/2} + \Psi^-_{1/2} = ia\frac{\cos kr \cdot e^{i\omega t}}{\sqrt{kr}}e^{-i(1/2\varphi+\varphi_0)}e^{-ik_z z} \ . \tag{4.36}$$

Simultaneously, the $\Psi_{1/2}$-wave is the traveling wave in the azimuth and axial directions, positive with respect to the Z-axis:

$$\left(\Psi_{1/2}\right)^+ = ia\frac{\cos kr}{\sqrt{kr}}e^{-i(1/2\varphi+\varphi_0)}e^{i(\omega t-k_z z)} \ . \tag{4.37}$$

The corresponding wave, traveling in the negative direction, is

$$\left(\Psi_{1/2}\right)^- = ia\frac{\cos kr}{\sqrt{kr}}e^{i(1/2\varphi+\varphi_0)}e^{i(\omega t+k_z z)} \ . \tag{4.37a}$$

Both waves form the standing wave in the radial and axial directions:

$$\Psi_{1/2} = ia\frac{\cos kr}{\sqrt{kr}}e^{i(1/2\varphi+\varphi_0)}\cos k_z z \cdot e^{i\omega t} \ . \tag{4.38}$$

However, in the azimuth direction, it is the traveling wave along the electron orbit. If we are uninterested in the description of the axial wave, we can omit the axial component and to consider only the radial-azimuth subspace:

$$\Psi_{1/2} = ia\frac{\cos kr}{\sqrt{kr}}e^{i(1/2\varphi+\varphi_0)} \cdot e^{i\omega t} \ . \tag{4.39}$$

Returning back to the traveling wave (4.34), let us write it as

$$\Psi^{+}_{1/2} = Ai\frac{e^{-i(kr+k_z z+\frac{1}{2}\varphi+\varphi_0)}e^{i\omega t}}{\sqrt{kr}}. \quad (4.40)$$

A *surface of the equal phase* of the traveling wave is defined by the equality

$$\phi = kr + k_z z + \tfrac{1}{2}\varphi + \varphi_0 = const \; . \quad (4.41)$$

If $r = const$, surfaces of the equal phase represent by themselves the spiral lines with the parameters:

$$k_z \frac{dz}{dt} = -\frac{1}{2}\frac{d\varphi}{dt} = -\frac{\omega_e}{2} = -\frac{2\pi}{2T_e} = -\frac{2\pi}{T_\varphi} = -\omega_\varphi \,, \quad (4.42)$$

where ω_e and T_e are the circular frequency and period of the electron's orbiting, ω_φ is the circular frequency of the azimuth wave of the fundamental tone and $T_\varphi = 2T_e$ is its period. Thus, we have

$$k_z v_z = -\omega_\varphi = -k_\varphi v_\varphi \text{ and } T_z = T_\varphi \,, \quad (4.42a)$$

where υ_z and υ_φ are, correspondingly, the axial and azimuth wave speeds. If the azimuth speed is negative, the clockwise motion occurs and the axial speed is directed in the positive direction along *Z*-axis.

At that, the wave radius $\bar{\lambda}_z$ defines the axial wavelength:

$$\lambda_z = 2\pi\bar{\lambda}_z = -v_z T_\varphi \, . \quad (4.43)$$

Such a wave should be called, strictly speaking, the *azimuth axial wave* because it is determined by the azimuth period. Because the azimuth axial wave (4.43) is related to two orbital turns, it is equal to the double wave screw pitch.

Under $z = const$, surfaces of the constant phase (4.41) are spiral and the following equality is valid:

$$k\frac{dr}{dt} = -\omega_\varphi \, . \quad (4.44)$$

At the cosmic levels, these are represented by the spiral galaxies, and in the microworld, they express the same wave structure.

When $r = const$ and $z = const$, the *constant phase* of the spatial-time wave $\phi = \omega t - kr - k_z z - \frac{1}{2}\varphi - \varphi_0 = const$, at the circular orbit, is connected with the frequency of the fundamental tone by the equality: $\omega dt - \frac{1}{2}d\varphi = 0$. From this we arrive at

$$\omega = \omega_\varphi . \tag{4.45}$$

During the one wave period of the fundamental tone, the electron twice runs the azimuth orbit, therefore, $\Delta\varphi = \dfrac{2\pi}{l} = 4\pi$, where $l = m = \frac{1}{2}$ is the order of the function and m is the azimuth number. When the order of the wave function is equal to $l = m = p$, the constant phase of the azimuth wave, at $r = const$ and $z = const$, is represented by the equality

$$\phi = \omega t - kr - k_z z - p\varphi - \varphi_0 = const . \tag{4.46}$$

Then, during the one wave period, the electron's azimuth displacement will be of

$$\Delta\varphi = 2\pi / p . \tag{4.47}$$

The amplitude of Ψ-function of displacement, in the cylindrical electron wave, obviously, takes the following form

$$A_m = A / \sqrt{\rho} . \tag{4.48}$$

In such a case, the amplitude speed of the wave will be equal to

$$\mathrm{v}_m = \omega A_m = \frac{\omega A}{\sqrt{\rho}} . \tag{4.49}$$

Because the circular motion is the sum of two mutually perpendicular potential-kinetic waves, hence, like the case the string (see Eq. (2.61)), the amplitude energy of an orbiting electron is

$$E = m\mathrm{v}_m^2 = m\omega^2 A_m^2 = \frac{m\omega^2 A^2}{kr} = \frac{m\omega^2 A^2}{\dfrac{\omega}{\mathrm{v}_0} r} = \frac{mA^2 \mathrm{v}_0}{r}\omega = \hbar_e \omega . \tag{4.50}$$

As we see, the electron energy is proportional to the *circular frequency*, i.e., to the *time wave number* of the time field of the orbit, ω. We will rewrite it as

$$E = mv_m^2 = \hbar_e \omega = h_e \nu = h_e \frac{v_0}{\lambda_e}, \tag{4.51}$$

where λ_e is the electron wave of *H*-atom space,

$$\hbar_e = \frac{mA^2 v_0}{r} \text{ and } h_e = \frac{2\pi m A^2 v_0}{r} \tag{4.51a}$$

are the orbital, radial and azimuth, electron actions. In the space of the stationary field of standing waves, we have the same (4.51) relations, however, here

$$\hbar_e = \frac{ma^2 v_0}{r} \text{ and } h_e = \frac{2\pi m a^2 v_0}{r}. \tag{4.52}$$

On the other hand, the electron is in the spherical field of *H*-atom, where its action is the constant value. Hence, at $r = r_0$, $a = r_0$ (see Eq. (4.25a)) and we have

$$\hbar_e = m v_0 a. \tag{4.53}$$

The wave atomic space, with the wave frequency $\omega = \upsilon_0 / \lambda\!\!\!^{-}_e$, induces outside the atomic space the external waves of matter-space-time of the same frequency, but with the speed c and wavelength λ, so that

$$\omega = \frac{\upsilon_0}{\lambda\!\!\!^{-}_e} = \frac{c}{\lambda\!\!\!^{-}}. \tag{4.54}$$

Therefore, the electron energy can be presented also as (see Eq. (4.23a))

$$E = \hbar_e \omega = h_e \nu = h_e \frac{c}{\lambda} = \frac{1}{2} h_\lambda \frac{c}{\lambda}. \tag{4.55}$$

With that, the electron's wave energy is equal its kinetic energy on the orbit:

$$E = \hbar_e \omega = m \upsilon_0 r_0 \omega = m \upsilon r \omega = \frac{1}{2} m \upsilon r \omega_{orb} = \frac{1}{2} m \upsilon^2, \tag{4.55a}$$

where $\omega = (1/2)\omega_{orb}$ is the circular wave frequency of the fundamental tone and ω_{orb} is the circular frequency of electron's revolution along the orbit, for which $\upsilon = r\omega_{orb}$.

Consequently, the energies $E = h\nu$ and the above-considered actions h, has no relation to the mystic massless quanta of Einstein's pure energy (which move in "vacuum" with light speed and have the wave features). These are the fundamental parameters of the electron orbit and electron itself and, hence, of the H-atom on the whole.

How can a mystic photon have physical properties, if its expansion (size), at least in the direction of the propagation of the wave beam, is equal to zero (according to the "great relativity theory")? The mystic photon is an object being outside of real space. It is a fruit of a headlong fantasy of abstractionism and machism, the fig-leaf covering the mechanical model of the wave field (proposed by Einstein) and nothing more.

Now, we will turn to the Ψ-function of the electron. Let us assume that the electron orbit is in the plane $z = 0$. Because the electron is the node of the wave orbit, hence, the boundary orbital conditions at the instant $t = 0$ must express the equality to zero of potential azimuth displacements in the node during one revolution:

$$\operatorname{Re} e^{-i(\frac{1}{2}\varphi+\varphi_0)}\Big|_{\varphi=0} = \operatorname{Re} e^{-i(\frac{1}{2}\varphi+\varphi_0)}\Big|_{\varphi=2\pi} = 0 . \tag{4.56}$$

These conditions are realized for the traveling electron wave in the positive direction if, e.g., $\varphi_0 = \pi/2$. In such a case, Ψ-function of the electron takes the form:

$$\Psi^{+}_{\frac{1}{2}} = Ai\frac{e^{i(\omega t-kr)}}{\sqrt{kr}}e^{-i(\frac{1}{2}\varphi+\frac{\pi}{2})}e^{-ik_z z} \tag{4.57}$$

or $$\Psi_{\frac{1}{2}} = \frac{A}{\sqrt{\rho}}e^{i\frac{1}{2}\varphi}e^{ik_z z}e^{i(\omega t-\rho)} . \tag{4.57a}$$

the function (4.57) describes the wave of the fundamental tone of the electron λ_e. Its length, as was revealed earlier, is equal to the doubled length of the electron orbit of the Bohr radius r_0:

$$\lambda_e = 4\pi r_0 . \tag{4.58}$$

The wave motion of the fundamental tone occurs in the nearest layers of the wave atmosphere of the *H*-atom, almost at its surface (we will discuss it later).

The equilibrium wave interchange of energy takes place between the *H*-atom and the surrounding field of matter-space-time. However, under the perturbations, the electron wave (4.58) can reiterate itself in the cosmic wave of the same frequency (see Eq. (4.54):

$$\lambda_c = \frac{4\pi r_0}{v_0} c . \tag{4.59}$$

The inverse quantity of this wave is the Rydberg constant:

$$R = \frac{1}{\lambda_c} = \frac{v_0}{4\pi r_0} \frac{1}{c} = \frac{1}{T_0 c} . \tag{4.60}$$

This formula has been obtained earlier at an analysis of oscillations of a string and we arrived at it again on the basis of the accurate wave calculation of the *H*-atom's space.

Let us turn now to the energetic calculation of electron transitions in the *H*-atom. The electron realizes the transitions of the *H*-atom from the *n*-th into *m*-th energetic state; it is the wave motion with the energy of transition (4.55). The law of conservation of energy, at such an extremely fast "quantum" transition, can be presented by the equality:

$$E_m + h\frac{c}{\lambda} = E_n . \tag{4.61}$$

Taking into account the potential energy of the electron in the spherical field of the *H*-atom (4.26), we obtain the following equation of the energetic balance:

$$h\frac{c}{\lambda} = E_n - E_m = \frac{mv_0^2}{2}\left(\frac{1}{m^2} - \frac{1}{n^2}\right). \tag{4.62}$$

Hence, we have $\frac{1}{\lambda} = \frac{mv_0^2}{2hc}\left(\frac{1}{m^2} - \frac{1}{n^2}\right).$ (4.63)

Thus, in the strict correspondence with the wave theory, without any hypotheses and formal fittings, we arrive at the spectral formula of *H*-atom (4.63) and the Rydberg constant

$$R = \frac{mv_0^2}{2hc} = \frac{v_0}{4\pi r_0 c} . \tag{4.64}$$

Note once more, in conclusion, the wave solutions for $l \neq 0$ do not concern the *H*-atom. They describe the shell structure of the others atoms of Mendeleev's periodic table.

5. HYDROGEN ATOM BY SCHRÖDINGER'S EQUATION

Let us remember some facts from history. E. Schrödinger published his "*For Lectures on Wave Mechanics*" in 1928. Three years later, H.E. White presented cross-sections of polar components of the Ψ-function[6]. Up to that time, scientific works in the decadence spirit were widespread. For example, the Pauli exclusion principle (a result of a free game of notions) was put forward in 1925. A whole series of theoretical constructions was carried out on the basis of the Rutherford-Thomson nuclear model, where the key role was attributed to electrons, etc. In these circumstances, Schrödinger was, to a definite degree, a hostage of the pressure of physical-mathematical decadence, which widely publicized quantum mechanics, as the highest achievement of human thought.

In this connection, it is appropriate to quote some discourses of *Joseph Clapper*, a specialist on free information[7]. He has stated that there is the sphere, where the mass media is extremely effective with the opinions of new problems. Under "new problems", he meant the problems about which neither an individual nor his friends (or members of the same group) have no opinion. The reason for the efficiency of the mass media, in the creation of views on new problems, is obvious. An individual is unable to protect oneself and, hence, information moves to unprotected ground. And, as soon as the opinion was formed, it becomes a new opinion, which can be strengthened easily, but changed in a very difficult way. Especially, such a process of formation of a definite view is the most efficient, when a man has no other sources of information. Thus, he depends on the mass media, accessible for him, yet more.

In 1933, *Max Born* wrote in his book on atomic physics[8] (All presented here and further quotations were taken from Russian editions of references): "Primarily, Schrödinger undertook an attempt to regard the corpuscles, in particular electrons, as wave packets. Although Schrödinger's formulae are quite correct, his interpretation could not hold out against the critique. Actually, on the one hand, the wave packets run throughout the course of time. However, on the other hand, the description of the interaction between two electrons, as the interaction of two wave packets in the ordinary three-dimensional space, encounters enormous difficulties.

The current accepted interpretation was proposed by Born. From the points of this interpretation, the probabilistic laws define a whole course of events in a physical system. A probability, defined by de Broglie's wave (associated with the state of a particle), corresponds to a location of a particle in space. Thus, the mechanical process is conjugated with the wave process, i.e., with the propagation of a probabilistic wave. This wave, submitted to Schrödinger's equation, defines the probability of any variant of course of

---

[6] H.E. White, *Phys. Rev.*, 37, 1416, 1931.

[7] *Modern Communication and Foreign Policy*, Washington, 1967, pp. 60-61.

[8] Max Born, *Atomic Physics*, Blackie and Son Limited, London-Glasgow, seventh edition, 1963; Mir, Moscow, 1965.

events in the mechanical frame of references. If for example, a wave of probability has the zero amplitude, in a point of space, it means that the probability to find an electron in this point is disappearingly small" (page 117).

Furthermore, on pages 172-173, we find: "In conclusion, we should consider once again the sense of the wave function itself... It should be expected that in wave mechanics as well, the wave function ψ, or rather its modulus squared, must gain in importance, because an instant value itself of the oscillating function, of course, cannot play a role because of the high frequency of oscillations. *The reason for taking the square of the modulus is that the wave function itself (because of the imaginary coefficient of the time derivative in the differential equation) is a complex quantity, while quantities susceptible of physical interpretation must of course be real*". (Underlined by the authors)

These judgements call up Newton's and Euler's views. *L. Euler*, in his "*Algebra*" (1770), has asserted: "*Square roots of negative numbers* are not equal to zero, are not less than zero, and are not greater than zero. From this it is clear that the square roots of negative numbers cannot be among the possible (actual, real) numbers. Hence, we have no another way except to acknowledge these numbers as impossible ones. This leads us to the notion of numbers, impossible in essence, which are usually called *imaginary (fictitious)* numbers, because they exist only in our imagination."

In connection to the statistical interpretation of the wave function, M. Born has continued (pages 174-175): "We speak often about the distribution of *electron density* in the atom or about the *electron cloud* around the nucleus. We mean by these words the *charge distribution*, which is obtained under the multiplication of the function of probability $|\psi_n|^2$ by the electron charge e. According to the statistical interpretation, the sense of this distribution is clear: in the photos of Figure 16, it is shown how this distribution can depict. The image represents projections (shadows) of electron clouds in different states."

A.K. Timiryazev gave the critique estimation to quantum mechanical concepts, which are based on positivism. His book[9] contains a critical analysis of the status quo in physics at that time (which has the actual meaning so far), concerning the acceptation of the statistical approach to the description of the atomic structure. Below are some passages from it.

"Beginning from 1923, the Bohr theory experienced deep crisis. Great difficulties have appeared during attempts to extend the theory ... for other atoms... Further, vast obstacles have arisen when it appeared necessary to calculate time, during the course of which the electron radiates energy, jumping from one orbit to another.

[9] A.K. Timiryazev, *Introduction in Theoretical Physics*, State Publishing House, GTTI, Moscow-1933-Leningrad, pp. 334-336.

This crisis was "*resolved*" in 1925, when Heisenberg's paper on the abstract mathematical theory of quanta appeared, where any hints of a model were deliberately removed... Thus, the new theory represents by itself a step back....

It is very characteristic that neither Heisenberg nor Schrödinger... deny the fact of the existence of atoms and electrons. But because of the fear of difficulties, which they encountered in the Bohr Theory, they rejected ideas about motion of electrons in the atom. In new theory, there is nothing about an orbit of an electron...

... Let us cite... the characteristic of Heisenberg's theory given by H. Thirring (1888-1976, Austrian-born physicist)[10]: "Heisenberg's theory represents a naked calculation scheme... The nakedness and unobviousness of this scheme are deliberate and desired. Responding to the question about the appearance of an atom, Heisenberg could have answered approximately in the following way: 'fool, your question is senseless like the question of a child: was the infant Christ a boy or a girl?' An atom is a thing, which should not look like something, in such a degree, similar to...who has signs of the sex and nationality. Atomic features are exhibited for our world of senses through the spectra, energetic levels, which are found under the excitation of atoms. And these quantities are uniquely defined by the laws of quantum mechanics".

...It is interesting to note that Professor Sommerfeld, assuming that new theory of quanta represents a great achievement, nevertheless, very unambiguously speaks about its dark sides... Sommerfeld's views are so interesting that we cite their here completely[11]: "The clearly expressed intention, which was in the first Heisenberg work on quantum mechanics, is the development of the method, which will be based exceptionally on the relations between the observable values in principle. Such notions as, for example, the location of an electron, the time of its revolution, and the form of the path of its motion, must be rejected from consideration. This limitation, in framework of only principally observable quantities, is based after all on Mach's philosophy. 30 years ago, leaning directly upon Mach, the popularization of the so-called energetics took place. Energetics has proposed to recognize only values of energy in the capacity of physically observed quantities".

Lois de Broglie wrote, in the Preface to D. Bohm's book[12]: "Classical physics has allowed the description of the evolution of events in nature, as causally unfolded in space and time ... and, consequently, the creation of the clear and exact models (for the imagination of a physicist). On the contrary, modern quantum physics forbids any presentation of such kind and makes it impossible. It accepts only such theories, whose foundation is made up of quite abstract formulae, and cast doubts on an idea related to the causal run of atomic and corpuscular phenomena. It recognizes only the laws of probability

---

[10] *Die Grundgedanken der neueren Quantentheorie*, H. Thirring, *Ergebnisse der exakten Naturwissensschaften*, VII Band, 1928, S. 410.

[11] A. Sommerfeld, *Atombau und Spektrallinien. Wellenmechanischer Ergänzungsband*, 1929, 44.

[12] David Bohm, *Causality and Chance in Modern Physics*, L., 1957; IL, Moscow, 1959 (in Russian.

and regards these laws as the laws, having the primary character and forming the final cognizable reality. It does not permit explaining these laws as a result of a causal evolution that occurs at the deeper level of the physical world..."

Lois de Broglie also noted in this work that physicists-theorists "under the influence of preconceived ideas, originated from the positivistic doctrine, assume..., that the indefinite and incomplete character of knowledge about the real processes in physics of the microworld (obtained from the experiment at its modern stage) is a result of the real indeterminacy of physical states and their changes. Hence, such an interpretation, evidently, is not a justified one. Perhaps, in the future, during the process of the cognition of the deeper level of physical reality, we will be able to interpret the probabilistic laws and quantum mechanics as a statistic result of the manifestation of the entire determined values of variables, which, in our time, are hidden".

In the conclusion of the Preface, Lois de Broglie wrote: Professor Bohm showed "that quantum mechanics has no right to regard its own modern views as final. And it is unable to forbid explorers to suppose an existence of the domains of reality, which are deeper than those, already investigated".

Note one more important work by *L.V. de Broglie*[13]. A manuscript of this work, published in 1982, is dated of 1950-1951, i.e., Lois de Broglie decided to publish it only 30 years later. In Preface, "*The evolution of Lois de Broglie's ideas concerning interpretation of wave mechanics*", *de Georges Lochak* (an editor of a French edition) focused attention on de Broglie's doubts, concerning the right understanding of quantum mechanics bases:

"His doubts became apparent through a great number of remarks, glued in between the manuscript pages, corrections, and small but essential additions, which are in the book... A great deal changed during the 30 years, but at that time, an unexpected turn (change) of the views of one of the prominent physicists of the 20th century was sensational and in some degree, even scandalous. In the corridors of Henri Poincaré's Institute, all have spoken about Lois de Broglie in a low voice, as if a terminal disease had suddenly been discovered and as if it should reasonable be aside from him... The debate has grown to a scale of religious wars and have obtained the world character."

After careful research of many problems related to the atomic theories, "he came to the conclusion that the *difficulties of the nuclear theory, like the difficulties of quantum electrodynamics, are unmovable in the framework of generally accepted ideas. These difficulties caused by the fundamental incapacity of all theory, in the large, to describe the space-time structures*. In his stronger growing persuasion, fully developed to 1952, he expressed by the following unequivocal phrase: *'On this day, all possibilities of*

---

[13] L. de Broglie, *Les Incertitudes D'Heisenberg et L'Interprétation Probabiliste de la Mécanique Ondulatoire*, Gauthier-Villars, Bordas, Paris, 1982. (Russian edition: L. de Broglie, *Heisenberg Uncertainty Principle and Probabilistic Interpretation of Wave Mechanics*, Mir, Moscow, 1986)

explanation of phenomena by wave mechanics, in that form, as it teaches, are, in a considerable extent, exhausted'."

Such are the historical estimations of quantum mechanics, as a theory, born on the basis of different philosophical trends of subjective character. Through these trends, one can refer to the different currents of the "philosophy of the blind". The formal mathematical guessing and fitting to the experiment, in the spirit of abstractionism, is the essence of this philosophy. Organs of eyesight and, hence, the right cerebral hemisphere of man are announced as unnecessary ones, even harmful. The left cerebral hemisphere and a free game of notions of learned blinds is accepted as praiseworthy. It is not necessity to prove that a blind man, from his birth, is not able to understand the World, like quantum mechanics, which represent by itself the theory of aggressively blind theorists. At the beginning of the 21st century, it is time to recognize this impartial truth.

It should also be noted that all humankind experience teaches that with the motion "up", in Cosmos, it increases *indeterminacy*. On the contrary, with the motion "down", in the microworld, it increases *definiteness* and *accuracy*, which are verified by modern electron technologies.

After general historical and philosophical remarks, let us turn now to the analysis of Schrödinger's radial functions, which are not of a suit with wave fields. We will begin from *s*-states of *H*-atom.

Schrödinger's radial functions of *s*-states define the electron density of probability, increasing the origin of coordinates. Hence, such solutions, unconditionally, point to the instability of the proton-electron dynamic system, more exactly, the impossibility of an existence of such a system.

About this disagreeable effect of Schrödinger's equation, N. Mott and I. Sneddon have gingerly noted as far back as the year, 1948[14]. They have written that "wave spherical functions of *s*-states are spherically symmetric ones and they *have the maximum in the origin of coordinates*, at $r=0$. That means that *probability of the penetration of the electron in an internal domain of the atom in such a case is too big*" (taken from the Russian version of the reference *).

Two physicists-theorists, Hans A. Bethe and Edwin E. Salpeter, have expressed quite opposite assertions.[15] We present the dependence $r^2 R_{n,l}^2(r)$ on *r*, i.e., the probability that the electron will be found in a spherical layer, between radii *r* and *r+dr*. It is clear that the *maximal charge density accounts for the bigger and bigger radii, according to the increase of the main quantum number n*". Let us discuss the above-quoted polar statements.

Schrödinger's radial functions for *H*-atom are determined by the expression

[14] N. Mott, I. Sneddon, *Wave Mechanics and its Application*, Oxford at the Clarendon Press, 1948; Russian translation by Nauka, Moscow, 1966, p.75.

[15] Hans A. Bethe, Edwin E. Salpeter, *Quantum Mechanics of One- and Two-Electron Atoms*, Springer-Verlag, Berlin–Göttingen–Heidelberg, 1957; Russian translation by Fizmatgiz, Moscow, 1960.

$$R_{n,l} = N_{n,l} e^{-\rho/(\kappa+1+l)} \left(\frac{2\rho}{\kappa+1+l}\right)^l (a_0 + a_1\left(\frac{2\rho}{\kappa+1+l}\right) + ... + a_\kappa \left(\frac{2\rho}{\kappa+1+l}\right)^\kappa), \quad (5.1)$$

where $n = \kappa + l + 1$ is the main quantum number, a_i are constant coefficients of Laguerre polynomial.

All radial functions of *s*-states, for which $l = 0$, have the form

$$R_{n,0} = N_{n,0} e^{-\rho/(\kappa+1)} (a_0 + a_1\left(\frac{2\rho}{\kappa+1}\right) + ... + a_\kappa \left(\frac{2\rho}{\kappa+1}\right)^\kappa). \quad (5.2)$$

The first extremum of the radial functions of *s*-states falls on the origin of coordinates and is equal to

$$(R_{n,0})_{\max} = a_0 N_{n,0}. \quad (5.3)$$

It means that the most probable place of localization of the electron is in the center of nucleus. Then, it is not clear, how the electron pierces the nucleus, whose density (according to the nuclear model) is about $2.5 \cdot 10^{14} g/cm^3$. It is impossible to imagine such a density. Most likely, *something is wrong with the nuclear atomic model.*

Thus, for *s*-states, the density of probability is turned out to be maximal, in the origin of coordinates, and extremely small, on the orbit. So that, the electron must leave the orbit and, penetrating the nucleus, be in its center. How the electron contrives to penetrate the nucleus, of unimaginably huge density, is a baffling question.

Hence, the main effect of solutions of the equation (5.1) is that *s*-states destroy the Bohr postulate on the stability of orbiting of the electron and show that the electron, certainly, will fall dawn on the nucleus and, what is more, permeate into its central part.

Here is a simple example. Under the condition $\kappa = 0$ and $l = 0$, an infinite series of the radial function is cut off at the zero member and the radial function $R_{n,l}(\rho)$ takes the form

$$R_{1,0} = 2e^{-\rho}. \quad (5.4)$$

Its root, $\rho_{1,0,\max 1} = 0$, corresponds to the maximum of this function which results in the infinite value of energy:

$$W = -\frac{e^2}{8\pi\varepsilon_0 a_0 (0)^2} = -\infty. \quad (5.5)$$

The absurdity is evident that does not require comments.

After getting rid of such disagreeable conclusions, the new wave radial function $P_{n,l} = R_{n,l}r$ was invented, frankly in the spirit of a free game of notions. How did they do it? Very simply. The real radial function $R_{n,l}$ was arbitrary multiplied by the radius *r*. Such a mathematical game allowed to "prove" the stability of the *H*-atom "in full agreement with quantum mechanics". Obviously, $P_{n,l} = R_{n,l}r = 0$ is always at $r = 0$.

In accordance with Born's interpretation, the density of probability to find the electron in an arbitrary point of space is defined by the modulus squared of Schrödinger's wave function. In spherical polar coordinates, the wave function has the form

$$\Psi = \psi_{n,l}(\rho,\theta,\varphi)e^{i\omega t}, \tag{5.6}$$

where

$$\psi_{n,l}(\rho,\theta,\varphi) = R_{n,l}(\rho)\Theta_{l,m}(\theta)\Phi_m(\varphi) \tag{5.6a}$$

is the amplitude (spatial) function.

The probability density of location of the electron in an elementary volume *dV* is determined by the differential

$$dw = \psi_{n,l}\psi^*_{n,l}dV = \rho_r\rho_\theta\rho_\varphi dV = \rho_w dV, \tag{5.7}$$

where $\rho_r = R^2_{n,l}$, $\rho_\theta = \Theta^2_{l,m}$, and $\rho_\varphi = |\Phi_m|^2 = \dfrac{1}{2\pi}$ are, correspondingly, the radial, polar, and azimuth components of the probability density; $\rho_w = \rho_r\rho_\theta\rho_\varphi$ is the total hypothetical probability density, which can be presented as

$$\rho_w = \psi_{n,l}\psi^*{}_{n,l} = \frac{1}{2\pi}R^2_{n,l}\Theta^2_{l,m}. \tag{5.8}$$

As follows from the equation (5.7), dimensionality of the wave function Ψ (if the le ngth of space is measured in centimeters) is

$$\dim \Psi_{n,l} = \dim \psi_{n,l} = 1/cm^{3/2}. \tag{5.9}$$

This dimensionality has no physical sense, because no physicist can show in reality the space with the dimensionality $cm^{3/2}$. In this sense, Schrödinger's wave function is a mathematical fantasy, far from reality. However, this drawback can be easily removed if the density of probability is determined according to the following expression:

$$\rho_w = \frac{1}{\Omega_0}\psi_{n,l}\psi^*_{n,l} = \frac{1}{2\pi\Omega_0}R^2_{n,l}\Theta^2_{l,m}, \tag{5.10}$$

where Ω_0 is an elementary volume of space, which is determined on the basis of the wave equation, or some physical conditions. From Schrödinger's equation it follows that

$$\Omega_0 = a^3 = \left(\frac{r_0}{Z}\right)^3, \tag{5.11}$$

where r_0 is Bohr radius, Z is the ordinal number of a hydrogen-like atom.

A hypothetical picture of the most probable electron density is defined by the extremes of the total density of probability in the three-dimensional space. But, because the azimuth component is constant, hence, the most probable distribution is determined, in essence, by two extremes of densities, $R^2_{n,l}$ and $\Theta^2_{l,m}$.

In spherical polar coordinates, an elementary volume $dV = r^2 \sin\theta d\varphi d\theta dr$ allows writing an elementary probability (which represents, strictly speaking, the third order differential) as

$$d^3w = \frac{1}{a^3}\left(\frac{1}{\sqrt{2\pi}}R_{n,l}\Theta_{l,m}\right)^2 r^2 \sin\theta d\varphi d\theta dr = \left(\frac{1}{\sqrt{2\pi}}R_{n,l}\Theta_{l,m}\right)^2 \rho^2 \sin\theta d\varphi d\theta d\rho \tag{5.12}$$

where $\rho = r/a$.

If we require the consideration of the probability within a spherical layer of an interval $(\rho, \rho + d\rho)$, then it is natural to perform the partial integration with respect to φ and θ. It leads to the second order differential

$$dw = \left(\int_0^{2\pi}|\Phi_m|^2 d\varphi \int_0^{\pi}\Theta^2 \sin\theta d\theta\right)R^2_{n,l}\rho^2 d\rho. \tag{5.13}$$

The coefficients of polar and azimuth functions are determined, in quantum mechanics, on the basis of the integrals of normalization of these functions:

$$\int_0^{2\pi} |\Phi_m|^2 d\varphi = 1, \int_0^{\pi} \Theta_{lm}^2 \sin\theta d\theta = 1. \quad (5.14)$$

In virtue of this, the equation (5.13) can be rewritten in the following form

$$dw = R_{n,l}^2 \rho^2 d\rho. \quad (5.15)$$

The coefficient of the radial function is defined by the integral

$$\int_0^{\infty} R_{n,l}^2 \rho^2 d\rho = 1. \quad (5.16)$$

The equation (5.16) allows the manipulation with the notion of the *radial density of probability* by introducing the new radial function $P_{n,l} = R_{n,l}\rho$ and, correspondingly, the pseudo density – the "radial probability":

$$D_{n,l} = P_{n,l}^2 = R_{n,l}^2 \rho^2. \quad (5.17)$$

The pseudo density *D* creates an illusion as if the probability of location of the electron in the origin of coordinates, i.e., in the center of *H*-atom, is equal to zero. Here is an example (taken from the Russian version of the book[16]) of technology of formation of the radial component of pseudo Ψ-function:

"The most probable value *r* is obtained from the expression $P(\mathbf{r},t) = \Psi^*(\mathbf{r},t)\Psi(\mathbf{r},t)$. The probability of location of the electron within a small domain between *r* and *r*+Δ*r* is equal, at $\Delta r \to 0$, to

$$P_{n,l}(r)dr = \lim_{\Delta r \to 0} \left[\int_0^{2\pi}\int_0^{\pi}\int_r^{r+\Delta r} \psi_{n,l,m}^* \psi_{n,l,m} r^2 dr \sin\theta d\theta d\varphi \right] = \eta_{n,l}^2(r) r^2 dr,$$

where $P_{n,l}(r)$ is the *function of the radial distribution* for a hydrogen-like atom (the probability calculated per unit length).... Returning to the expression for the radial distribution of probability density $P_{n,l}(r)$, we can write

[16] W.H. Flygare, *Molecular Structure and Dynamics*, Prentice-Hall, Inc., Englewood Cliffts, New Jersey, 1978; Russian translation by Mir, Moscow, 1982, V.1, pp. 180-181.

$$P_{n,l}(r) = [r\eta_{n,l}(r)][r\eta_{n,l}(r)] = [R_{n,l}(r)]^2 ,$$

where $R_{n,l}(r)$ is determined as the radial amplitude function".

Analyzing the above-presented material, we observe an excellent game of symbols in the required direction. (For definiteness, let us designate the pseudo radial function by the symbol $\widetilde{R}_{n,l}(r)$). At first, the standard symbol of the radial function $R_{n,l}(r)$ is substituted for the nonstandard symbol $\eta_{n,l}(r)$. Then, using the standard symbol of the radial function $R_{n,l}(r)$, the pseudo radial function is introduced in accordance with the equality

$$R_{n,l}(r) = r\eta_{n,l}(r) . \qquad (5.18)$$

Such a logical tightrope walking means only one thing: "the radial function does not suit us because it shows that the orbiting electron is unstable; this is why we substitute the symbol, and the notion itself, of the radial function $R_{n,l}(r)$ according to the following correspondence:

$$R_{n,l}(r) \quad \rightarrow \quad \widetilde{R}_{n,l}(r) = rR_{n,l}(r) \text{''}. \qquad (5.18a)$$

After such manipulations, there is no choice, as only to present the diagrams of the "radial amplitude function $R_{n,l}(r)$", i.e., the pseudo radial function $\widetilde{R}_{n,l}(r)$. But now we have the absolutely other picture of the distribution, where the "probability" to find an electron in the center of a nucleus, in all cases, will be equal to zero.

In actual fact, if we will lean upon the real radial function, we must recognize that at $l = 0$ motion of the electron around the nucleus is, to the higher degree, the unstable one and the electron must "bore" the nucleus and prove to be in its center. However, such a conclusion was undesirable for creators of quantum mechanics.

Thus, removing the troublesome radial function $R_{n,l}(r)$ (or that is the same $R_{n,l}(\rho)$) from the probabilistic description of the electron motion and substituting it for the pseudo radial function $rR_{n,l}(r)$ (or $\rho R_{n,l}(\rho)$), the "*explanation*" of an existence of stationary states of the hydrogen atom, in accordance with Bohr theory, "*turns out well*" in some cases.

Of course, W. Flygare had no relation to the birth of such a mathematical fitting. He plainly tried to substantiate by his own way the evident substitution of Schrödinger's radial function for the pseudo radial function.

On the basis of formulae (1.9), (1.9a), and (5.1), Schrödinger's Ψ-function for *H*-atom can be written as

$$\Psi_{n,l} = \frac{N_{n,l}}{2\pi} e^{-kr/(\kappa+1+l)} \left(\frac{2kr}{\kappa+1+l}\right)^l (a_0 + a_1\left(\frac{2kr}{\kappa+1+l}\right) + \ldots$$
$$\ldots. + a_\kappa \left(\frac{2kr}{\kappa+1+l}\right)^\kappa)\Theta_{l,m}(\theta) e^{-im\varphi} e^{i\omega t}, \tag{5.19}$$

where $k = 1/a$ is the number, inverse to Bohr radius a.

Let the coefficient of the variable r will be presented in the following form:

$$k_r = \frac{ik}{\sigma + l}, \tag{5.19a}$$

where $\sigma = \kappa + 1$ is the number of members of Laguerre polynomial. Then the Laguerre polynomial can be written as

$$f(-2k_r ir) = (a_0 + a_1\left(\frac{2kr}{\kappa+1+l}\right) + \ldots + a_\kappa\left(\frac{2kr}{\kappa+1+l}\right)^\kappa). \tag{5.19b}$$

As a result, Schrödinger's Ψ-function will be presented in the following form

$$\Psi_{n,l} = \frac{N_{n,l}}{2\pi} f(-2k_r ir) e^{i(\omega t + k_r r)} \Theta_{l,m}(\theta) e^{-im\varphi}. \tag{5.20}$$

The radial component of the wave function

$$R_{n,l} = N_{n,l} f(-2k_r ir) e^{i(\omega t + k_r r)} \tag{5.21}$$

represents by itself the wave with the imaginary radial wave number and the amplitude (5.19b). At that the frequency ω is also the "imaginary" one:

$$\omega = k_r v_0 = \frac{ikv_0}{\sigma + l}, \tag{5.21a}$$

where v_0 is the wave speed.

Since the wave number k_r of Schrödinger's space is an imaginary quantity, therefore, Schrödinger's Ψ-functions do not describe the spatial waves; they express the aperiodic perturbations of space independently of an interpretation of Ψ-function.

Accordingly, because of this reason as well, solutions of Schrödinger's equation do not correspond to the real wave picture of the microworld. They exclude the formation of waves of H-atom in the surrounding space that is in full contradiction with reality.

As was already mentioned in Section 1 of this chapter, according to Born's interpretation, the extremes of radial functions $R_{n,l}$ define the radii of shells of the most probable states: $r = a\rho_{n,l,\max i}$, where i is the number of an extremum. However, in an overwhelming number of cases, relative values of radii of extremal shells are not equal to integers squared: $\rho_{n,l,\max i} \neq m^2$ and, hence, these negate the condition of cutting off (1.5). Such shells define the energetic levels, non-existent in nature:

$$W = -\frac{e^2}{8\pi\varepsilon_0 a\rho_{n,l,\max i}}. \quad (5.22)$$

Of all pseudo functions, only pseudo functions of the kind $(R\rho)_{n,n-1} = const\,\exp(-\rho/n)\cdot\rho^{n-1}$ have extremes corresponding to the integers squared: $\rho = n^2$; hence, they "confirm" Bohr's orbits: $r = a_0 n^2$. In this connection, let us consider in detail the radial and pseudo radial solutions. We will denote the radii of shells of extremes of radial and pseudo radial functions, correspondingly, as $\rho_{n,l,\max i}$ and $\rho_{n,l,\max is}$.

The following pseudo radial function $P_{1,0}$ corresponds to the radial function $R_{1,0} = 2e^{-\rho}$:

$$P_{1,0} = 2e^{-\rho}\rho. \quad (5.23)$$

Its root is $\rho_{1,0,\max 1s} = 1$. The energy, corresponding to it,

$$W = -\frac{e^2}{8\pi\varepsilon_0 a_0 (1)^2},$$

satisfies H-atom spectrum. However, the pseudo function itself (5.23) is not a solution of Schrödinger's equation. So that, we have no right to apply its roots to the formula (5.22), which is related to the other (radial) function.

The rest of radial and pseudo radial functions, for $n = 2, 3, 4, 5$, have the form as follows.

$\underline{n = 2}$, $W = -\frac{e^2}{8\pi\varepsilon_0 a_0 2^2}$ *(Energy by Bohr)*

By Schrödinger:

$l=0$, $\kappa=1$: $R_{2,0}=\frac{1}{2\sqrt{2}}e^{-\rho/2}(2-\rho)$:

$\rho_{2,0,\max 1}=0$, $\rho_{2,0,\max 2}=2^2$ → $W=-\frac{e^2}{8\pi\varepsilon_0 a_0(0)^2}=\infty$, $W=-\frac{e^2}{8\pi\varepsilon_0 a_0 2^2}$.

Although the last energy satisfies the cutting off condition, but simultaneously the first root negates it.

$P_{2,0}=\frac{1}{2\sqrt{2}}e^{-\rho/2}(2-\rho)\rho$ (The pseudo radial function):

$\rho_{2,0,\max,s}=0.7639320225\quad 5.2360679774\,99$

→ $W=-\frac{e^2}{8\pi\varepsilon_0 a_0\rho_{2,0,\max,s}}$ → $R=\infty$,

----

$l=1$, $\kappa=0$: $R_{2,1}=\frac{1}{2\sqrt{6}}e^{-\rho/2}\rho$:

$\rho_{2,1,\max 1}=2$ → $W=-\frac{e^2}{8\pi\varepsilon_0 a_0 2}$ → $R=\infty$.

$P_{2,1}=\frac{1}{2\sqrt{6}}e^{-\rho/2}\rho^2$ (The pseudo radial function):

$\rho_{2,1,\max 1s}=4$ → $W=-\frac{e^2}{8\pi\varepsilon_0 a_0 2^2}$. The formal "conformity" with the Bohr energy takes place here (the fitting happens). In this case ($l=1$, $\kappa=0$), the maximum density of the hypothetical probability $R_{2,1}$ is defined by the root $\rho_{2,1,\max 1}=2$, which eliminates the cutting off condition and, hence, the value of the radial function becomes infinite $R=\infty$. This unpleasantness has been "removed" by means of the pseudo radial function $P_{2,1}$ the maximum of which $\rho_{2,1,\max 1s}=2^2$ gives the energy, corresponding formally to the Bohr energy of the electron. But, as was elucidated above, this root has no relation to the radial function and, correspondingly, energy (5.22). This absurdity is inherent for all other functions.

$\underline{n=3}$, $W=-\frac{e^2}{8\pi\varepsilon_0 a_0 3^2}$ *(Energy by Bohr)*

By Schrödinger:

$l=0$, $\kappa=2$: $R_{3,0}=\frac{2}{81\sqrt{3}}e^{-\rho/3}(27-18\rho+2\rho^2)$:

$\rho_{3,0,\max 1}=0 \rightarrow W=-\frac{e^2}{8\pi\varepsilon_0 a_0 \rho_{3,0,\max 1}}=\infty$

$\rho_{3,0,\max 2,3}=3.531370333$, 11.46862697 ; $W=-\frac{e^2}{8\pi\varepsilon_0 a_0 \rho_{3,0,\max}} \rightarrow R=\infty$,

$P_{3,0}=\frac{2}{81\sqrt{3}}e^{-\rho/3}(27-18\rho+2\rho^2)\rho$ (The pseudo radial function):

$\rho_{3,0,\max,s}=0.7400370804 \quad 4.185930091 \quad 13.074032828$

$\rightarrow W=-\frac{e^2}{8\pi\varepsilon_0 a_0 \rho_{3,0,\max,s}} \rightarrow R=\infty$.

$l=1$, $\kappa=1$:

$R_{3,1}=\frac{4}{81\sqrt{6}}e^{-\rho/3}\rho(6-\rho)$,

$\rho_{3,1,\max}=1.75735931 \quad 10.242640687 \rightarrow W=-\frac{e^2}{8\pi\varepsilon_0 a_0 \rho_{3,1,\max}} \rightarrow R=\infty$.

$P_{3,1}=\frac{4}{81\sqrt{6}}e^{-\rho/3}\rho^2(6-\rho)$ (The pseudo radial function):

$\rho_{3,1,\max,s}=2.227998127 \quad 10.7720018726 \rightarrow W=-\frac{e^2}{8\pi\varepsilon_0 a_0 \rho_{3,1,\max,s}} \rightarrow R=\infty$.

$l=2$, $\kappa=0$:

$R_{3,2}=\frac{4}{81\sqrt{30}}e^{-\rho/3}\rho^2$:

$\rho_{3,2,\max 1}=6 \rightarrow W=-\frac{e^2}{8\pi\varepsilon_0 a_0 6} \rightarrow R=\infty$.

$P_{3,2}=\frac{4}{81\sqrt{30}}e^{-\rho/3}\rho^3$ (The pseudo radial function):

$\rho_{3,2,\max 1s}=9 \rightarrow W=-\frac{e^2}{8\pi\varepsilon_0 a_0 3^2}$. The formal "conformity" with Bohr energy takes place – the fitting happens.

$\underline{n=4}$, $W=-\frac{e^2}{8\pi\varepsilon_0 a_0 4^2}$ (*Energy by Bohr*)

By Schrödinger:

$l=0,\ \kappa=3$:

$R_{4,0}=\frac{1}{768}e^{-\rho/4}(192-144\rho+24\rho^2-\rho^3)$:

$\rho_{4,0,\max 1}=0 \rightarrow W=-\frac{e^2}{8\pi\varepsilon_0 a_0 0^2}=\infty$,

$\rho_{4,0,\max 2-4}=3.419589235 \quad 9.90409441 \quad 22.673316353$

$W=-\frac{e^2}{8\pi\varepsilon_0 a_0 \rho_{4,0,\max 2-4}} \rightarrow R=\infty$.

$P_{4,0}=\frac{1}{768}e^{-\rho/4}(192-144\rho+24\rho^2-\rho^3)\rho$ (The pseudo radial function):

$\rho_{4,0,\max 1s}=0.732343789 \quad \rightarrow W=-\frac{e^2}{8\pi\varepsilon_0 a_0 \rho_{4,0,\max 1s}} \rightarrow R=\infty$,

$\rho_{4,0,\max 2s}=4 \rightarrow W=-\frac{e^2}{8\pi\varepsilon_0 a_0 2^2} \rightarrow R=\infty$

(The last energy corresponds to the H-atom spectrum, but it does not satisfy the cut-off condition by the reason which were elucidated above and leads to the divergence of the radial function, $R=\infty$)

$\rho_{4,0,\max 3,4s}=10.649564176 \quad 24.618092034 \rightarrow W=-\frac{e^2}{8\pi\varepsilon_0 a_0 \rho_{4,0,\max 3,4s}} \rightarrow R=\infty$.

$l=1,\ \kappa=2$:

$R_{4,1}=\frac{1}{256\sqrt{15}}e^{-\rho/4}\rho(80-20\rho+\rho^2)$:

$\rho_{4,1,\max}=1.69690610\ ,\ 8.7491349\ ,\ 21.55395899 \rightarrow W=-\frac{e^2}{8\pi\varepsilon_0 a_0 \rho_{4,1,\max}} \rightarrow R=\infty$.

$P_{4,1}=\frac{1}{256\sqrt{15}}e^{-\rho/4}\rho^2(80-20\rho+\rho^2)$ (The pseudo radial function):

$\rho_{4,1,\max,s}=2.8303936591 \quad 9.589180277 \quad 23.580426063$

$\rightarrow W--\frac{e^2}{8\pi\varepsilon_0 a_0 \rho_{4,1,\max,s}} \rightarrow R=\infty$.

$l=2,\ \kappa=1$:

$R_{4,2}=\frac{1}{768\sqrt{5}}e^{-\rho/4}\rho^2(12-\rho)$:

$\rho_{4,2,\max 1,2} = 5.071796769 \quad 18.928203230 \quad \rightarrow W = -\dfrac{e^2}{8\pi\varepsilon_0 a_0 \rho_{4,2,\max 1,2}} \rightarrow R = \infty$.

$P_{4,2} = \dfrac{1}{768\sqrt{5}} e^{-\rho/4} \rho^3 (12-\rho)$ (The pseudo radial function):

$\rho_{4,2,\max, s} = 6.788897449 \quad 21.2111025509 \quad \rightarrow W = -\dfrac{e^2}{8\pi\varepsilon_0 a_0 \rho_{4,2,\max, s}} \rightarrow R = \infty$.

$l = 3, \ \kappa = 1$:

$R_{4,3} = \dfrac{1}{768\sqrt{35}} e^{-\rho/4} \rho^3$:

$\rho_{4,3,\max 1} = 12 \rightarrow W = -\dfrac{e^2}{8\pi\varepsilon_0 a_0 \rho_{4,3,\max 1}} \rightarrow R = \infty$.

$P_{4,3} = \dfrac{1}{768\sqrt{35}} e^{-\rho/4} \rho^4$ (The pseudo radial function):

$\rho_{4,3,\max 1s} = 16 \rightarrow W = -\dfrac{e^2}{8\pi\varepsilon_0 a_0 4^2}$.

The formal "conformity" with Bohr energy takes place here - the fitting has happened)

$\underline{n = 5}, \ W = -\dfrac{e^2}{8\pi\varepsilon_0 a_0 5^2}$ (*Energy by Bohr*)

By Schrödinger:

$l = 0, \ \kappa = 4$: $R_{5,0} = \dfrac{2}{46875\sqrt{5}} e^{-\rho/5} (9375 - 7500\rho + 1500\rho^2 - 100\rho^3 + 2\rho^4)$:

$\rho_{5,0,\max} = 0 \quad 3.3730382 \quad 9.46156165 \quad 19.4953468 \quad 37.6700533$

$P_{5,0} = \dfrac{2}{46875\sqrt{5}} e^{-\rho/5} (9375 - 7500\rho + 1500\rho^2 - 100\rho^3 + 2\rho^4)\rho$:

$\rho_{5,0,\max, s} = 0.7288877086 \quad 3.9261159207 \quad 10.0897481038 \quad 20.3321188417$
39.9231294250

$l = 1, \ \kappa = 3$:

$R_{5,1} = \dfrac{4}{46875\sqrt{30}} e^{-\rho/5} \rho(3750 - 1125\rho + 90\rho^2 - 2\rho^3)$,

$\rho_{5,1,\max} = 1.594239454 \quad 8.172289247 \quad 18.80470089 \quad 36.4287704077$

$$P_{5,1} = \frac{4}{46875\sqrt{30}} e^{-\rho/5} \rho^2 (3750 - 1125\rho + 90\rho^2 - 2\rho^3):$$

$\rho_{5,1,\max,s} = 2.7638777872 \quad 9.0359270606 \quad 19.3000556408 \quad 38.9001395111$

$l = 2 \quad \kappa = 2:$

$$R_{5,2} = \frac{4}{46875\sqrt{70}} e^{-\rho/5} \rho^2 (525 - 70\rho + 2\rho^2):$$

$\rho_{5,2,\max} = 4.804896151 \quad 15.9559233 \quad 34.2391805$

$$P_{5,2} = \frac{4}{46875\sqrt{70}} e^{-\rho/5} \rho^3 (525 - 70\rho + 2\rho^2):$$

$\rho_{5,2,\max,s} = 6.3157051741 \quad 16.990494945 \quad 36.693799880$

$l = 3, \ \kappa = 1:$

$$R_{5,3} = \frac{4}{46875\sqrt{70}} e^{-\rho/5} \rho^3 (20 - \rho): \ \rho_{5,3,\max 1} = 10 \quad \rho_{5,3,\max 2} = 30$$

$$P_{5,3} = \frac{4}{46875\sqrt{70}} e^{-\rho/5} \rho^4 (20 - \rho): \ \rho_{5,3,\max 1,2s} = 12.192235935 \quad 32.807764064$$

$l = 4, \ \kappa = 0:$

$$R_{5,4} = \frac{4}{140625\sqrt{70}} e^{-\rho/5} \rho^4, \ \rho_{5,4,\max 1} = 20 \rightarrow W = -\frac{e^2}{8\pi\varepsilon_0 a_0 \rho_{5,4,\max 1}} \rightarrow R = \infty.$$

$$P_{5,4} = \frac{4}{140625\sqrt{70}} e^{-\rho/5} \rho^5, \ \rho_{5,4,\max 1s} = 5^2 \rightarrow W = -\frac{Z^2 e^2}{8\pi\varepsilon_0 a_0 5^2}$$

(The formal conformity with Bohr energy takes place here – the fitting happens)

Let us once more analyze the radial and pseudo radial functions.

An introduction of the pseudo radial function means the substitution of Schrödinger's $\psi_{n,l}$-function (5.19) for the new (pseudo) function $\Xi_{n,l} = r \cdot \psi_{n,l}$, which does not satisfy Schrödinger's wave equation,

$$\Xi_{n,l} = r \cdot \frac{N_{n,l}}{2\pi} e^{i(\omega t - k_n r)} f(2k_n r) \Theta_{l,m}(\theta) e^{-im\varphi}, \tag{5.24}$$

where $k_n = \dfrac{k}{k + l + 1}$. At that the tacit transformation of the three-dimensional space of events into the one-dimensional space took place. Now, the H-atom is represented by a linear formation.

Disregarding all above-stated and the fact that Ξ-function is not a solution of Schrödinger's equation, let us continue the discussion: "is an introduction of this function, *as the probabilistic function*, rightful? The modulus squared, $\Xi_{n,l}\Xi^*_{n,l}$, defines the density of "*probability*" (compare with (5.8)):

$$\rho_w = \Xi_{l,m}\Xi^*_{l,m} = \frac{1}{2\pi}r^2 R^2_{n,l}\Theta^2_{l,m} = \frac{1}{2\pi}P^2_{n,l}\Theta^2_{l,m}. \tag{5.24a}$$

Everything is revealed through a comparison; therefore, we will consider the classical problem, where we will deliberately make the same error, like the one accomplished at the introduction of the radial pseudo function $P_{n,l}$.

Let us assume that some material points-events are randomly distributed in three-dimensional space, with the increasing density to the origin of coordinates, according to Gaussian distribution:

$$\rho_w = \frac{1}{\sqrt{(2\pi)^3 \cdot \sigma^3}} e^{-\frac{r^2}{2\sigma^2}}, \tag{5.25}$$

where σ is some parameter of distribution.

The probability, that the material points-events are located within a small domain, between r and $r+\Delta r$, is

$$\Delta w = \int_r^{r+\Delta r} \rho_w dV = \int_r^{r+\Delta r} \rho_w 4\pi r^2 dr = \int_r^{r+\Delta r} \frac{1}{\sqrt{(2\pi)^3 \cdot \sigma^3}} e^{-\frac{r^2}{2\sigma^2}} 4\pi r^2 dr. \tag{5.26}$$

If $\Delta r << 1$, then the integral (5.26) can be substituted for the finite differential

$$\Delta w = \rho_w \Delta V = \frac{1}{\sqrt{(2\pi)^3 \cdot \sigma^3}} e^{-\frac{r^2}{2\sigma^2}} 4\pi r^2 \Delta r, \tag{5.26a}$$

where $\Delta V = 4\pi r^2 \Delta r$ is an elementary volume of space, presented in the form of a spherical layer of the thickness Δr.

For the comparison of probabilities of events in different elementary domains of three-dimensional space, we must take only *equal* volumes ΔV. Otherwise, the picture of probability will be distorted.

Probabilities of events in two arbitrary equal spherical layers-volumes,

$$\Delta w(r_1) = \frac{1}{\sqrt{(2\pi)^3} \cdot \sigma^3} e^{-\frac{r_1^2}{2\sigma^2}} 4\pi r_1^2 \Delta r_1 ,$$

$$\Delta w(r_2) = \frac{1}{\sqrt{(2\pi)^3} \cdot \sigma^3} e^{-\frac{r_2^2}{2\sigma^2}} 4\pi r_2^2 \Delta r_2 ,$$

are defined by the following relation

$$\frac{\Delta w(r_1)}{\Delta w(r_2)} = e^{-\frac{r_1^2 - r_2^2}{2\sigma^2}} = \frac{\rho_w(r_1)}{\rho_w(r_2)} . \tag{5.26b}$$

The partition of the space of events into equal volumes, in which, directly or indirectly, frequencies of random events are counted up, is an indispensable condition. It is the indispensable procedure for the probabilistic estimations of events. Accordingly, it is inadmissible to violate this axiom if we wish to see the objective picture of events, but not its distorted image.

For example, we are interested in the probabilistic trajectory of the movement of a Mr. "N", within the bounds of a small town. For this purpose, we must divide the town into parts of equal areas. Then, it is necessary to count up the frequency of appearances of "N" there (in every part of the town) during a long enough period of time. As a result, we will obtain the field of probability of the events. Obviously, in this field of events, some line of extremal values of probability, which defines the probabilistic movements of "N", will be found. The probabilistic trajectory will coincide, mainly, with the route to his work, and it also will reflect his definite business and private interests.

Taking into consideration all of the aforementioned, we can return to our problems. What will happen if the formula of differential probability (5.26a) will be rewritten as follows?

$$dw = \rho_w dV = \left(\frac{4\pi r^2}{\sqrt{(2\pi)^3} \cdot \sigma^3} e^{-\frac{r^2}{2\sigma^2}} \right) dr . \tag{5.27}$$

Following *formal logic*, the change of positions of cofactors in a formula does not change its product; so that, no errors were committed at rewriting of the equation (5.26a) in the form of the equation (5.27). *Dialectical logic* asserts that at the permutation of cofactors (which resulted in the formation of new cofactors) there is no quantitative error, but a chance exists for qualitative error. The latter leads to gross distortions of the described events and to the erroneous quantitative effects.

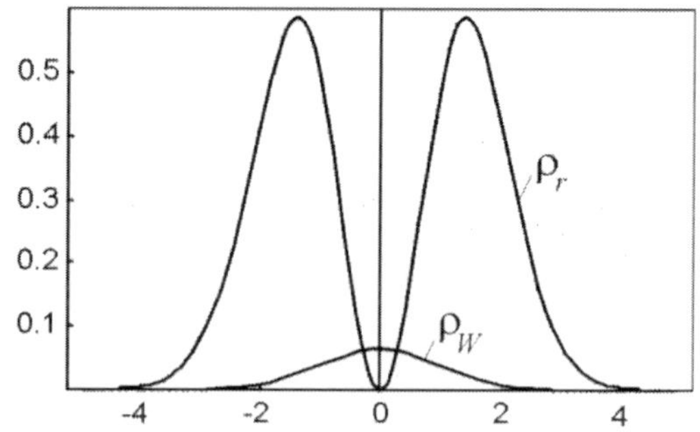

Figure 3.8. Plots of the density ρ_w and the pseudo density ρ_r.

In the expression (5.27), the first step, on the way to distortions of the objective field of probability, was made. Namely, the differential volume $dV = 4\pi r^2 dr$ was substituted for the new differential *dr*, which *takes the constant value* in all points of the space of events. In such a case, the former equal differentials *dV* will be related between themselves as

$$\frac{dV_2}{dV_1} = \frac{r_2^2}{r_1^2}\,. \tag{5.28}$$

Now, it is impossible to speak about a comparison of events. The density of pseudo probability can be called the "radial probability" or the "linear density". It must be written, according to the equation (5.27), in the following form

$$\rho_r = 4\pi r^2 \rho_w = \frac{4\pi r^2}{\sqrt{(2\pi)^3}\cdot\sigma^3} e^{-\frac{r^2}{2\sigma^2}}\,. \tag{5.29}$$

For comparison, the plots of functions of the density ρ_w and the pseudo density ρ_r, for $\sigma = 1$, are presented in Figure 3.8.

The graph of the real density of probability ρ_w, in the *three-dimensional space*, shows that the probability of distribution of material points-events increases in the direction of the origin of coordinates. Quite the contrary, the pseudo probability, related to the *linear space* (where there are no material points-events) points out that the "probability" of finding the events falls to zero in the origin of coordinates.

It is a direct evidence of absurdity. Figuratively saying, operating with mathematics in the spirit of a free game of notions, the "black" (the maximal distribution in the origin of coordinates) is announced as the "white" (the zero distribution in the origin of coordinates). This absurdity has happened because the *equal differential volumes* of space, $dV = 4\pi r^2 dr$, were deliberately substituted for the *differentials-functions*

$$dV(r) = 4\pi r^2 d(\text{const}), \tag{5.30}$$

where $d(\text{const}) = dr$ is the *constant differential* in all space of events.

As an example of inadmissibility of such erroneous transformations, let us imagine a spherical galaxy, in which the density of stars is expressed by the function (5.25). The above mentioned manipulation with the linear density will lead to the laughable conclusion, as if the probability of finding of stars falls to zero, according as we approach to the center of the galaxy.

We can supplement the problem, in question, with an example of the two-dimensional distribution of material points-events by the Gaussian distribution with the density of probability

$$\rho_w = \frac{1}{\sqrt{(2\pi)^2 \cdot \sigma^2}} e^{-\frac{r^2}{2\sigma^2}} \tag{5.31}$$

and the differential probability

$$dw = \rho_w dS = \frac{1}{\sqrt{(2\pi)^2 \cdot \sigma^2}} e^{-\frac{r^2}{2\sigma^2}} 2\pi r dr. \tag{5.32}$$

Here, the falsification of events is also possible through an introduction of the "radial probability"

$$\rho_r = \frac{2\pi r}{\sqrt{(2\pi)^2 \cdot \sigma^2}} e^{-\frac{r^2}{2\sigma^2}} = re^{-\frac{r^2}{2\sigma^2}}, \tag{5.33}$$

which announces the zero probability in the origin of coordinates.

Let us imagine an ant-hill in a forest. Its external contours correspond to the graph of the density of probability (5.31) (Figure 3.9a).

Obviously, the probability of distribution of ants in the ant-hill follows the similar formula of density. Now, we decided to prove that, as approaching to the center of the ant-hill, the probability of finding of ants there falls. To this end, it is necessary to take advantage of the "radial probability" (5.33) (Figure 3.9b). We propose to theorists of such probabilities to poke a hand in the center of an ant-hill in order to examine the correctness of their "theory".

Thus, in positivism, the notion the *truth* is a subjective notion. The true one is that it satisfies "practical needs" of a theory. And the correspondence of a theory to the real picture of nature has, most often, the speculative character, because the world, by positivism, is the subjective reality.

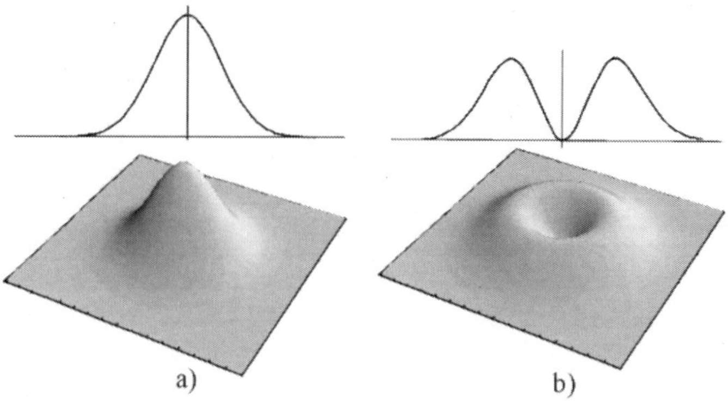

Figure 3.9. The transformation of the structure of a studying object (a), described by the density of probability ρ_w, similar to (5.31), into the completely other object (b), described by the different, in principle, "radial probability" ρ_r (5.33).

For the completeness of the above analysis, we should touch the case of the linear distribution of probability of material points-events by Gauss, with the density of probability

$$\rho_w = \frac{1}{\sqrt{2\pi}\cdot\sigma} e^{-\frac{r^2}{2\sigma^2}} \tag{5.34}$$

and the differential probability

$$dw = \rho_w dr = \frac{1}{\sqrt{2\pi}\cdot\sigma} e^{-\frac{r^2}{2\sigma^2}} dr \,. \tag{5.35}$$

In this situation, the actual probability and the "radial probability" are equal (coincide); accordingly any distortions are impossible here.

In virtue of all above-stated, an introduction of the pseudo radial function in the three-dimensional space of events of the H-atom is unlawful and represents the method of the coarse fitting to the required results. Such formal constructions obtained the wide dissemination in physics.

Let us consider now the discrepancy of the radii of stationary shells, defined by Schrödinger's radial functions, with the radii, defined by Bohr postulates. As an example, we will take the radial function $R_{3,0}$, corresponding to the numbers $\kappa = 2$ and $l = 0$ ($n = 3$):

$$R_{3,0} = \frac{2}{81\sqrt{3}} e^{-\rho/3} (27 - 18\rho + 2\rho^2) \,. \tag{5.35}$$

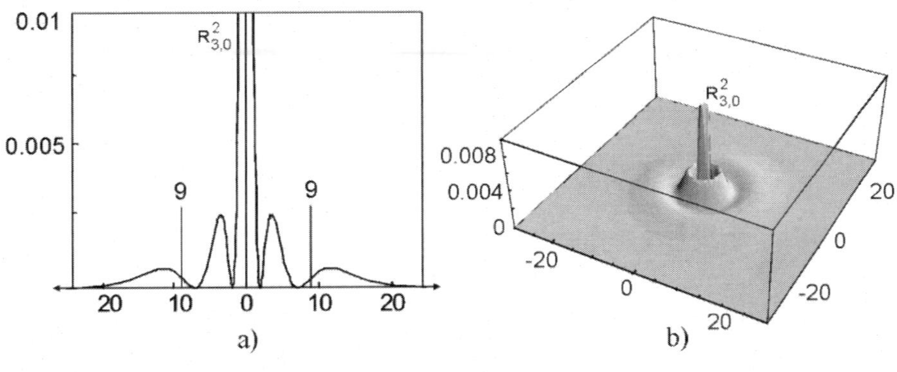

Figure 3.10. The density of hypothetical probability of s-state, $R_{3,0}^2$, for Schrödinger's Ψ-function with the parameters, $n = 3$ and $l = 0$.

One- (a) and two-dimensional (b) graphs of change of the radial component of density of probability $R_{3,0}^2$, in the dependence on the distance, along the radius ρ, are presented in Figure 3.10.

The radial function squared, $R_{3,0}^2$, has the maximum in the origin of coordinates. There are also two smaller maxima, defining the two shells of the most probable localization of the electron (if we will strictly follow the quantum mechanical interpretation of Ψ-function). Extremes of the radial function are as follows:

$$\rho_{3,0,\max 1} = 0, \ \rho_{3,0,\max 2} = 3.531370333, \ \rho_{3,0,\max 3} = 11.46862697.$$

At the same time, according to the cut-off condition, only the radius $\rho = 3^2 = 9$ defines the stationary shell of the electron, corresponding to this function. Two vertical lines in Figure 3.10a, at the distance equal to 9 from the origin of coordinates, indicate its location. As we see, there is no maximum (shell) of such radius among extremes of $R_{3,0}^2$! It is no wonder, the radial function $R_{3,0}$ is "ignorant". It does not "know" that it represents by itself the reduced function (obtained as a result of the cut off operation) and, therefore, it cannot defines anything here, including the most probable localization of the electron.

Let us assume that the radial solutions for the H-atom in quantum mechanics are correct. Then, it is necessary to pay attention to the next myth, concerning an absence of trajectories (orbits) in the solutions of Schrödinger's equation. It is easily to verify that Schrödinger's Ψ functions define orbits, along which (in accordance with his equation) electrons move.

The natural wave motion occurs along wave trajectories. For example, the trajectories of charged particles are observed experimentally, but they are called, by the modest word, "*tracks*" in order to confirm the "*absence*" of trajectories. Here, we meet with the practical application of the method of "*a free game of notions*", when the indisputable fact of an existence of trajectories is linguistically concealed. A trace of a charged microparticles, for example in the Wilson cloud chamber, is similar to a trajectory of a jet plane, which left a trace on its way, which can be also called a "track". But such a verbal trick cannot change the reality of an existence of trajectories of planes. An analysis of trajectories of charged particles in the magnetic field lies in the base of mass spectrographs. Why, if they exist outside, cannot trajectories be inside atoms?

Schrödinger's equation was built on the basis of an introduction the energy of an electron (orbiting around a proton) in the wave equation. Namely, it was constructed on the basis of Kepler's laws, which uniquely presume an existence of trajectories. Hence, solutions of Schrödinger's equation must contain the probabilistic trajectories.

The calculations of the three-dimensional distribution of maxima of the modulus of the wave function $\psi_{n,3,1}$ are presented in Figure 3.11. These maxima point to an existence of three probabilistic circular trajectories (Figure 3.11d).

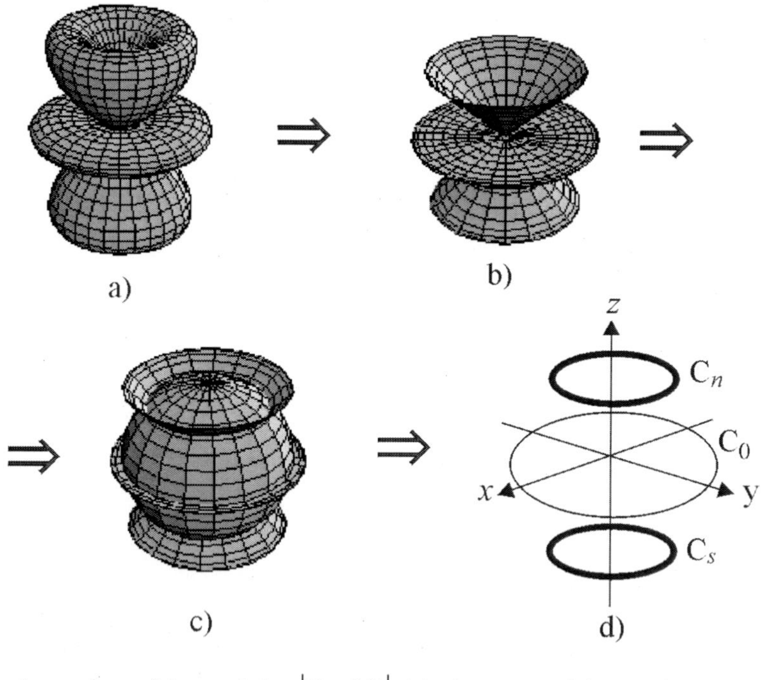

Figure 3.11. The polar surface of the modulus $|\Theta_{3,1}(\theta)|$ (a); the cones of the maximal density of the polar probability (b); the combination of the sphere of the maximal radial probability density and the cones of the maximal polar probability density (c); the three-dimensional distribution of maxima of the modulus of the amplitude function $|\psi_{n,3,1}| = |R_{n,3}(\rho)\Theta_{3,1}(\theta)\Phi_1(\varphi)|$, or $|\psi_{n,3,1}| = |R_{n,3}(\rho)\Theta_{3,1}(\theta)|$ (d).

An existence of probabilistic trajectories is quite natural, since Schrödinger's equation realizes the transition from an orbit to the probabilistic field of motion. Extremes of the wave function in such a field must coincide with these trajectories. The data, presented in Figure 3.31, confirm this.

In the above-presented example, the modulus squared $|\psi_{n,3,1}|^2$ defines three circular trajectories: two polar, C_n and C_s, and one equatorial, C_0. Two polar trajectories are defined by principal extremes of the wave function; the equatorial trajectory – by the collateral extremes.

On such orbits, as was discussed above, the total energy of an electron does not conform to the hydrogen atom spectrum. It is impossible also to co-ordinate the most probable polar trajectories with the usually accepted, quantum mechanical, interpretation of Schrödinger's equation solutions.[17]

In the spirit of traditions of some physicists of the first half of the 20th century, M. Clyde and J. Selbin have regarded, in their book[18] Schrödinger's equation as a particular case of the general wave equation. It is impossible to disagree with their view:

"The wave function is presented as $\Psi = \psi(x,y,z)e^{-i\omega t}$. Then, the general wave equation is transformed, by form, into Schrödinger's wave equation by the following way:

$$\nabla^2\Psi = \frac{1}{\upsilon^2}\frac{\partial^2\Psi}{\partial t^2} \Rightarrow \nabla^2\Psi + \frac{\omega^2}{\upsilon^2}\Psi = 0\text{, but because } \upsilon = \lambda\omega = \frac{\hbar\omega}{p} = \frac{E}{p},$$

then $\nabla^2\Psi + \dfrac{p^2}{\hbar^2}\Psi = 0$ and $\Delta\Psi + \dfrac{2mE}{\hbar^2}\Psi = 0$, or $\Delta\Psi + \dfrac{2m(W-U)}{\hbar^2}\Psi = 0$".

Because the wave equation describes the plural mass processes, the value of energy in Schrödinger's equation should be regarded as a fundamental measure of the wave field of the studying object. Hence, there are no bases to assert that the orbits (Figure 3.11) must belong, without fail, to the electron.

G.C. Pimentel and R.D. Spratley, describing the basic concepts of quantum chemistry*, have noted the following. The standard requirements (to the properties of Ψ-function) have "*stipulated an appearance, in the wave equation solution, of the linear spectrum, which exactly coincides with the hydrogen atom spectrum*" (the Russian version[19] p. 30). As follows from the above considered, all these "exact" coincidences is the myth and nothing more.

Another widespread opinion, contained in their work (p. 31), is that "ψ does not contain information about a trajectory of motion of an electron. Actually, in the microworld, the notion of the "*electron trajectory*", in an atom and molecule, *loses a sense*. Instead of the

---

[17] L.G. Kreidik and G.P. Shpenkov, *Foundations of Physics*, Bydgoszcz, 1998.

[18] M. Clyde, Jr., Joel Selbin, *Theoretical Inorganic Chemistry*, Reinhold Dublishing Corporation, N-Y, Chapman & Hall, LTD, London; Russian translation by "Himiya", Moscow, 1969, p. 49.

[19] G.C. Pimentel and R.D. Spratley, *Chemical Bounding Clarified Through Quantum Mechanics*, Holden-Day Inc., San Francisco-Cambridge-London-Amsterdam, 1970; Russian translation by "Mir", Moscow, 1973.

trajectory, the modulus squared, $|\psi|^2$, allows one to determine only the *probability of location* of the electron in the volume. The $|\psi|^2$-value is bigger, when the probability of location of the electron is higher, and smaller, when the electron appears seldomly. At the nodal surface, where ψ-value is equal to zero, the electron could never be."

All these assertions, which have been repeated for tens of years, do not correspond to reality (see, once more, the solutions of Schrödinger's equation, presented in Figure 3.11). Apart from the considered above, the next question arises. How has it happened that the field of all Ψ-functions is created by one electron, which is "spread" over all space? Maybe it is the field of possible positions of many electrons, belonging to different *H*-atoms. In the last case, Schrödinger's equation describes some "averaged probabilistic" *H*-atom, but this picture does not agree with *H*-atom spectrum.

Regarding the Bohr postulates, they wrote * (p. 46): "... a fault of the atomic planetary model, able to call headache, is the demand that the motion of a charged electron along a curvilinear trajectory will occur without radiation and a loss of energy. This difficulty does not arise in quantum mechanics, since the notion of the "trajectory" is absent there. The wave description of the electron does not give any information about an existence of its trajectory. Therefore, the question about a lost of energy, caused by the radiation, in such a model, generally, does not arise. Moreover, it is impossible to say that a trajectory exists. Quantum mechanics plainly does not accept it. At the transition from the classical equation of motion to the equation of quantum mechanics, a member, corresponding to the kinetic energy, was changed by definite means. We can render these changes in the form of the following order: Break off think about trajectories – we cannot approach the nature and describe the atomic properties by the equation of motion, which contains the notion of the "trajectory". And, actually, we must not regret about it. If there is no trajectory, then the electron must not emit energy. As we know, the latter is really so".

As follows from the solution, shown in Figure 3.11, all above cited quotations are, in essence, emotional assertions. They remind us the literary stories, which do not rely on the solutions of Schrödinger's equation, but only repeat the mythology of formal suppositions. They are not supported by proofs, because, throughout the existence of quantum mechanics, the *three-dimensional distribution of extremes of Schrödinger's Ψ-functions has never been presented*. If only this would be done, a myth on the absence of orbits (trajectories) would disappear.

Finally, let us briefly touch upon the radiation of potential-kinetic waves emitted by the *H*-atom. The natural motion in the Universe is the longitudinal-transverse motion-rest, the main form of which is the spiral-like motion. This motion represents by itself a superposition of two mutually perpendicular plane-polarized waves. It is the natural vortical motion-rest. And all natural motions in nature are based on the dynamic equilibrium of exchange of matter-space-motion-rest, which forms the "zero" equilibrium radiation.

If some disturbance of the equilibrium happens, then the wave perturbations of matter-space arise. In particular, the orbital motion of an electron, as the natural motion, takes place arbitrarily for a long time, until the dynamic equilibrium of the exchange of matter-space-motion-rest of the H-atom with the ambient field of matter-space, will not be upset.

When we set a charged macroobject in a uniform rotation, the definite perturbation in the surrounding space arises as a result. Such a motion is unnatural (artificial) and, therefore, accompanied always with the wave perturbation.

On the contrary, the orbital motion of an electron is the natural motion in H-atom space. It is a part of the wave motion of matter-space at the subatomic level with its own dynamic equilibrium "zero" radiation. Such a state of motion of the electron is continued so long, until some wave perturbations will transfer it in a new stationary state. And again, the electron transition, in the new state, is accompanied with the wave perturbation of the ambient field of matter-space, which raises over the zero exchange (radiation).

SHORT BIOGRAPHY OF DR. GEORGE SHPENKOV

George P. Shpenkov, retired professor, the author of *Friction Surface Phenomena* (Elsevier Science, 1995) and one of the discoverers of *Hydrogen Wear of Metals phenomenon* (registered in former USSR, No. 378). He was born in Minsk, Belarus, in 1937 and received there his B.A. in physics from Belarusian University in 1965. He then went to St. Petersburg (Leningrad), Russia, to work with Nina Aleksandrovna Gorunova and Solomon Meerovich Ryvkin at Ioffe Physico-Technical Institute, under whose supervision he completed the D.Phil. in physics of semiconductors in 1968. He taught at Gomel University (Condensed Matter Sub-Faculty) in Belarus from 1969 to 1972, when he moved to Minsk, where he has spent years researching into applied physics problems at Radiation Introscopy Laboratory and teaching physics at Substance Structure Sub-Faculty in Belarusian University.

He works in Poland from 1992: at Silesian University in Katowice (Institute of Technological Problems) to 1996, then at University of Technology and Life Sciences (Institute of Mathematics & Physics) in Bydgoszcz, where he has taught up to 2008.

His first book, *Physical Chemistry of Friction*, appeared in 1978 (in Russian), followed by its extended and revised version (1991), *Friction Surface Phenomena* (1995). He is the co-author of *Alternative Picture of the World* (1996), *Foundations of Physics*: 13.644. Collected Papers (1998), and *Atomic Structure of Matter Space* (2001) with Leonid Kreidik.

He is the author of over a hundred scholarly articles on various aspects of physics, published in different journals, proceedings, etc. Working at Ioffe Institute, he discovered laser radiation in A2B4C52 semiconductor compounds and generation of the second harmonic in them. One of his Ph.D. thesis referees was Zhores Ivanovich Alferov 2000 Nobel Laureate in Physics.

He defended also D.Sc. degree in Solid State Physics in Tomsk (at Institute of Firmness Physics and Material Science, Siberian Division of Academy of Sciences, Russia, 1990). University lecture classes he has taught include all sections of General Physics, Quantum Mechanics and Quantum Chemistry, Modern Physics (for philosophers), Selective Chapters of Surface Science (a monograph course), etc.

Subjects of current interest include an analysis of basic concepts of physics and their reconsideration to conformity with dialectical logic and philosophy on the basis of new ideas first put forward in the book *Alternative Picture of the World* (1996). Main achievements of this analysis are a series of the discoveries (expecting their universal recognition) related to primordial problems of physics such as the nature of mass and charge of elementary particles, the shell-nodal (molecule-like) structure of atoms, microwave background radiation of atoms, the nature of Lamb shift, dynamic model of elementary particles, and etc.

REFERENCES

[1] L.G. Kreidik and G.P. Shpenkov, *Foundations of Physics*, Bydgoszcz, 1998.

[2] L.G. Kreidik and G.P. Shpenkov, *Alternative Picture of the World*, V.2, Bydgoszcz, 1996.

[3] *Dictionnaire de la abstraite* [*Dictionary of the abstract*], Paris, 1957, pp. 2-3.

[4] Waller M.D., *Cladni Plates*, London, Staples Press, 1960.

[5] L. de Broglie, *Les Incertitudes D'Heisenberg et L'Interprétation Probabiliste de la Mécanique Ondulatoire* [*Heisenberg Uncertainties and the Probabilistic Interpretation of Wave Mechanics*], Gauthier-Villars, Bordas, Paris, 1982. (Russian edition: L. de Broglie, Heisenberg Uncertainty Principle and Probabilistic Interpretation of Wave Mechanics, Mir, Moscow, 1986)

[6] M. Clyde, Jr., Joel Selbin, *Theoretical Inorganic Chemistry*, Reinhold Dublishing Corporation, N-Y, Chapman & Hall, LTD, London; Russian translation by "Himiya", Moscow, 1969, p. 49.

[7] G.C. Pimentel and R.D. Spratley, *Chemical Bounding Clarified Through Quantum Mechanics,* Holden-Day Inc., San Francisco-Cambridge-London-Amsterdam, 1970; Russian translation by "Mir", Moscow, 1973.

[8] W.H. Flygare, *Molecular Structure and Dynamics*, Prentice-Hall, Inc., Englewood Cliffts, New Jersey, 1978; Russian translation by Mir, Moscow, 1982, V.1, pp. 180-181.

[9] N. Mott, I. Sneddon, *Wave Mechanics and its Application*, Oxford at the Clarendon Press, 1948; Russian translation by Nauka, Moscow, 1966, p.75.

[10] Hans A. Bethe, Edwin E. Salpeter, *Quantum Mechanics of One- and Two-Electron Atoms*, Springer-Verlag, Berlin, Göttingen,Heidelberg, 1957; Russian translation by Fizmatgiz, Moscow, 1960.

[11] *Die Grundgedanken der neueren Quantentheorie* [*The basic ideas of the newer quantum theory*], H. Thirring, Ergebnisse der exakten Naturwissensschaften, VII Band, 1928, S. 410.

[12] A. Sommerfeld, *Atombau und Spektrallinien*. Wellenmechanischer Ergänzungsband [*Atomic structure and spectral lines. Wave mechanical supplement tape*], 1929, 44.

[13] David Bohm, Causality and Chance in Modern Physics, L., 1957; IL, Moscow, 1959 (in Russian)

[14] A.K. Timiryazev, *Introduction in Theoretical Physics*, State Publishing House, GTTI, Moscow-1933, Leningrad, pp. 334-336

[15] Max Born, *Atomic Physics*, Blackie and Son Limited, London-Glasgow, seventh edition, 1963; Mir, Moscow, 1965

[16] Gentill K., *Acoustics*, 4, 58, 1957.

In: Old Problems and New Horizons in World Physics ISBN: 978-1-53615-430-6
Editors: Volodymyr Krasnoholovets et al.

Chapter 3

THE WAVE BEHAVIOR AND SUBMICROSCOPIC CONCEPT OF THE MICROWORLD: BEYOND QUANTUM MECHANICS

Victor Christianto[1,*], Volodymyr Krasnoholovets[2,†] and Florentin Smarandache[3,‡]

[1]Satyabhakti Advanced School of Theology – Jakarta Chapter, Indonesia
[2]Institute of Physics, Kyiv, Ukraine
[3]Deartment of Mathematics and Sciences, University of New Mexico, Gallup, US

ABSTRACT

In some recent papers ('Hooft and others), it has been argued that quantum mechanics can arise from classical cellular automata. Nonetheless, G. Shpenkov has proved that the classical wave equation makes it possible to derive a periodic table of elements, which is very close to Mendeleyev's one, and describe also other phenomena related to the structure of molecules. Hence the classical wave equation complements Schrödinger's equation, which implies the appearance of a cellular automaton molecular model starting from classical wave equation. The other studies show that the microworld is constituted as a *tessellation* of primary topological balls. The tessellattice becomes the origin of a submicrospic mechanics in which a quantum system is subdivided to two subsystems: the particle and its inerton cloud, which appears due to the interaction of the moving particle with oncoming cells of the tessellattice. The particle and its inerton cloud periodically change the momentum and hence move like a wave. The new approach allows us to

* Corresponding Author's Email: victorchristianto@gmail.com; http://researchgate.net/profile/Victor_Christianto.
† Author's Email: krasnoh@iop.kiev.ua.
‡ Author's Email: florentin.smarandache@laposte.net.

correlate the Klein-Gordon equation with the deformation coat that is formed in the tessellatice around the particle. The submicroscopic approach shows that the source of any type of wave movements including the Klein-Gordon, Schrödinger, and classical wave equations is hidden in the tessellattice and its basic exciations – inertons, carriers of mass and inert properies of matter.

Keywords: Schrödinger equation, Klein-Gordon equation, classical wave equation, Shpenkov's model, periodic table of elements, molecule, cellular automata, submicroscopic concept, tessellettice, inertons

INTRODUCTION

Elze [1] wrote about possible re-interpretation of quantum mechnics (QM) starting from classical automata principles. This is surely a fresh approach to QM, initiated by some authors including Gerard 't Hooft [3]. In the mean time, in a series of papers Shpenkov [2, 3-14] draws attantion to the fact that the spherical solution of Schrödinger's equation says nothing about the structure of molecules; the spherical solution and its comparison with experimental data are hardly discussed properly in textbooks, with an excuse that it is too complicated.

According to Shpenkov [2, 3-14], the classical wave equation is able to derive a periodic table of elements, – which is close to Mendeleyev's periodic table, – and also other phenomena related to the structure of molecules. Hence it seems Shpenkov's interpretation of classical wave equation can complement Schrödinger equation.

However, the Schrödinger equation is a quantum equation that describes the motion of the appropriate particle-wave since all quantum objects manifest charecteristics of both particles and waves. Considering Shpenkov's results, one can ask: *why do the particle's characteristics dissapear and what exactly is the subject of purely wave behaviour in a quantum system?*

In order to answer these questions, we involve recent studies of Krasnoholovets (see, e.g., Ref. [15]), in which a submicroscopic concept has been developed. Namely, it has been shown how the motion of a canonical particle occurs in physical space constructed as a cellular structure named the tessellattice.

In this work we carry out studies of the Schrödinger equation and wave equation and show how they both are related to the tessellatice that forms their essence and fills with physical content.

SCHRÖDINGER EQUATION VS. CLASSICAL WAVE EQUATION (OF SOUND)

Shpenkov's work is based on [13]: (1) Dialectical philosophy and dialectical logic; (2) The postulate on the wave nature of all phenomena and objects in the universe. He uses the classical wave equation

$$\Delta\hat{\Psi} - \frac{1}{c^2}\frac{\partial^2\hat{\Psi}}{\partial t^2} = 0. \tag{1}$$

This equation is also known as the wave equation of sound or string vibration (see, e.g., Refs. [16, 17]).

Shpenkov shows that the classical wave equation is able to:

1. Derive a periodic table of elements (slightly different from but close to the Mendeleyev's periodic law) based on spherical solution of his standing wave equation [10];
2. Give a dynamical model of elementary particles [6];
3. Derive binding energy of deuterium, tritium, helium and carbon [9];
4. Derive the atom background radiation of hydrogen which corresponds to the observed COBE/CMBR (Cosmic Microwave Background Radiation) [4];
5. Derive the shell-nodal model of atoms and molecules [7];
6. Explain anisotropy of graphene [13];
7. Describe the shell-nodal picture of carbon and grapheme [12];
8. Describe electron "orbitals";
9. Describe electron "spin";
10. Derive neutron magnetic moment;
11. Derive proton magnetic moment;
12. And other things [11].

Therefore, it seems that Shpenkov's wave model of particles and molecules may be a promising alternative to complement the standard quantum/wave mechanics.

Besides, Shpenkov points out to some serious weaknesses associated with (spherical solution of) Schrödinger's equation of a quantum system, which is in the potential field of one or another nature:

1. Its spherical solution is rarely discussed completely, though the complete spherical solution of Schrödinger's wave equation does not agree with any experiment.

2. The Schrödinger equation is able only to arrive at hydrogen energy levels, and it has to be modified and simplified for other atoms. For example, physicists are forced to use an approximate approach called *Density Functional Theory* (DFT) in order to deal with N-body system [18].
3. The introduction of variable wave number k into the Schrödinger equation, depending on electron coordinates, and the omission of the azimuth part of the wave function, were erroneous [19]. Schrödinger's variable wave number should be questioned because the potential function cannot influence the wave speed or consequently the wave number.
4. Introduction of the potential function V into the wave equation, which results in dependence of the wave number k on the Coulomb potential, *generates divergences* that do not have a physical justification. They are eliminated in an artificial way [5, p. 27].
5. Modern physics erroneously interprets the meaning of polar-azimuthal functions in Schrödinger's equation, ascribing these functions to atomic "*electron orbitals*" [14, p. 5].
6. Schrödinger arrived at a correct result of hydrogen energy levels using only a radial solution of his wave equation, with one major assumption: the two quantum numbers found in the solution of his wave equation were assumed to be the same with Bohr's quantum number [8].
7. Quantum mechanics solutions, in their modern form, contradict reality because on the basis of these solutions, the existence of crystal substances-spaces is not possible [5, p. 26].
8. Schrödinger's approach yields abstract phenomenological constructions, which do not reflect the real picture of the micro-world [8].
9. Schrödinger himself in his 1926 paper apparently wanted to interpret his wave equation in terms of vibration of string [20, 21]. This is why he did not accept Born's statistical interpretation of his wave equation until he died.

In the initial variant, the Schrödinger equation had the following form [8]:

$$\Delta\Psi + \frac{2m}{\hbar^2}\left(W + \frac{e^2}{4\pi\varepsilon_0 r}\right)\Psi = 0 \tag{2}$$

The wave function satisfying the wave equation (2) is represented as

$$\Psi = R(r)\,\Theta(\theta)\,\Phi(\varphi)\,T(t) = \psi(r,\theta,\varphi)\,T(t) \tag{3}$$

where $\Psi(r, \theta, \varphi) = R(r)\,\Theta(\theta)\,\Phi(\varphi)$ is the complex amplitude of the wave function because

$$\Phi_m(\varphi) = C_m e^{\pm im\varphi} \tag{4}$$

For standard method of separation of variables to solve the spherical Schrödinger equation (see e.g., Refs. [22, 23]).

Equations for the functions Φ, Θ and *T* are known in the theory of wave fields. Hence these equations presented nothing new. Only the *R* was new. Its solution turned out to be *divergent*. However, Schrödinger together with H. Weyl (1885-1955), contrary to the logic of and all experience of theoretical physics, artificially cut off the divergent power series of the radial function $R(r)$ at a κ-th term. This allowed them to obtain the radial solutions, which, as a result of the cut off operation, actually were the fictitious solutions [8].

Furthermore, it can be shown that the time-independent Schrödinger equation [22]

$$\nabla\Psi + \frac{2m}{\hbar^2}(E - V)\Psi = 0 \tag{5}$$

can be written in the form of standard wave equation [8]

$$\nabla\Psi + k^2\Psi = 0 \tag{6}$$

where

$$k = \pm\sqrt{\frac{2m}{\hbar^2}(E - V)}\,. \tag{7}$$

Comparing equations (2), (5) and (6), we obtain [8]

$$k = \pm\sqrt{\frac{2m}{\hbar^2}\left(W + \frac{e^2}{4\pi\varepsilon_0 r}\right)}, \tag{8}$$

which means that the wave number *k* in Schrödinger's radial wave equation is a quantity that *varies continuously in the radial direction*. Is it possible to imagine a field where the wave number, and hence the frequency, change from one point to another in the space affected by the field? Of course, it is not possible. Such wave objects do not exist in Nature.

SHPENKOV'S INTERPRETATION OF CLASSICAL WAVE EQUATION

Shpenkov suggested starting the study of the wave equation in the form

$$\nabla\Psi + \frac{\omega^2}{c^2}\Psi = 0 \tag{9}$$

instead of the standard classical wave equation (1). In eq. (9) the wave number $k = \omega / c$ is invariable [5]. Here, ω denotes a fundamental carrying frequency of the wave field at the corresponding level of space, and c denotes the speed of light. In order to correct the faults of wave mechanics, it is necessary to write down the above wave equation that meets the conditions: (a) the wave number is constant, and (b) the azimuth factor must be taken into consideration along with radial $R(r)$ and polar factor of the wave-function [5].

In this case, the differential equation for the radial factor $R(r)$ is:

$$\rho\frac{d^2R_l}{d^2\rho} + 2\rho\frac{dR_l}{d\rho} + \left[\rho - l(l+1)\right]R_l = 0 \tag{10}$$

where $\rho = kr$.

Shpenkov suggests a kind of a fundamental wavelength corresponding to expression (9) equal to

$$\lambda_e = c / \omega_e = 1.603886998 \times 10^{-10}\ \text{m},\ \omega_e = 1.86916197 \times 10^{18}\text{s}^{-1} \tag{11}$$

because it is one-half of a mean value of the interatomic distance in a solid.

The detailed analysis to find the spherical solution of equation (9) is discussed in Shpenkov's other papers [7, 10]. Some consequences of the solution of Shpenkov's interpretation of the classical wave equation are [5]:

1. As masses of atoms are multiple of the neutron mass (or hydrogen atom mass), following Haüy's ideas makes it reasonable to suppose that any atom, like the elementary Haüy's molecule, is the neutron (H-atom) molecule.
2. Therefore, atoms should be considered as neutron (H-atom) quasi-spherical multiplicative molecules. The word 'multiplicative' means that strong bonds must couple particles constituted of these elementary molecules, which we call multiplicative bonds.
3. Potential polar-azimuthal nodes of spherical shells in stable atoms (nucleon molecules) contain by two coupled nucleons.

4. Polar potential-kinetic nodes (not filled with nucleons in the most abundant and stable atoms) are ordered along the z-axis of symmetry (in spherical coordinate system) of the atoms.
5. Exchange (interaction) between completed nodes inside (strong) and outside (electromagnetic) of the atoms is realised by exchange charges of with the fundamental frequency (11).
6. Principal azimuth nodes of the wave space of atoms are marked by ordinal numbers. These numbers coincide with the ordinal numbers of elements of Mendeleyev's periodic table. The quantity of neutrons, localized in one node, is equal to or less than two.
7. Arranging atoms with the same or similar structure of outer shells one under another, one arrives at the *periodic-nonperiodic law of spherical spaces* that constitutes periodic table, slightly differing from the conventional one of Mendeleyev.

There are of course other researchers who use the classical wave equation to study atoms and particles (see e.g., Mills [24] and Close [25]).

Correspondence between Classical Wave Equation and Quantum Mechanics

A connection between classical and quantum mechanics has been studied at least by several researchers (see e.g., Refs. [26-28]). Ward and Volkmer [29] discussed a relation between the classical electromagnetic wave equation and Schrödinger equation. They derived the Schrödinger equation based on the electromagnetic wave equation and Einstein's special theory of relativity. They began with electromagnetic wave equation in one-dimensional case:

$$\frac{f^2 E}{f x^2} - \frac{1}{c^2}\frac{f^2 E}{f t^2} = 0. \tag{13}$$

This equation is satisfied by plane wave solution:

$$E(x,\, t) = E_0 e^{i(kx-\omega t)}, \tag{14}$$

where $k = 2\pi / \lambda$ and $\omega = 2\pi\nu$ are the spatial and temporal frequencies, respectively. Substituting equation (14) into (13), then we obtain

$$\left(\frac{\partial^2}{\partial x^2}-\frac{1}{c^2}\frac{\partial^2}{\partial t^2}\right)E_0 e^{i(kx-\omega t)}=0, \tag{15}$$

or

$$\left(k^2-\frac{\omega^2}{c^2}\right)E_0 e^{i(kx-\omega t)}=0, \tag{16}$$

which arrives us to a dispersion relationship for light in free space: $k=\omega/c$. This is similar to the wave number k in eq. (8).

Then, recalling from Einstein and Compton that the energy of a photon is $\varepsilon = h\nu = \hbar\omega$ and the momentum of a photon is $p = h/\lambda = \hbar k$, which allows us to rewrite eq. (14) using these relations:

$$E(x,\,t)=E_0\,e^{\frac{i}{\hbar}(px-\varepsilon t)}. \tag{17}$$

Substituting expression (17) into eq. (13) we find

$$-\frac{1}{\hbar^2}\left(p^2-\frac{\varepsilon^2}{c^2}\right)E_0 e^{\frac{i}{\hbar}(px-\varepsilon t)}=0, \tag{18}$$

which results in the relativistic total energy of a particle with zero rest mass

$$\varepsilon^2 = p^2c^2. \tag{19}$$

Following de Broglie, we may write the total relativistic energy for a particle with non-zero rest mass

$$\varepsilon^2 = p^2c^2 + m_0^2c^4. \tag{20}$$

Inserting expression (20) into eq. (18), it is straightforward from (15) that we get

$$\left(\nabla^2-\frac{m_0^2c^2}{\hbar^2}\right)\Psi=\frac{1}{c^2}\frac{\partial^2\Psi}{\partial t^2}, \tag{21}$$

which is the Klein-Gordon equation [30, 31] for a free particle [29]. Now we want to obtain Schrödinger equation, which is non-relativistic case of eq. (21). The first step is to approximate $\varepsilon^2 = p^2c^2 + m_0^2c^4$ as follows

$$\varepsilon = m_0c^2\sqrt{1+\frac{p^2}{m_0^2c^2}} \cup m_0c^2 + \frac{p^2}{2m_0} \cup m_0c^2 + | \quad . \tag{22}$$

After some approximation steps, Ward and Volkmer [29] arrived at the Schrödinger equation

$$-\frac{\hbar^2}{2m}\nabla^2\phi = i\hbar\frac{\partial\phi}{\partial t} , \tag{23}$$

where the non-relativistic wave function ϕ is also constrained to the condition that it can be normalized to unit probability.

In the meantime, Hilbert and Batelaan [32] explored equivalence between the quantum and acoustic system. A simple physical system was discussed, which mirrored the quantum mechanical infinite square well with a central delta well potential. They find that the analytic solution to the quantum system exhibits level splitting, as does the acoustic system. They compare the acoustic resonances in a closed tube and the quantum mechanical eigen-frequencies of an infinite square well and showed that the acoustic displacement standing wave is

$$\xi(x) = \xi_{\max}\sin\left(n\pi x/(2a)\right) \tag{24}$$

for the nth resonance. Eq. (24) has the same shape as the quantum mechanical wave function.

So we can conclude that there exists formal connection between the classical wave equation and Schrödinger equation, but it still requires some assumptions and approximations. Shpenkov's interpretation of classical wave equation looks as more realistic for atomic and molecular modelling.

TWO ROUTES TO CELLULAR AUTOMATA MODEL OF WAVE EQUATION

In the previous section we have argued that Shpenkov's interpretation of classical wave equation looks as more realistic for atomic and molecular modeling. Now we shall outline a cellular automata model of classical wave equation.

A plausible method to describe cellular automata model of wave equation was depicted by Yang and Young [33]. For the 1D linear wave equation, where c is the wave speed they presented a scheme

$$\frac{u_i^{n+1} - 2u_i^n + u_i^{n-1}}{(\Delta t)^2} = c^2 \frac{u_{i+1}^n - 2u_i^n + u_{i-1}^n}{(\Delta x)^2}. \tag{25}$$

After some steps eq. (25) can be rewritten in a generic form (by choosing $\Delta t = \Delta x = 1, t = n$) as follows $u_i^{t+1} + u_i^{t-1} = g(u^t)$, which is reversible under certain conditions. This property comes from the reversibility of the wave equation because it is invariant under the transformation: $t \rightarrow -t$.

O'Reilly has shown that the coupled Maxwell-Dirac electrodynamic system can be implemented in an analog cellular-automaton operating within a 3D regular face-centered cubic lattice [34]. The result of this approach can be expressed in terms of a second order wave equation, namely: $s_i^{t+1} = s_i^t + \dot{s}_i^{t+1}$. He concludes that the second order wave equation is arguably one of the simplest possible continuous-valued cellular automata update equations that do anything physically interesting, though all of electrodynamics can be built of elaborations of this one fundamental interaction.

Thus, cellular approach allows one to construct equations that describe physical systems without using second order equations.

THE TESSELLATTICE AS THE SOURCE FOR THE FORMALISM OF CONVENTIONAL QUANTUM MECHANICAL

A detailed theory of real physical space was developed by Bounias and Krasnoholovets starting from pure mathematical principles (see e.g., Ref. [35]). A submicroscopic theory of physical processes occurring in real physical space was elaborated by Krasnoholovets in a series of works (see e.g., monograph [15]). Those studies show that our ordinary space is constructed as a mathematical lattice of primary topological balls, which was named a tesselllattice.

In the tesselllattice, balls play the role of cells. This is a physical vacuum, or aether. Matter emerges at local deformations of the tessellattice when a cell (or some cells) changes its volume following a fractal law of transformations. Such a deformation in the thessellatie can be associated with the physical notion of mass.

The motion of a fractal-deformed cell, i.e., a mass particle, is occurred with the fractal decomposition of its mass owing to its interaction with ongoing cells of the tesselllattice. Such an interaction generates a cloud of a new kind of spatial excitations named 'inertons.' This means that "hidden variables" introduced in the past by de Louis Broglie, David Bohm and Jean-Pierre Vigier have acquired a sense of real quasiparticles of space.

Thus in monograph [15] it has been shown that inertons are carriers of a new physical field (the inerton field), which appears as a basic field of the universe. Inertons as quasi-particles of the inerton field are responsible for quantum mechanical, nuclear and gravitational interactions of matter.

A particle moving in the tessellatice is surrounded with its inerton cloud. This picture can easily be compared with the formalism of quantum mechanics because the particle wrapped with its inerton cloud can be projected to the particle's wave ψ-function determined in an abstract phase space. In such a pattern, the overlapping of ψ-functions of nearest particles means that the particles' inerton clouds overlap and thus we obtain real carriers of the quantum mechanical interaction, which provide a short-range action between the particles studied.

The particle's de Broglie wavelength λ plays the role of a section in which the moving particle emits its inerton cloud and in the next odd section λ these inertons come back to the particle passing on the momentum to it.

How can we write the interaction of a moving particle with its inerton cloud? The interaction can be written between the particle and an ensemble of inertons, which accompany the particle. The ensemble is presented as one integral object, an inerton cloud. The speed υ_0 of the particle the particle satisfies the inequality $\upsilon_0 << c$. At such presentation, our study is significantly simplified and is reduced to the consideration of a system of two objects: the particle and its cloud of inertons, which the particle periodically emits and adsorbs when moving along its path. In this case the Lagrangian (2.1) is transformed to the following one written in two-dimensional Euclidean space

$$L = \tfrac{1}{2} m_0 \dot{x}^2 + \tfrac{1}{2}\mu_0 \cdot \left[(\dot{\chi}^{\parallel})^2 + (\dot{\chi}^{\perp})^2 \right] - \frac{2\pi}{T}\sqrt{m_0 \mu_0}\, x \dot{\chi}^{\perp}. \tag{26}$$

In the Lagrangian (26) (2.49) the first term describes the kinetic energy of the particle with the mass m and the velocity $\dot{x}$, which moves along the axis X; the second term depicts the kinetic energy of the whole inerton cloud whose mass is μ_0 and its center-of-mass has the coordinate $\chi^{\parallel}$ along the particle's path and $\chi^{\perp}$ is the transverse coordinate; the third term is the interaction energy between the particle and the inerton cloud where $1/T$ is the frequency of their collisions.

By using the substitution

$$\dot{x}^{\perp} = \dot{\chi} + 2\pi\sqrt{m_0 / \mu_0}\, x / T, \tag{27}$$

we carry out a kind of a canonical transformation that leads to the following Lagrangian

$$\tilde{L} = \tfrac{1}{2} m_0 \dot{x}^2 - \tfrac{1}{2}(2\pi / T)^2 m_0 x^2 + \tfrac{1}{2}\mu_0 \cdot \left(\dot{\chi}^2 + (\dot{\chi}^{\parallel})^2\right). \tag{28}$$

We can see from the effective Lagrangian (28) (2.51) that in such a presentation the particle's behavior is described as a classical harmonic oscillator and the accompanying inerton cloud moves by its own hidden principle (though it does not disturb the particle).

The Hamiltonian function according to the definition

$$H = \sum_i \dot{Q}_i \partial L / \partial \dot{Q}_i - L.$$

In our case the Hamiltonian is

$$H = \dot{x}\partial L / \partial \dot{x} + \dot{\chi}\partial L / \partial \dot{\chi} - \tilde{L}. \tag{29}$$

The effective Hamiltonian based on the Lagrangian (28) (2.51) of the oscillating particle in the system of the center-of-mass of the particle and its inerton cloud in the explicit form becomes

$$H = p^2 / (2m_0) + m_0 (2\pi / T)^2 x^2 / 2. \tag{30}$$

Solutions of the equations of motion given by the Hamiltonian (30) are well known for different presentations. In particular, the function (30) allows one to derive the Hamilton-Jacobi equation

$$(fS_1 / fx)^2 / (2m_0) + m_0 (2\pi / T)^2 x^2 / 2 = E \tag{31}$$

from which we obtain the equation for a shortened action

$$S_1 = \square_{x_0}^{x} p\, dx = \square_{x_0}^{x} \sqrt{2m_0 [E - (2\pi / T)^2 x^2 / 2]}\, dx. \tag{32}$$

The function (32) (2.55) enables the solution x as a function of t in the form

$$x = \frac{\sqrt{2E / m_0}}{2\pi / T} \sin(2\pi t / T). \tag{33}$$

Now we can calculate the increment ΔS_1 of the action (32) (2.55) of the particle during the period T; in terms of the action-angle variables

$$\Delta S_1 = \oint p\, dx = \oint \sqrt{2m_0\left(E - m_0(2\pi / T)^2 x^2\right)}\, dx$$
$$= \oint \sqrt{2m_0\left(E - E\sin^2(2\pi t / T)\right)}\sqrt{2E / m_0}\cos(2\pi t / T)\, dt$$
$$= 2E\int_0^T \cos^2(2\pi t / T)\, dt = 2E\left(\frac{t}{2} + \frac{\sin(4\pi t / T)}{4(2\pi / T)}\right)\Bigg|_{t=0}^{t=T} = ET. \tag{34}$$

The final result (34) can be rewritten as follows

$$\Delta S_1 = E \cdot T = E / \nu \tag{35}$$

where the notation $\nu = 1/T$ is entered.

Since the constant E is the initial energy of the particle, i.e., $E = \frac{1}{2} m_0 \upsilon_0^2$, the increment of action (35) can also be presented in the form

$$\Delta S_1 = \tfrac{1}{2} m_0 \upsilon_0^2 \cdot T = m_0 \upsilon_0 \cdot \tfrac{1}{2} \upsilon_0 T = m_0 \upsilon_0 \cdot \lambda \tag{36}$$

where the parameter λ is the spatial amplitude of oscillations of the particle along its path.

If we equate the increment of the action ΔS_1 to the Planck constant h, we immediately arrive at the two major relationships of quantum mechanics introduced by de Broglie for a particle:

$$E = h\nu, \ \lambda = h / (m_0 \upsilon_0). \tag{37}$$

Thus the amplitude of special oscillation of a particle is exactly the particle's de Broglie wavelength.

Having obtained the relationships (37), we can present the complete action for a particle

$$S = S_1 - Et = \int^{x} p\,dx - Et \tag{38}$$

in two equivalent forms:

$$S = m_0 \upsilon_0 x - Et \tag{39}$$

and

$$S = h \times (x / \lambda - \nu t). \tag{40}$$

The relationships (39), (40) and (37) allow the derivation of the Schrödinger equation. If in a conventional wave equation

$$\Delta \psi - \frac{1}{(\upsilon_0 / 2)^2} \frac{\partial^2 \psi}{\partial t^2} = 0 \tag{41}$$

(where $\frac{1}{2}\upsilon_0$ is the average velocity of the particle in the spatial period λ) we insert a wave function, whose phase is based on the action (40),

$$\psi = a \exp\left\{ i 2\pi [x / \lambda - \nu t] \right\}, \tag{42}$$

and set $\upsilon_0 = \lambda \cdot 2\nu$, we get the wave equation in the following presentation:

$$\Delta \psi + \left(2\pi / \lambda \right)^2 \psi = 0. \tag{43}$$

Then putting $\lambda = h / p$ and extracting the momentum p from the function (32) (i.e., $p^2 = 2mE$) we finally obtain a conventional time-independent Schrödinger equation

$$\Delta \psi + \frac{2 m_0 E}{\hbar^2} \psi = 0. \tag{44}$$

Thus, we can see that the moving system of a particle and its inerton cloud obeys the Schrödinger equation.

THE PARTICLE'S DEFORMATION COAT AND THE KLEIN-GORDON EQUATION

As we discussed above, Ward and Volkmer [29] demostrated the derivation of the Klein-Gordon equation (21) for a mass particle starting from its total relativistic energy $\varepsilon^2 = p^2c^2 + m_0^2c^4$ (20). They also showed that a non-relativistic approximation of the same energy (20) results in the Schrödinger time-dependent equation (23).

Usually the Klein-Gordon equation [30, 31] is applied for the description of an abstract relativistic particle that does not possess spin. However, the submicrscopic concept of physics presented in monograph [15] makes *it possible to relate the Klein-Gordon equation to a real object,* namely, a deformation coat that is developed around the mass particle created in the tessellatice.

In fact the creation a particle means the appearance of a local deformation, i.e., a volumetric fractal deformation of the appropriate cell of the tesselllatice. The local deformation must induce a tension state in ambient cells, which may extend only to a definite radius *R*. So behind the radius *R*, the tessellattice does not have any distorsion, it is found here in a degenerate state.

The study [15] shows that in the microworld such fundamental physical parameters as mass and charge vary at the motion. Namely, in a section (the even section) equal to the particle's de Broglie wavelength λ the mass m is transferred to a tension ξ and the charge e changes to the magnetic nomopole g. In the odd section λ the mass and charge are restored. The same happened with cells that form the particle's deformation coat. When the particle is moving, it pulls its deformation coat as well, i.e., ambient cells adjust to state of the particle. In the deformation coat the state of cells osciallates between the tension ξ and mass m. A collective oscillating mode of the deformation coat is specified by the energy [15] $E = \hbar\omega$, which in turn equals the total energy of the particle mc^2.

The discussed oscillations can be described by a plane wave mode $E(x,t) = E_0 e^{i(kx - \varepsilon t)}$ (17). Then following the arguments (17) – (21), we immediately derive the Klein-Gordon equation (21). Note that in our case the particle that obeys the Klein-Gordon equation is the deformation coat that accompanies the moving particle. This deformation coat is specified with the radius equal to the particle's Compton wavelength [15] (see p. 57).

If the speed υ of a particle satisfies the inequality $\upsilon << c$, we following reasoning (22) and (23) will arrive at the Schrödinger equation (23).

DISCUSSION AND CONCLUSION

We have discussed some weaknesses of the Schrödinger equation for description of atom and molecules. Then we have debated Shpenkov's wave model of atom and molecules based on classical wave equation. His model is able to arrive at a periodic table of elements, which is close to Mendeleyev's periodic law. We also have reviewed a plausible cellular automaton molecular model for classical wave equation, as an alternative to Cellular automaton quantum mechanics (by Elze, Gerard't Hooft etc.).

At last we have considered the submicrosicopc concept that allows one to easily derive the Schrödinger and Klein-Gordon equations starting from first submicrosopcic principles. It is interesting that for the first time we now can identify the Klein-Gordon equation with a real object that is described by this equation – it is the particle's deformation coat that is induced in the tessellattice at around the appropriate created canonical particle.

The submicroscopic concept, which is based on space constituted as the tessellattice of primary topological balls, introduces a new physical field, namely the inerton field, which appears as a fundamental field of the universe. Inertons emerge at any motion of particles; in particular, they arise in atoms and around owing to uninterrupted motion of electrons, nuclei and nucleons.

Thus the motion of a quaatum system is characterised by its separation to two joined subsystems: the particle itself and its inerton cloud. Their oscillation dynamics exhibits obvious features of the wave motion. Although the deformation coat that accompanies the moving particle behaves in a special way, it is described by the Klein-Gordon equation, which also manifests the wave properties.

Our analysis shows that oscillations of inertons are present in any movement of a material object. Inertons clearly demonstrate wave behavior. This means that inerton oscillations appear in atoms and molecules. Hence inerton oscillations justify Shpenkov's model [4–14], which applies a classical wave equation of sound to atoms and molecules: the wave funcation Ψ used by Shpenkov describes oscillations of an inerton field and the location of the corresponding nodes in the oscillating wave studied.

Thus, quantum mechanical models, cellular automata, and a cellular automaton molecular model that uses a wave equation can be covered by studies originated from the tessellattice and the submicroscopic behaviour of quantum systems, which invloves an inerton field that binds canonical particles with the tessellattice and between themselves. Further investigations in this direction are recommended, which will shed light on the cornerstones of the microworld.

ACKNOWLEDGMENTS

One of the authors (V. C.) thanks George Shpenkov for replying to some questions regarding his wave model of the periodic table of elements and sending a dozen of his works. The first author (V. C.) also would like to express his gratitude to Jesus Christ who always encouraged and empowered him in many occasions. He is the Good Shepherd. *Soli Deo Gloria!*

REFERENCES

[1] Elze, H.-T. (2015). Are nonlinear discrete cellular automata compatible with quantum mechanics? *J. Phys.: Conference Series 631*, 012069.

[2] Elze, H.-T. (2017). Quantum models as classical cellular automata. *J. Phys.: Conf. Series 845*, 012022.

[3] 't Hooft, G. (2014). *The cellular automaton interpretation of quantum mechanics*, arXiv:1405.1548 [quant-ph].

[4] Shpenkov, G. P., and Kreidik, L. G. (2002). Microwave background radiation of hydrogen atoms, *Revista Ciências Exatas e Naturais* [*Exact and Natural Sciences*] Journal *4*(1), 9-18, http://www.unicentro.br/pesquisa/editora/revistas/exatas/v4n1/Microwave.pdf.

[5] Kreidik, L. G., and Shpenkov, G. P. (2002). Important results of analyzing foundations of quantum mechanics, *Galilean Electrodynamics & QED EAST 13*(2), 23-30, http://shpenkov.janmax.com/QM-Analysis.pdf.

[6] Shpenkov, G. P., and Kreidik, L. G. (2004). Dynamic model of elementary particles and fundamental interactions. *Galilean Electrodynamics*, Special Issue – GED East *15*(2), 23-29.

[7] Shpenkov, G. P. (2005). The nodal structure of standing spherical waves and the periodic law: What is in common between them? *Physics Essays 18*(2), 196-206.

[8] Shpenkov, G. P., and Kreidik, L. G. (2005). Schrödinger's error in principle. *Galilean Electrodynamics 16*(3), 51-56, http://shpenkov.janmax.com/Blunders.pdf.

[9] Shpenkov, G. P. (2005). *The binding energy of helium, carbon, deuterium and tritium in view of shell-nodal atomic model and dynamic model of elementary particles*, http://shpenkov.janmax.com/stronginteraction.pdf.

[10] Shpenkov, G. P. (2006). *An elucidation of the nature of the periodic law, Chapter 7 in The mathematics of the periodic table*, Rouvray, D. H. and King, R. B., ed., Nova Science Publishers, New York, pp. 119-160.

[11] Shpenkov, G. P. (2006). *A new theory of matter-space-time: Evidences in support of an advantage over the modern theory accepted in physics and the perspective to be*

of use. A lecture delivered in Military Academy, Warsaw, Poland, October, 20, http://shpenkov.janmax.com/Theory-DM-English.pdf.

[12] Shpenkov, G. P. (2007). *The shell-nodal structure of the carbon atom and graphene*, http://shpenkov.com/pdf/talk2017Berlin.pdf.

[13] Shpenkov, G. P. (2010). *Anisotropy of unstrained pristine graphene*. http://shpenkov.janmax.com/GrapheneAnisotropy.pdf.

[14] Shpenkov, G. P. (2013) *Dialectical view of the world: The wave model (selected lectures). Vol. I: Philosophical and mathematical background*. http://shpenkov.janmax.com/Vol.1.Dialectics.pdf.

[15] Krasnoholovets, V. (2017). *Structure of space and the submicroscopic deterministic concept of physics*. Apple Academic Press, Oakville and Waretown.

[16] Pain, H. J. (2005). *The Physics of Vibrations and Waves, 6th ed*. J. Wiley & Sons, Ltd. ISBN: 0-470-01295-1(hardback); 0-470-01296-X (paperback).

[17] Rienstra, S. W., and Hirschberg, A. (2014). *An introduction to acoustics. Eindhoven University of Technology*, www.win.tue.nl/~sjoerdr/papers/boek.pdf.

[18] Anonymous, *Density Functional Theory for Beginners*. (2014). http://newton.ex.ac.uk/research/qsystems/people/coomer/dft_intro.html.

[19] Huang, X. (2012). *How did Schrödinger obtain the Schrödinger equation*?, http://vixra.org/abs/1206.0055.

[20] Schrödinger, E. (1926). *Quantisation as a Problem of Proper Values. Part I. In Collected papers in Wave Mechanics*. Providence, Rhode Island: AMS Chelsea Publishing, http://einstein.drexel.edu/~bob/Quantum_Papers/Schr_1.pdf.

[21] Schrödinger, E. (1926). An undulatory theory of the mechanics of atoms and molecules. *Phys. Rev*., Second series *28*(6), 1049-1070.

[22] Anonymous, *Atomic Spectra*, p. 19-21, http://astrowww.phys.uvic.ca/~tatum/stellatm/atm7.pdf.

[23] Fowler, M. (2007). *Classical wave equations*, p. 10-12. http://galileo.phys.virginia.edu/classes/252/Classical_Waves/Classical_Waves.pdf.

[24] Mills, R. L. (2008). *The grand unified theory of classical physics. Vol. I. Atomic physics; Vol. II*. Molecular physics. Blacklight Power. Printed by Cadmus Communications, A Canveo Company Richmond, UA.

[25] Close, R. A. (2008). *The classical wave theory of matter*, http://www.classicalmatter.org/ClassicalMatterWaves.html.

[26] Christianto, V. (2014). A review of Schrödinger equation & classical wave equation. *Prespacetime J*. *5*(5), http://www.prespacetime.com.

[27] Landsman, N. P. (2005). *Between classical and quantum*, arXiv:quant-ph/0506082.

[28] Rienstra, S. W., and Hirschberg, A. (2014). *An introduction to acoustics*. Eindhoven University of Technology, www.win.tue.nl/~sjoerdr/papers/boek.pdf.

[29] Ward, D. W., and Volkmer, S. (2006). *How to derive the Schrödinger equation*. arXiv:physics/0610121.

[30] Klein, O. (1926). Quantentheorie und fiunfdimensionale Relativitíatstheorie [Quantum Theory and Five-Dimensional Relativity Theory], *Zeit. fiur Phys.* 37, 895-906.

[31] Gordon, W. (1926). Der Comptoneffekt nach der Schríodingerschen Theorie [The Compton effect according to the Schriodinger theory], *Zeit. fiur Phys*. 40, 117-133.

[32] Hilbert, S. A., and Batelaan, H. (2007). Acoustic analog to quantum mechanical level splitting. *Am. J. Phys. 75*(11), 1003; also in Faculty Publications, Department of Physics and Astronomy, University of Nebraska - Lincoln. Paper 103, http://digitalcommons.unl.edu/physicsfacpub/103.

[33] Yang, X.-S., and Young, Y. (2010). *Cellular automata, PDEs, and pattern formation*, arXiv: 1003.1983.

[34] O'Reilly, R. C. (2006). *An analog cellular automaton model of Maxwell-Dirac electrodynamics.* https://pdfs.semanticscholar.org/e214/f1e82db00c12d8266db878f3287691750623.pdf.

[35] Bounias, M., and Krasnoholovets, V. (2004). The universe from nothing: A mathematical lattice of empty sets. *International J. Anticipatory Computing Systems 16*, 3-24; arXiv:physics/0309102.

In: Old Problems and New Horizons in World Physics ISBN: 978-1-53615-430-6
Editors: Volodymyr Krasnoholovets et al.

Chapter 4

THE INNER WORKINGS OF REALITY

***Robert Neil Boyd*[1,*] *and Adrian Klein*[2]**
[1]Princeton Biotechnology Corporation, Barrington, NJ, US
ECAO Aff., Israel

"We have to be guided by the facts, and if the facts seem incredible, as they often do, we have first of all to assure ourselves that they are facts, and then explore what those facts actually mean, without intellectual prejudice."

Sir Oliver Lodge

ABSTRACT

The Subquantum microscope constructed in Serbia has made images of entities as small as 10e-95cm., well smaller than the Planck length. From the evidence provided by various direct instrumented observations, there is no lower bound to the small. The entirety of all the sciences, including all of the physics, are described by and are originating from agglomerations and activities of Subquantum Infinitesimals. Subquantum Infinitesimal-based physics unifies all the sciences under one "umbrella", correcting the anomalies and paradoxes of QM, leading to a QM that is deterministic and alive, and restores Consciousness to physics. Physics is no longer based on "inert" random processes. The Subquantum eliminates the *mathematical fantasies* of Einstein's version of relativity, while restoring Galilean relativity as valid up to an infinite velocity. All matter at all scales, all fields, all forces, the life sciences and Consciousness physics are contained by the understandings inherent in the Subquantum Infinitesimals, which are the smallest unit of matter, and at the same time, the smallest unit of Consciousness. This implies that everything has Consciousness and as well, memories of all that the given entity has ever experienced. That each infinitesimal has a memory, is implied by experimental results

* Corresponding Author's Email: rnboydphd@comcast.net.

which have physically proved that the "vacuum" (actually a Subquantum plenum) has memory, and by the experimentally observed vast arrays of information which are conveyed by the various vector potentials, which information is conveyed by the groups of infinitesimals which are causing the vector potentials. These experimental results, and others, imply that everything has some form of Experiential Consciousness and experiential memory. In addition, it has been shown that infinitesimals can traverse vast distances with velocities ranging up to an infinite velocity. This is the basis for all experimentally proven behaviors of quantum mechanics, including "non-local" phenomena, such as the quantum information field. The infinitesimals form a bipartite ether, composed of free ether entities traveling with velocities up to infinity and captive ether entities which are constrained to an "atmosphere" which adheres to, and interpenetrates existing matter and by Subquantum entities which are caught up in the local conveyances of the various forces, at light speed or less. All existing matter constantly radiates and absorbs information-bearing infinitesimals, which process results in local and non-local experiential and emotional communications between all observables. Empirically, experiential and emotional information forms the entire basis of an undivided and Intelligent Universe.

Keywords: SubQuantum infinitesimals, Fractional Quantum Hall Effect, Mobius transformations, infinite velocity propagations, Liénard-Wiechert potential, vacuum memory, quantum mechanics, time "substance", time force, SubQuantum (SQ) aether, SQ Information Storage, SQ Information Transport, SQ Unification, SQ physics, information physics, information correlated, holographic universe, ambient intelligence, consciousness, SQ Being, FPU Lattice, brain, Kolmogorov Limit, quantum cosmology, Coulomb field, experiential information, Weinberg Angle, Cabibbo Angle, Weak Force, Strong Force, Yang-Mills field, absolute distance, relativity, SQ Microscope, continuous creation, implicate order, self, qualia, personality information, unity consciousness, Eidetic Information, monad, sentience, pre-physical, SU2, bio-fields, Froehlich oscillations, EEG, ZPF, sympathetic resonance, information resonance, consciousness resonance, La Place gravitation, PEAR, L-Field, DNA, signature field, morphogenic, palyngenesis, Golden Mean, Phi Ratio, Neural Networks, Cramer's conjecture, quantum entanglement, interdimensional resonator, OrchOR, formative world

INTRODUCTION

Nobel Laureate (Physics) Richard Feynman once said: "If your theories and mathematics do not match the experiment, then they are wrong."

We extend foundational understandings regarding the physics of Consciousness into a vastly more comprehensive and novel approach, regarding the inner workings of what is actually Sentient Reality. The Subquantum approach to Consciousness has been developed by the authors of this chapter, through a decade of research-team efforts, experiments and observations. The resulting Subquantum-based peer-reviewed works are all rooted in empirical, reproducibly observable, actual facts (rather than unsupported assumptions, beliefs, or speculative fantasies, or imaginings, or theories) (see: http://worlds-within-worlds.org/publications-and-research.php, for example) [1].

We are exploring together the unexplored frontiers of the Subquantum constituents of Reality, which obviously exert a "pre-energetic" (pre-field) function at their upper level of complexity (the Quantum), thus originating and controlling all of physical reality as we know it, including the Brain, which is one of the most complex physical-energetic objects known to man.

Our physical world has long ago been demonstrated to boil down to vibrating interactive energy patterns, which operate in various Force and Field domains which are described in the forms of carefully checked mathematical expressions. Energy is considered to have a quantified nature, implying discrete (separate equally-sized bundles) energetic behaviors in its dynamic properties. Orbital jumps in atomic structures are such discrete events, which are resulting in the appearance of "sudden" energy Quanta. Until now, these Quanta have been treated as irreducible units of energy, which appear to pop up in zero time, as "random" events.

Exclusively Quantum-based conundrums and paradoxes began to change after the advent of a Nobel Prize winning experiment, called "The Fractional Quantum Hall Effect" (FQHE, Laughlin, Stormer, Tsui, 1998) with results which introduced several new kinds of quasi-particles, all of which exhibit fractional elementary charge and fractional statistics, which are expressing themselves in the form of Quantum Hall transition states. Based on these physical results, the observed presence of a set of "hierarchical states", originating from condensing quasi-particles, led to an extension of the original Laughlin-Stormer-Tsui plasma model, into a new hierarchical Subquantum understanding, of the origins of particles and quanta [2].

These conditions are displayed inside Quantum "limiting states", and are going towards the infinitely small, well beyond the previously assumed measurement limits of the small in particle physics (i.e., the Planck length), to the much smaller Subquantum level, wherein composite fermions and fractional states of electrons can be described. This easily extends to the infinitely small, since it is easily shown that any measurement of distance can be infinitely subdivided. This new set of infinitesimal physical domains, with new inherent mathematical particularities strongly reminds us of previously rejected concepts of the Ether, as being the binding and conveying media of physical entities, structures, and fields.

The Subquantum ether is comprised of two main parts, one motional, which is responsible for all known quantum behaviors, and the other "static", where SQ particles are constrained by existing matter and comprise an "atmosphere" which is visible to the unaided eyes at the edges of physical objects, as a sort of fog which appears black or white, depending on the local circumstances. Related to this, we have admired the discoveries of Prof. V. Krasnoholovets as related to "inertons" as he calls them. He has developed an instrumentation system that is capable of imaging the "atmosphere" of "inertons" that surround an electron, which partly verifies the "static" portion of our version of "ether" [3]. His discoveries provide verifications for the static portion of our SQ ether, as well as supporting the superluminal attribute of the two part SQ ether, as Prof. Krasnoholovets has been able to apply his inerton technology so as to image stars in real time, to image where they are actually located at this instant. This recording method is not constrained to light speed limitations [4].

This team is in the process of constructing a different type of "static ether" force-detection apparatus based on the rare mineral "shungite" which is found in the Karelia region of Russia. The better grades of shungite are ~ 95% fullerines with a mixture of C_{60} and C_{70} molecules. Shungite from the Karelia region constantly emanates a measurable, usually repulsive force when the material is moved into close proximity to other materials, which force especially noticeable at protrusions. This force can be experienced by the human musculature when the Shungite emanation interacts with various materials. Unfortunately, human muscular sensitivity varies from person to person, and muscular sensitivity is modified by the emotional circumstances of the individual.

Because of this, we are going to attach very sensitive load cells to a strongly emanating cube of shungite such that the force which arises during interactions of the shungite emanation with other materials, will be picked up by the load cells and fed into a circuit which will send the measurement data out to a software program which will map the forces in a 2D color-coded array, which will be put into correspondence with color photographs of the surface(s) being measured.

This measured force is not electric and it is not magnetic and it is not gravitation. So what is it? We think it is a continuous flux of SQ ether of the "static" variety. There is a polarity to interactions between the shungite cube and other materials, as there are locations where an attraction can be measured. We may be measuring an interaction with what is classically known as "Vril", which interpenetrates all of space and comprises a portion of the SQ ether which is active during Creation Processes. We expect we will find information is transported by these forces, when we are able to arrive at a position where we can record eidetic information by some means.

During the last decade, Boyd and Smith discovered 14 non-linear projective Mobius transformation solutions to the Maxwell equations, in addition to the original Mobius (bi-linear) transformation discovered by V. Fock, of Germany, during 1948. The Mobius transformation solutions to the Maxwell equations allow that entities of these classes can exhibit propagation velocities from zero, to an infinite velocity (and everything in between). The 15 classes of Mobius propagations are generators of the Conformal Group Spin (2, 4) = SU (2, 2), consisting of: 6 rotations and boosts; 4 special conformal transformations; 4 translations; and 1 dilation. (See: http://www.valdostamuseum.com/hamsmith/SegalConf.html#conformalEMG) [5].

The Mobius (bi-linear) transformation solutions thus behave in a manner completely different from our accustomed Lorentz transformation propagations, which are constrained by the various measurable physical properties of the plenum, such as permittivity and permeability, to propagate at or near the speed of light. (*c* is not a constant due to the fact that the physical properties of the media can vary in space and time, both locally and non-locally, so c is now considered as a reliable average velocity for transverse electromagnetic propagations.)

The fact that there do exist superluminal propagating entities, has allowed us to realize many new vistas in physics research, especially as related to Quantum and Subquantum mechanics. All the known behaviors of Quantum systems are caused by, controlled by, and preceded by Subquantum (SQ) activities.

When we ignore relativistic considerations as physically unprovable, as according to Einstein, we find we can rely on experimentally proven Quantum considerations to develop a cosmology that develops continuously throughout an infinite volume universe, with no beginning and no end [6]. From topological thermodynamics considerations we know that entropy and negentropy in Quantum Cosmology have equal averaged sustained values when examining the entire of the infinite volume, which means there can never be any "heat death" of the infinite volume universe.

Starting from these novel understandings, these writers decided long ago to engage on a long and tortuous path, exploring how "normal reality" emerges from the Subquantum (SQ). We are also studying the various SQ control mechanisms which are related to the brain's Quantum activity. In our conceptual framework, a brain is described as a highly evolved resonant transduction system, able to process Information-guided Subquantum flux patterns of various complexities, which eventually become our sensory, emotional, and intuitive experiences.

This perspective implies considering Experiential Information as a foundational constituent of Reality, where experiential information is actually conveyed internal to all infinitesimal units. From this point of view, we are able to generate an infinite number of variations and combinations, of many types of information, according to propagation originated SQ flux information interchanges, which partially result from the infinite range of velocities with which information-bearing entities can propagate.

In addition, as infinitesimals slow down below infinite velocity, they start experiencing time. (Time does not exist, for an infinite velocity entity.) Then at various superluminal velocities, they occasionally can combine with other infinitesimals. When this happens, the experiential information contained in each infinitesimal is combined with the information inherent in all the other infinitesimals in the given Subquantum entity, which live in the given scale of smallness.

The primary types of information conveyed by Subquantum entities are emotions, and "mental pictures", but many other varieties of information are also remembered by them, such as magnetic field intensities, electric field intensities, gravitational influences (if any), and so on (see: Bruce DePalma's "memory of space" experiments, for more on this) [7]. The memory capacity of the individual infinitesimal is quite vast, an extraordinary fact we are still having trouble coming to grips with. (Reference, Poponin and Gariaev's "DNA Phantom" Experiments [8]). Also reference the astounding quantity of information which is conveyed by all the varieties of Vector Potentials [9]. Our team has postulated that infinitesimals are multi-dimensional entities. This speculation is the most prominent possible explanation for the memory capacity of infinitesimals.

These behaviors, and the information conveyed by the infinitesimals, naturally exhibit informational and emotional resonances. The total velocity range that Subquantum structures operate in, extends from zero to infinity. For practical purposes the velocity regimes can be divided into zero velocity, subluminal, luminal, superluminal, and infinite velocity domains, as related to psycho-physical correlates which result in, what we consider to be "normal reality".

Information interchanges take place at all scales of size, and at all velocities, and are involved with every kind of material and field. (Except when entities are at zero velocity. Then, they cannot transmit.) Due to the fact that everything stores and/or carries information, the fact of the matter is that everything has, and conveys, the qualities of uniqueness and personality. So stars and rocks and clouds have personality and personality energy. All forces and all fields, including light, convey eidetic information regarding all possible varieties of experience, especially personality and emotional-state information, and mental pictures. Everything is unique, without exception, in the Natural World.

As an example of a practical use of this fact, the Cherokee People have used stars as sort of "cosmic email" information exchanges, for thousands of years. When they go out hunting for deer to eat, at night (when most deer are out eating), before the hunting party goes on their separate ways, they agree on which star they are going to use as their communications hub. When everyone agrees, they go out their separate ways, and look for deer to feed their families with. Every once in a while, each member of the hunting party stops and gets very involved with the agreed-on star, seeking information which will be conveyed by way of that star, by other members of the hunting party.

The star receives and stores and transmits experiential information, personality information, mental pictures, and emotions, to and from each member of the party who unifies their consciousness with the star, by way of focused attention on/with the star. So the star is their "news channel". But the star is way better, because it conveys, not words, but actual personal experiences, in their entirety, to all the people who are using that particular stellar communications vehicle, in real time. There is no "speed of light" delay involved, whatsoever.

So Joe finds a deer and shoots it and then finds out it is too heavy for him to manage by himself. So he looks at the agreed on star and becomes consciously unified with it, and then sends out the message, using feelings and pictures, that he needs some help with this huge deer and gives picture information regarding his location. Before too long, other members of the hunting party start showing up, exactly where Joe is, and they help him carry the deer back to their village. During the day, they use mountains, trees and rocks, and even air, to store and convey their messages to one another. No radios or phones or internet devices are needed to accomplish such Consciousness-based communications. At one time, such communications were commonplace.

This leads us to a new understanding of velocity-encoded and time-encoded Information transportation systems, which are acting throughout the infinite volume of the Universe, conveying information at up to an infinite velocity. Non-local information transports are the mechanism which directly results in the Quantum Field, the quantum potential, quantum "entanglement", quantum correlations, and quantum coherence and decoherence. In addition, local and non-local informational interactions with matter at the scale of the Planck length, and larger, are the determining factors in unitary quantum evolution U (the Schrodinger equations), and the origination of so-called "quantum state reductions" R (also known as "quantum jumps", which are quantum-information-originated "avalanches", which result in local changes of state).

Subquantum information transports and correlations, result in an endless hierarchy of super-implicated orders of complexity [9, 10] and interacting Informational structures, which are evolving in the several velocity regimes which are less than an infinite velocity, as mentioned above. Information flows which are arising from the overarching Cosmic Harmony (both as a local, and a non-local Information Source) arise at all velocities. This can be followed, as extending from the lowest-velocity states, which are typically corresponding to ponderable physical matter, up to the infinite velocity Information-controlled Quantum Potential. Source-Originated information flows also exist at all scales, and partially comprise the Ambient Intelligence.

The Time Force is caused by a "time substance", inherent in the same ether flux which causes gravitation. However, the Time Force is many orders of magnitude smaller than the force due to gravitation, which is in turn, 10^{39} times weaker than the force due to an electric field. The fact that time is a force cannot be denied, simply due to the fact that all things inevitably age (see: Kozyrev's experiments with the "Time Substance") [11].

Information-creating influences start at the highest implicate (potential possibilities) layers of Reality. Infinite velocity infinitesimals carry information, non-locally. Infinite velocity information transport is the starting point of all non-local quantum behaviors, such as the quantum information field. When the velocity of an entity is infinite, there is no time for that entity. Information is carried in this timeless condition, along a nonlinear path of velocity decreases, which decreases are due to interactions with other entities. These interaction events result in deterministic outputs which influence the various levels of manifestation.

In a huge synoptic overall perspective, we can follow the ubiquitous presence of Information from the lowest Sub Quantum Unit (the "Bhutatma" in the Vedic tradition), through the structures formed in the physical Quantum domain and up to the cosmic scale, at the same time contemplating the way various Informational patterns that form at higher velocity regimes sequentially require appropriate expression modalities in subjacent implication levels. This requirement is the fundamental organizing, structuring and controlling principle of the Universal Information-guided All, and involves endless interdimensional feedback loops of causation, acting in super-positioned reciprocal time vectors.

The fact that Subquantum understandings are based on empirical facts, rather than being yet another theoretical mathematical fantasy, has made it difficult for people to really register the significance of these expressions, and what they mean. Recent multitudinous Subquantum (SQ) physics results, require major shifts in our foundational understandings regarding the true nature of Mind and Consciousness. The Subquantum Information Field and its activities, unifies all of the sciences, and describes and explains every kind of phenomena so far encountered by human beings.

Here is a listing of influences, activities, and effects that can be directly attributed and/or correlated to the common grounds of Subquantum (SQ) ether activities, SQ information storage, and SQ information transports. Namely,

- SQ ether flux is the origination and cause of:
  - Newton's Laws (especially action and reaction),
  - Gravitation,
  - Gravitational variations,
  - Gravitational "anomalies",
  - Gravitational periodicities,
  - Information-correlated superluminal variations in measured rest mass,
  - Time variations (Variations in the pace of time).
- SQ ether flux is the origination and cause of:
  - Quantum Mechanics
  - Quantum Non-locality,
  - Quantum potential,
  - Quantum field,
  - Planck's "constant",
  - Fine structure "constant",
  - Quantum "entanglement",
  - Quantization,
  - Quantum state reduction (quantum jumps) [quantum avalanches],
  - Quantum unitary evolution (the Schrodinger equations),
  - Quantum fluctuations (Zitterbewegung, or quantum jitter),
  - "Zero Point Energy",
  - And so on.

- SQ ether fluxes are the origination and cause of:
  - The Maxwell equations (which assume, and prove the existence of, an ether),
  - Electromagnetic fields,
  - Electrical fields,
  - Magnetic fields,
  - Electromagnetic fields,
  - "Spin fields" and their information transport effects,
  - Torsion fields (primarily theoretical, at this point),

  Variations in the radiation patterns of normal, complex, and symplectic electromagnetic transmission antennas, resulting from:
  - Operator's attention,
  - Operators' intention,
  - Operators' emotional states,

  Resulting in:
  - Divergences in the quantum potential,
  - Variations in human consciousness factors (attention, intention, cognition, emotionals),
  - Variations in the measured speed of light.
- SQ ether flux is the origination and cause of:
  - Chemistry,
  - Avogadro's number,
  - Distance-Coupled variations of pH in information-correlated solutions and materials,
  - Variations in the interaction rates between chemicals,
  - Variations in mixing rates, especially in nonlinear, chaotic, or bifurcating systems,
  - "Self-Organization" per Prigogine,
  - Homeopathy,
  - Physical Properties of the atomic elements and the periodic chart of the atomic elements,
  - And so on.
- Time Force:

  SQ ether flux is the origination and cause of the Time Force, as in Kozyrev's instrumented observations of "time waves". When considered in a matter = "ether sponge" analogy, variations in the flow, or waves in the "Time Substance" are resulting from:
  - The deformation of any physical object (This also results in information storage in the object),
  - The encounter of an air, or water jet, with an obstacle,

  - The operation of an hourglass filled with sand,
  - The absorption or emission of light,
  - Friction,
  - Burning,
  - The actions of an observer, such as a movement of the head,
  - The heating or cooling of an object,
  - Phase transitions in substances (solid to liquid, liquid to vapor, etc.),
  - Dissolving and mixing rates of chemical substances,
  - The "fading" death of plants,
  - "Non-light radiation" from astronomical objects,
  - (Sudden?) changes in states of human consciousness and the emotions,
  - Coupled variations in temperature,

  (See: "The Source Field Investigations" by D. Wilcock) [12], (also see: "The Ether Model as Result of the New Empirical Conception" by A. Mishin) [13] (also see: "The Field" by L. McTaggart) [14].

- SQ ether flux is the origination and cause of:
  - Non-Local Thermodynamics!
  - Van Der Waals gas,
  - Non-conservation of energy in open, far from equilibrium, thermodynamic systems (such as this Universe) (See: The Category Theory of Topological Thermodynamics) [15],
  - Negentropy (organizing, ordering, restorative, and energy-gaining behaviors, as opposed to entropy, which produces chaos and loses energy).
- SQ ether information-carrying flows are the origination and cause of:
  - "Esoteric", "Paranormal", and "Subtle Energy" Fields,
  - Telluric Currents,
  - Vril and its information-connective and energetic behaviors in Creation Processes,
  - "Inertons" (V. Krasnoholovets) and their physical attributes [16],
  - Chi,
  - Huna,
  - Orgone and its effects (Reich),
  - Od and Odic emanations, and their effects (Baron von Reichenbach),
  - Telekinesis,
  - Telepathy,
  - Empathy,
  - Clairvoyance,
  - Healing touch,

  - Remote healing,
  - "Remote Viewing"
  - Precognition,
  - ITC,
  - Premonitions,
  - Sheldrake's "Morphogenic Field" and morphic resonance (http://www.sheldrake.org/) [17],
  - Mesmer's Battery (Anton Mesmer, France) and its effects [18],
  - Galvanic Battery and Galvanic Current (Luigi Galvani, Italy) and their effects [19],
  - Radionics and its effects,
  - Psychotronics and its effects,
  - Psionics and its effects.
- Information-Bearing SQ ether flux, is the origination and cause of:
  - DNA sequence information,
  - DNA resequencing information,
  - "Life Force",
  - Qi,
  - Nature's Harmony (Ambient Intelligence),
  - Remote Healing,
  - Homeopathy,
  - Flower Remedies,
  - And so on.
- SQ ether flux is the origination and cause of:
  - Continuous Creation,
  - Miracles,
  - Divine Intervention.
- SQ ether flux is the origination and cause of:
  - All the attributes and effects of Consciousness. This is what we are studying now.

We need to know factually, not theoretically, what is the actual relationship of Mind (Self/Being) to the physical. And we need to understand the functioning of the interfaces between Self and all levels of experienceable Reality.

In essence, the brain interfaces directly into the SQ Information/Being realms, which is where the Self actually resides, when it is not connected with the so-called Physical Universe. No manner of Being (Self) actually needs a brain to have direct experiences, but the Being uses a specific physical vehicle for interfacing the Self with the given Physical Universe, to have specific types of experiences and to accomplish various types of activities. The vehicle of Consciousness can be biological, non-biological, or even non-physical (e.g., Beings which are coherent SQ clouds of semi-material form, with temporarily material attributes and the ability to directly influence matter, such as Angels or "Orbs") [20].

The Consciousness Itself is inherently in another realm and has selected to interface with this particular Reality through one or another type of various possible vehicles. But everything is all connected to everything else, at the level of Pure Awareness, while selected experiences are had through an inherent ability of the Consciousness, the ability known as Attention. What the Attention is on determines what will be experienced.

SQ information inputs to the brain, from the Being, and inputs from the Ambient Intelligence, are part of an Active, Aware, and Intelligent, Living Universe. Information inputs cause changes in the quantum functions which are found internal to the brain cells. These quantum behaviors arise in the form of the activities of single "isolated" electrons captured inside "quantum shells" which are internal to "micro-tubes", physical structures which are internal to cells [21].

The observed quantum-originated functional changes internal to brain cells, produce the electrical activities seen in the nervous system. These activities then feed back into the Being, in a process known as "cognition". In other words, our experiences of Being Alive and getting sensory and intuitive information from our surroundings, are cognitions. We can observe then, a bi-directional experiential feed-back loop between Being and the brain.

The idea that All Reality is Holographic, is quite appealing, and is increasingly mentioned in the scientific and philosophic literature of our time. But what does the term "Holographic Universe" actually mean?

Is the Brain structured according to some hyperdimensional blueprint, able to produce a consciousness different from its own physical nature by some still totally elusive way? Or is it, by its conjectured hyperdimensional nature, a unique resonant system, which is able to tune into an external (non-physical) field made of complex information, such as the Quantum potential?

Is the brain an Information generator, or is it instead the receiver of information, transducing, and then processing, our sensory and intuitive experiences? Is the existence of the brain a prerequisite for high-level Subjective Information processing? Overwhelming empirically tested, experimental and experientially based, compelling evidence has appeared in recent times, which supports the understanding that the Brain is the receptor, not the origination, of Consciousness. These important results arise from the entire spectrum of modern scientific developments, pointing us towards a non-materialistic worldview.

The non-reductionist perspective is quickly replacing old mechanistic reductionist dogmas and fixated unquestioned assumptions. However, some basic ingredients seem to be still missing from most of these new non-reductionist concepts. Our hope is to provide public access to the various missing understandings, thus offering a completely new horizon, so as to alleviate existing global academic struggles which aim to find a self-consistent, unified and comprehensive expression of Information's physicality and the fundamental behaviors of Information which are directly influencing the foundations of Physics, as illustrated in the list above.

What is this Reality primarily made from? Matter, Energy, Information, Consciousness, or all of these? This is the crucial point we decided to challenge, in our meta-conceptual framework which is addressing Space, Time and Information on an equal footing, while carefully detecting various global interaction patterns and mutual implication levels – on a scale expanding from infinitesimal entities, (the primal Subquantum Units), which live far below the Plank (10^{-33} cm) or even Kolmogorov (10^{-54} cm) limits, to infinitely large scales of Being, where metric descriptive terms vanish.

In this no-beginning and no-end Universe continuum, creation is running in a continuous flow, as emergent processes from the very essence of Being, the all-contending overarching Cosmic Harmony, usually equated with a deterministic Divine source.

The basis of Continuous Creation concepts underlying this perspective result in a non-paradoxical, perfectly logical whole, which makes the "Big Bang" and the related "expanding Universe" obsolete, physically untenable conjectures. (Quantum Cosmology) [6].

For our current purpose, it will suffice to consider the perpetual creation/dis-creation processes which are universally active and endless. There was never any "beginning" of the Universal Subquantum Plenum that manifests as "Space" and Matter and Energy and Life and Consciousness. As mentioned above, entropy and negentropy are balanced in an infinite volume universe. The SQ is the limitless foundational framework of all manifestation. Dynamic events and patterns are governed by Informational presets which evolve in a not less infinite range of velocities the Sub Quantum Units and their combinations are demonstrably engaged in. We address ultimately the Universal Information storage and carrier systems, which are both the source and result of the quite vast incomprehensible "everything".

"Every living being is an engine geared to the wheel-work of the universe. Though seemingly affected only by its immediate surroundings, the sphere of external influences extends to infinite distance." ~Nikola Tesla (New York American, February 7, 1915).

This novel approach is rooted in an overwhelmingly wide pool of information, originating in a comprising, objective, experimental, and multidisciplinary extended database. This database includes in its explanatory range, both "normal" and "anomalous" information-related phenomena, in a far-reaching, paradox-free, and coherent epistemological framework.

Novel Subquantum concepts (which differ from previous SQ concepts, published by respected authors such as DeBroglie, Bohm, Pearson, LaViolette, Valentini, and Zazerskiy, to name but a few), allow for superluminal velocity propagations in an Information-Controlled Physical Reality.

Subquantum expressions can describe the originations of the experiential windows which comprise our sensory awareness systems, as they are made available to the (Coherent SQ) Being (also known as a Biologically-Involved Consciousness). This is originating in the physical and energetic dynamics of our particular kind of Reality.

When the electric field at a given point, is evaluated according to Liénard–Wiechert potential, the result is identical to the result which obtains when one hypothesizes that the Coulomb field propagates with an infinite velocity. Experiments performed during the past several years have confirmed this hypothesis to be valid: The Coulomb Field does propagate with an infinite velocity.

For example, see:

- http://www.pandualism.com/c/coulomb_experiment.html [22];
- http://gsjournal.net/Science-JournalsPapers/Author/1624/Allen,%20Robnett [23];
- https://link.springer.com/article/10.1140/epjc/s10052-015-3355-3 [24];
- http://prola.aps.org/abstract/PR/v39/i4/p616_1 [25].

That magnetic fields may also propagate at an infinite velocity, is implied in some of the more recent results. It has already been expressed by many, such as Newton, LaPlace, Eddington, Mach, Jefimenko, Van Flandern, Podkletnov, and Boyd, that gravitation also propagates with an infinite velocity. We already know that quantum physics experiments have proven quantum "non-locality" (infinite velocity) as an irrefutable fact. A Subquantum infinitesimal particle, with an infinitely small mass, can easily propagate at an infinite velocity, as demonstrated by the several Mobius (bilinear) transformation solutions to the Maxwell equations found by V. Fock, R.M. Kiehn, R.N. Boyd, and T. Smith [5].

So it appears that infinite velocity is the most common velocity in the Universe, with light speed being an exception to the otherwise consensus infinite velocity dynamics of the Universe, with less than light speed events being more common than light-speed events. Infinite velocity SQ dynamics is the basis which unifies all the sciences.

Adding to the list from above, the following items are relevant to SQ dynamics:

- Collider results are intermediated by SQ particles [26].
- The Weinberg angle is due to SQ intermediary interactions.
- The Cabibbo angle is due to SQ intermediary interactions.
- Weak force is a SQ particle group behavior interaction.
- "Strong force" – Penrose agrees with us that there is no strong force [27].

There are no so-called "virtual particles" [as conjectured by QED]. There are intermediary SQ particle interactions, comprised of actual SQ particles [28].

The mass of the Yang-Mills gauge field is infinitesimal, and it is intermediated by SQ particles.

Planck's constant is not a constant (experiments in this regard were performed in 2002) [29].

Distance is an absolute. That means volume is an absolute. (The universe does not and cannot "expand".)

Einstein's version of "relativity" produces a false identity between distance and time, which is like saying that a meter is identical to a second.

This false identity was required in order to appear to validate another false identity, which was created between light speed and time, which is like saying a given velocity is identical to a second. Velocity is measured in terms of distance traveled per unit time.

However, a 60 MPH velocity is not the same as a minute. To compound this irrational concept, it was further stated that distance is a variable ("contracting") where light speed is concerned, an experimentally unprovable proposition.

In actual fact, no matter how rapidly a given distance is traversed, that distance remains constant. Also in actual fact (not theoretically), the force due to time, and the force due to gravitation are both variables. In addition, it has been experimentally repeatedly demonstrated that the speed of light is not a constant, but varies according to the local values of permittivity and permeability of the SQ media, and the polarization angle of the photons relative to the source.

Any measure of distance can be infinitely subdivided.

Relativistic electrodynamics fails [28].

Heisenberg uncertainty is refuted experimentally, in several situations. [Dehmelt, Nobel Laureate, and Boyd, Nobel Nomination] [30].

A sentient, cognizant, active, aware, and deterministic Quantum Mechanics is the result of Subquantum activities.

Properties of "Quantum Matter" Lie Mid-Way Between Classical Matter and Mind [31].

Eidetic Information is additive and cumulative.

Kolmogorov vortices (10^{-58} cm) naturally result from extending Kolmogorov's studies of turbulence towards the small [32].

A SQ microscope has been designed by Boyd, tested for proof of principle in Germany, and constructed in Serbia, which has imaged entities as small as 10^{-95} cm [33].

The Fermi-Pasta-Ulam lattice (FPU lattice) is comprised of group behaviors of SQ particles [34].

Given the vast memory capacity of the infinitesimals, we have postulated that they may be inherently hyper-dimensional entities, responsible for many or all of our hyperdimensional observations as related to the physics [35, 36].

There are many more items which should be in the list, but this listing, combined with the listing above, should be sufficient for now.

Infinite velocity SQ propagations in an overarching Cosmic Harmony, makes perfect sense in our Continuous Creation/Dis-Creation model, where atoms and subatomic particles are dissociated or annihilated by photonuclear interactions with gamma rays, and by SQ interactions. SQ information transport mechanisms, which are involved in all known fields and all forces, contend an infinite array of potentially structured informational combinations. Thus any Informational structure of a given complexity may be traced back, as an implementation aspect of the Source, where it exists Beyond Time in its "virtual" form.

Dynamic events and patterns are governed by Informational presets which evolve in a not less infinite range of velocities the Sub Quantum Units and their combinations are demonstrably engaged in. We address ultimately the Universal Information storage and carrier systems, which are both the source and result of the quite vast incomprehensible "everything".

Information patterns cluster in various Qualia (sensory items) defined structures, fall in the range of the Being's ability to process information. Among biological entities with brains, this happens by way of genetic activations of the Quantum connectivity of brain, which are resulting from environmental observables, which was studied by Gariaev [35]. In some situations, information is circumventing conscious anchorage functions by direct resonance modalities operating between the Self and higher-implicated orders [36] of SQ-mediated Informational stuff.

This second variety of information input (transcendent and intuitive SQ information) is a range-transcending one, which works in perfect coordination with lower (Quantum mediated) processes, which are normal to the brain's Quantum efficiency. As a result of combined clusters of additive Information contents, new stable SQ Information structures are formed, operating in different regimes of SQ flow speeds, and showing diverse resonant aggregation tendencies, according to the complexity levels of the information. This process occurs along the lines of the interactive principles we explained earlier.

Moreover, it seems reasonable to assume that total Information capacity of SQ information combinations reflects the sum of the components, though not in a linear way (applying no metric), but in a manner that is exponentially magnified, and selected, along non-linear vectors of non-quantifiable experiential increments.

As amply emphasized earlier, everything in physical manifestation reflects by its Unique Personality, a personality structure with an infinite variety of accompanying energetic signatures (such as emotional information) which can convey interactively to other Beings, the Personality Information they store and carry at any given point in linear time. (See the above description of "Cherokee hunting group" communications.)

This kind of communication channel might be described in such instances, as an information bearing etheric string that connects the hunting group's individuals, to the star's current celestial position. (The distance to the given star, in terms of "light years" has no relevance, but the actual location of the star at the time of the experience is relevant.)

The group of hunters is "tuned in" to the star by way of a direction-specific etheric string which is carrying eidetic information. Thus the hunters are able both to transmit and receive eidetic information and states of being, as transmitted and received by the star, by receptively integrating the information thus provided by the star (upon accomplishing Unification with that star through unwavering Attention), into their own personal cognitive domains.

This kind of communication happens when one Consciousness becomes unified with another Consciousness, through the faculty of unremitting Attention (no thinking). The Consciousness faculty of Attention results in progressive emotional and informational resonance with the object of the attention. Ultimately, Unification of the Observer and the Observed occurs, to the degree that such Unification can be comfortably tolerated, by the individual. So, in the example given, the hunters unified with the Consciousness of the selected star, which acted as a communications terminal for the hunters, on their way to find deer to feed their families with.

Endless similar instances have been described for cases where birds, trees, mountains, other objects, or even thin air, mediate meaningful Information, which is stored internal to the SQ structure which interpenetrates all things, sometimes with accompanying energetic cues, such as visible discontinuities or strong "feelings", when such experiences are accepted into the experiences of the individual, without judgment or prejudice. Energy field conversions and Unifications, which act to convey Eidetic Information by progressive emotional and informational resonances due to the faculty of Attention, are amply documented in both experimental and experiential literature.

Such information storage and transmission occurs in SQ-controlled non-metric (subtle) domains, by the vehicles of the SQ particles, which are beyond the Quantum activity range and operate in higher time-resolution domains. In a higher time resolution, "Quantum Jumps" behavior is further divisible, ad infinitum, offering, from Informa-tional perspective, an endless chain of intermediate states corresponding to different implication levels [37] of subtle Information-based control.

With these understandings in mind, let's adventure into contemplating the real nature of Self and the self-conscious Information processing modalities as seen from our SQ perspective.

Individual self-conscious patterns are formed at the "monadic" ("monad", meaning "the Oneness, individuated") level of the upward/downward bi-vector Information-processing apparatus of Nature. These individualized structures reach interactive expression and integration ability in the pre-physical and physical domains while preserving their own complexity and coherence during interaction processes. The Informational Being descends into matter/energy, by ensuring – along a "downward" path of lower-complexity morphic resonances and pattern-generating effects – the setup of a fully functional biological form and its energy-transducing nervous system, which is the Being's biological interface with the Quantum-operated physical world.

Of course, the same behavior applies to Informational blueprints of any complexity, be them reactive, conscious, or self-conscious ones, which result in nature's various "kingdoms" which are biological, non-biological, and "pre-physical" domains. Everything, including rocks and clouds is sentient, and "alive", an understanding which disagrees with prevalent crude biological criteria. This view is easy to reach when one admits the definition of an infinitesimal as the smallest unit of Consciousness. Then, since everything is made from agglomerations of infinitesimals and their activities, Sentience is involved in all things and all activities, at all scales, down to the infinitely small and up to the infinitely large.

Everything is a dynamic Information interface, exhibiting the bidirectional flow of content running along interacting proxy velocity vectors and extending in upward and downward interference patterns inside given windows of coaxial homogeneity blueprints. These patterns in turn participate in endless sequences of increasing complexity as information combinations go up to (or down to) an undifferentiated proto-information field which comprises an infinite expression potential.

Our main challenge is to find out, in an experimentally provable range: Where are the human consciousness and its brain located on the overall map of Consciousness Domains? An acceptable understanding of this question implies, as a prerequisite, an accurate epistemological mapping of the various subtle energy biofields comprising the SQ essence. This ensures that a seamless integration of interactions between the person, and various environmental factors, can be correctly mapped.

Our approach to living systems is vastly transcending vitalist doctrines, which are limited to postulating a vital principle, as distinct from biochemical reactions.

According to J. White [38] who references to various bio-energy field control mechanisms, can be found in 97 different cultures. In common perspective, biofields have been defined as an endogenous coherent EM field, which is related to living organisms, as emergent from classical cellular fields, comprising some new features, or alternatively, as an SU2 symmetry-related complementary phenomenon related to classical EM fields. In various attempts at definition, biofields are expected to restore the long-range connection modalities which we observe between molecular and supra-molecular processes (of the kind we see in cytoplasmic functions).

Classical interpretations of biofields, as exclusive features of living systems have been repeatedly questioned by supporters, regarding its origin in symmetrical physical demands which may be applied to the whole of Nature. This view is ascribing to living systems a more highly complex reaction behavior regarding such requirements, than non-living matter is able to provide.

This approach is consistent with Froehlich endogenous coherent oscillations, invoked for the explicitly nonlinear sensitivity of living systems, to weakly ionized and non-ionized EM fields [39].

Nevertheless, the biofield's observed coherence, and its partial inability to be localized, suggest its enfolded position, as lying between energetic and Informational processes, a taxonomic requirement which is fully satisfied by our SQ model for sentient reality. The biofield's non-local essence was first confirmed by instantly-transferred cortical photo-stimulation experiments, performed during 1994 (Grinberg, Delanor & All) [40].

According to Thaheld (2005), [41] known EEG correlation patterns fall short from explaining the nonlocal bio-correlational distributions of neural ionic currents and the strength and densities of electric fields, operating in brain. These observations are consistent with a new approach (introduced by Driesh), [42] where subtle bio-regulatory processes are seen as being the result of "mind-like" organizational effectors, rather than physical energy fields (of the kind invoked by R. O. Becker & G. Selden [43] in their experimental results related to electrically de-differentiated cells).

How are the biofields coupled into both energetic and Informational fields?

In order to answer this question, a particular category of SEs ("subtle energies") (see our Nov. 2008 installment) [44] is constructively invoked. We remind our readers about Zero Point Fluctuations (ZPF) in Vacuum. ZPF and quantum jitter provide direct physical evidence of the existence of the SQ. These behaviors demonstrate the infinite energetic background of the physical world, by a supposedly ceaseless production of physical particles from the ZPE, which are subsequently dissolving onto the background by a process of hyperdimensional inter-plays, which are occurring between different enfolded orders [37].

This down-conversion of multidimensional SE fields, aimed at coupling into 3D force fields, requires a novel energy/Information transduction technology (see Clarus' sympathetic resonance technology) [45].

Sympathetic resonators are capable of SQ interactions with the electromagnetic spectrum, ordering its fundamental magnetic vector properties, and affecting, through the Quantum field, interactions and physical processes which are occurring in the subatomic range. From this perspective, the bio-energetic human chakra system, is conveniently placed at the crossroads of energetic and Informational SQ interactions, where the dynamics of perception, attention, and intention, determine our biological reality. Such processes occur by way of ether flux information, which flux is known in Vedantic teachings as "Prana", which corresponds to the Chinese notion of "Qi".

Early experimental protocols related to the physical aspects of the biofield, were performed by B. Payne, in 1981, [46] pointing to a biofield strength which was observed to be on the order of over 100 million times greater than the magnitude of the body's magnetic field. Biofield deflections have also been correlated to synchronous geomagnetic index fluctuations, pointing to physically originated bio-energetic components.

At the same time, non-physical correlations of biofield intensity values, are consistently related to various emotional states, in experimental subjects. Such analytic parameters seem to attest to an increasing complexity of responses in bio-systems, to environmental geomagnetic influences, of the kind reported in R. I. Jones' uniform daily rotation experiments in plants [47].

Vortex behaviors seem to be a foundational basis, and are evidently a primary and fundamental organization pattern, in our Universe. Kolmogorov vortices are at the smallest scales of turbulent flows. At the Kolmogorov scale, viscosity dominates and the turbulent kinetic energy is dissipated as

$$\boldsymbol{n} = (v^3/e)^{1/4},$$

where n is the size of the vortex, v is kinematic viscosity and e is the average rate of dissipation of turbulent kinetic energy, per unit mass [48].

We have discovered by direct derivation, that vortex formations exist all the way down to the 10^{-58} m Kolmogorov Limit, which is the smallest entity which is still influenced by gravitation. Entities smaller than this are the primary agencies causing gravitation, in our modernized version of the infinitesimal model of Gravitation published by the Marquis de LaPlace during 1870 [49].

Some have considered that torque forces may interact with EM, gravity and nuclear forces, modulating their respective effects, especially around biosystems. This concept, which is highly relevant for biofields, strongly supports the notion of an increased Information field involvement, of the kind our SQ model suggests, as regarding high-level functions which are evolving in the brain.

We think that similar control mechanisms are at work in the various lower-level bio-regulatory networks as well, such as splicing, methylation, glucositation and protein synthesis on ribosomes, as well as being involved in all other biological integration processes.

We are facing a different, non-chemical level of organization, one which is reminiscent of morphic causation principles, which have been proposed as early as 1944 by A. Girwitsch [50]. Girwitsch ontologically differentiated between a physical embryo and its organizing pre-patterning morphic fields, which call forth genetic responses.

From there, S. Savva's [51] concepts of the mind components of biofields (1997), serve as behavior controlling aspects in all fundamental biological programs. From there, further on to our SQ-driven Information control mechanisms, the way was widely open. As Savva pointed out, biofield control operates through four independent control subsystems, which are the nervous system, chemical interactions, coherent biophotonic EM (Van Wijk) [52] and Subtle Energy channels (Subtle Energy is an energetic which not reducible to any of the well-known fundamental physical forces, such as electricity).

This special feature has been experimentally proven by Bockris (1997) [53] in low temperature nuclear transmutations of lead into gold, in concentrations up to 300 ppm. These transmutations are only occurring in the presence of a "psi operator", a person who is capable of intentionally speeding up, and slowing down, by an act of will, the nuclear decay rate of americium (Yan Xin, 2002) [54]. Such experimental results are strongly consistent with the Princeton Engineering Anomalies Research (PEAR) laboratory publications of Princeton University which demonstrate predetermined expectation deviation values which are far from randomness, during Random Event Generator (REG) experiments [55].

As a consequence of the special interdimensional status of biofields, they cannot be significantly blocked by any physical screening. Their effects disobey all space constraints, of the kind that the known isotropic physical fields display.

Bio-information fields are composed in anisotropic, and net-like patterns, which do not attenuate with the square of the distance. These fields are seen acting along the exact lines of any pre-established emotional or intention bond. At lower levels of unfolding information controls, human-originated bio-information determinants have been proved to influence bacterial growth parameters (E. Rausher, B. Rubik, 1966) [56]. Such observations open wide perspectives of cytophysiological and cytogenetic control leading to perfect control of DNA-related systems [8].

More comprehensive data may be found in H. S. Burr's well-known concepts related to L-Fields (life organization units), [57] as well as the works of his collaborator's. L. Ravitz' works [58] are pointing at a direct relationship between the L-Field and a person's mental, physical, and emotional conditions. Ravitz reportedly has even been able to show that the L-Field, as a whole, disappears before physical death. Such observations have been correlated with anecdotal extrasensory monitoring of sudden disappearances of the normally observed auric fields, around those individuals who are facing an impending, and sudden, but unexpected death.

Quantum bio-holographic data are increasingly invoked in advanced modellings of sentient reality. In the light of modern investigations, DNA is seen as a self-calibrating antenna, which is able to process quantum holographic information stored in the SQ portions of photons, by phase conjugate adaptive resonances. In the brain's overall biofield, narrow spectral frequency-related interdimensional windows (e.g., 38-40 Hz) allow sharp frequency-adaptive couplings of organized enfolded Information matrix fields, which live in projective space, which are coupling into environmentally-originated, neurally supported, Quantum configurations.

At a different level of enfolded information, similar pre-geometrically organized information patterns, related to embryogenetic determinants, are able to set up stress gradients in the vacuum, these behaviors occur according to embryogenetic holography principles, thus guiding the implementation of the ontogenetic pathways, in accordance with the brain's anticipated task performances. From this novel perspective, brain structures and connectivities, organize under highly complex biofield determinants, as a Quantum sufficient energy-processing tool capable of shifting its own integration range, into higher complexities of information matrices, thus producing the Being's expression in the material world.

The common denominator for both the brain's quantum activity, and the Informational Being, is in their common hyperdimensional Subquantum background, which is able to accommodate asymmetrical monistic determinations, which we propose should replace obsolete dualistic and panpsychist models.

Perhaps the most promising recent development in bio-energy-backed technology, with an unprecedented potential in advanced health care programs, is introduced by UCLA Professor Emeritus Valerie Hunt [59]. Her Aurameter (TM) is able to detect, in a predictive way, anti-coherent frequency bands in individual "Signature Field" patterns of subtle energy, making possible their transactional shift into holographically balanced health patterns, by perfectly calibrated frequency transfers.

Most interestingly, Prof. Hunt strongly emphasizes that the emotional constellations which are surrounding stagnant memory fields, are originating in past "lifehoods" (a term she uses in a metempsychotic context). Such emotional constellations have persisting blockage effects on current-life subtle energy dynamics, which are directly related to health. Such anti-coherent frequency trends seem to be stored in fundamental SQ domains, and are composed of Information which is embedded internal to the overall biofield structures, where the brain's bio-energetic components are also embedded.

These findings strongly support reportedly successful hypnotic regression therapies, which tap into memory storage systems which originate from personality-related events from the remote past. Recalling of such memories is possible as a result of resonant alignments between various SQ conformal information maps, which are presented as superimposed information symmetry configurations in the brain.

As described elsewhere, in the monadic spectrum of Subquantum interferences, a relatively stable Subquantum condensate forms by ether flux convergence running in higher (transpersonal) fields. This is an individualized proto-conscious crystallization of mixed Informational content, acting as further downward inductor of lower-complexity aggregated states, in sequentially higher density ether fields.

During this process, the conscious monadic essence is enabled to extend into and interact with progressively lower reality domains. A seamless chain of morphogenic (shape-creating) effects are induced in adjacent domains along this "downward induction" pathway, which run under the coordination parameters which are originating in the monadic effector, responsible for its initiation.

Emotional and cognitive functions related to this progressive immersion of the Being into low-velocity Sub Quantum regimes are unfolding, at their respective levels of manifestation, according to the dynamic permeability of the respective media, down to the domains where the physical laws become effective.

Progressively, obeying these laws and constructively using them by downward Informational control, an appropriate interface between the Self and its physical environment is implemented by activating corresponding bio-morphogenic fields in subtle energy domains (etheric body) then normal matter, able to organize – and further on to perform stability maintenance requirements of the thus formed interface system.

This process allows for an optimal efficiency in the interaction between the Being and its rough chemical-physical vibration and Quantum-mechanical media, which it transiently inhabits:

One by one, subtle high-energy fields pertaining to morphogenic processes originate the set-up of energy centers and communication systems which are at the limit of physicality, according to well-organized blueprint presets (the Chakras and their nadic webs of connections), preparing in this way, the foundational physical mechanism, the ultimate tool, which anchors Being into matter (in our physical human situation), the brain.

Subquantum very-low velocity residual components of the morphogenic apparatus, engaged in stability preservation of the bio-system, are easily detectable as cloud-like etheric formations interpenetrating physical macrostructures. The reactive efficiency of these cloud-like structures, to changing conditions (health parameters at organismic, systemic, histological, cell and cell components levels) has been amply described in the relevant literature, as polychromatic auric events. Morphic (shape producing) resonance fields are observed at work, not only in downward influences (as just described), but also in upward synthesis processes of spatial distribution presets, which are activated by various energy supplies (palyngenesis, DNA-phantom etc.). So, a bi-directional feed-back loop is always active and available for physical manifestation.

Without addressing here, subtleties related to the chromosomal radiations which are carrying and applying DNA Information signals (of both local and cosmic origination and distribution) to "self-organizing" matter in biosystems (far beyond the scope of this contribution), let us explore the result of a physically accomplished brain structure, and its multitasking interfaces, which interact with the Being's various cognitive abilities.

It's quite obvious that at a biological level, all the observable structures (from cell constituents to organismic ensembles) serve the sole purpose of maintaining a functionally efficient brain structure, able to ensure the individual's accurate navigation in its environment.

Anatomically and functionally, the brain is obviously the most complex object we know. At the level of its multi-layered neuro-biomolecular network systems, there are obviously at work, distribution orders which arise from far beyond internal structural codes (DNA). This is possible due to the brain's hyper-dimensional essence, connecting it to non-physical Universal matrices (via the right temporal lobe and its associated limbic system). This also relates to the mechanism of the morphic resonance-mediated control of post-genetic ontogenetical phase correlations.

Nonlocal Subquantum Information matrices are brought into resonant alignment with existing neural connectivity presets, according to shared dynamic configuration maps. (In some poorly peer-checked references, a special task in Information storage, transport and coupling to matter has been suggested for inert Krypton configurations operating at the hypothalamic walls of the brain's tertium ventriculum.) Though this conjecture is not yet investigated enough, SQ Information content might easily be stabilized in chemically inert gases and then delivered into biosystems, at their higher range, thus controlling the thalami-hypothalamo-hypophyso-adrenal axis of global emotion-controlled endocrine regulation.

Extending the Rubner-Kleiber scaling computation laws, K. Togbey & All [60] performed a fractal analysis of the relevant allometric exponents of the Brain, reaching a 4/5 law. This implies that a 5th dimensional function is present in resonant Neural Network connectivities. This finding relates the brain's Information-coding abilities to the Golden Mean ratio, by equating EEG metrics to number of harmonics times 2φ. Wave packets, scaled in powers of the Golden Mean, exert a patterning action upon the time-lags separating neural firing events, over several orders of magnitude. Harmonic spectra below and above the brain's fundamental 2φ frequency matrix, are involved. Neural networks operate according to powers of 2φ/2, efficiently using them for Fibonacci series related Information coding, involving Hausdorff dimensions which are implementing the fractal distribution patterns of SQ information structures.

Log-normal and power-Law distributions of these variables are thus governing inheritances and aberrant processes in biosystems, where the power-law distribution accurately describes the Fokker-Plank equation, as applied to the discrete variables associated with SQ Informational Units and their combinatorial matrices.

The brain's Quantum sufficiency, as amply discussed in recent research data, allows for the ceaseless bidirectional flow of various Information patterns, displaying a wide range in their degrees of complexity (dependent on the underlying Sub Quantum configurations), across the aforementioned hyper-dimensional interface. These flow patterns run both in neuro-cybernetic linear Time and also along retro-causative non-physical signaling pathways, in the domain of hidden variables, violating Bell's theorem (P. V. Kurakin) [61] conforming to our infinite-velocity Subquantum information transport understandings, also known as the Quantum Potential.

Photon emission/absorption events are not random probabilistic frameworks, as current mainstream views suggest. Instead, electromagnetically conveyed information deterministically implements informationally enfolded guidance of the energetic (Quantum) phenomena as observed in the brain. Target-related conformal signal transmission blueprints reach various targets in the neural domain of Quantum activity in zero time. This is accurate, but the infinite velocity involved in such processes is poorly supported by Cramer's conjecture of "advanced and retarded time" in his transactional interpretation of Quantum Mechanics. Given the fact that experiment has proved that the motional portion of the Subquantum ether, in the form of the Coulomb field, is able to propagate and convey information at an infinite velocity [22], Cramer's "advanced and retarded time" is not the correct understanding [62].

Thus, a perfectly coordinated deterministic control is implemented upon the unfolding action templates involved in neural activity, which underpins the overall outcome of

Consciousness-related Quantum processes, and the elusive features of phenomenal continuities of Consciousness. Information subunits aggregate, in both directions along the Time vector, into an experiential "nowness", which is accessible to the Conscious Being. At the same time, this process provides the global reference frames for mnestic structures, which are stored in atemporal Information fields, which are actually required for setups of specific connection meanings. Such events involve Quantum non-local effects, originating in instances of space-time-free hyperdimensional couplings of the SQ Units, to Information fields, implementing in this way a much more efficient, effective, and inclusive communications system than the one predicted by standard Quantum entanglement schemes.

From this advanced perspective, it's quite reasonable to assume that brain operates in real physical time at the Quantum level, but it also operates as an interdimensional resonator, one which is capable of coupling hidden-time SQ signal processing events, into bio-molecular energy-related Quantum states.

Time-patterning Quantum events are only the statistical results of deterministic SQ processes, which are evolving in zero-time frameworks, which we have associated to a Grassman projective space, whose limiting singularity is the Quantum-embedding Euclidean space.

Our extended team reached by now a quite advanced stage of developing an infinite tetration and self-root related mathematical formalism able to express these kinds of interdimensional transitional geometries, involving non-integer hyper operations, which will highlight the many new connections hereby suggested, between Informational 3D spinor spaces, Grassman projective information space, the Pfaff dimensions, the Lie algebras of infinitesimals, and the Euclidean volume of brain, putting in a new light on the brain correlates of the activities of the Being, some of which have been previously addressed by the Penrose-Hameroff "OrchOR model" of Consciousness [21].

Being is, by its axiomatic essence, ontologically irreducible relative to the bio-molecular backgrounds which are required for setting up the various resonant Quantum maps which are involved in Consciousness. The internal coherence of Consciousness results from an Informational homogeneity at the most basic analytical level, and as an individualized monadic structure which is interconnected with the Universe as a whole and follows sequentially higher orders of guidance and coordination sources.

In order to reach the physical realm of energy, it projects a fragment of its holographic Universe-as-Consciousness structure into a previously morphogenically caused transactional anchorage system, the brain, thus establishing fundamental interaction patterns as briefly discussed here.

The Consciousness window of the biological Being is centered, during its brain-coupled condition, on fulfilling environmental integration tasks related to time-dependent fluctuations of information-space configurations, in their own causation and retro-causation range.

Ontogenetic compensation effects remove automatic physiological functions from the active attention field into autonomic neural network activities, at the lower limit of the Consciousness Window (remaining still accessible to its control, by specially trained mental procedures addressing the supposedly "subconscious" domain).

A similar fluctuation in Information transfer permeability into and from the Conscious domain operates at its upper limit, too, allowing for "supra-conscious" data intrusions into Quantum-sensitive processing spectra, implemented by intuitive and extrasensory channels.

The integral dynamic structure of the Being is engaged in a ceaseless increase of its complexity by the Information exchanges which are running in all of its effective domains, the brain-coupled one included, as long as this coupling is at work. The resulting integral complexity is preserved as the Being's overall cognitive spectrum disregarding its temporal and partial interaction with a transient physical brain.

Trans-personal connection abilities operate undisturbed, even after disconnecting from the physical energy anchorage (physical form) temporary used for enhancing its gain of understanding. This view is strongly supported by an overwhelming experimental database which has piled up for years, in modern scientific literature. This understanding is very hard to dismiss upon previous obsolete dogmatic backgrounds.

The brain is an ideal anchorage for organized Information fields which arise from ambient energy-information processes. In more complex biological organisms, the brain supplies the active transducing components for integration, implementing symmetrical Information flows between interconnected Quantum structures, which are involved in the ambient environmental coupling process, with the brain.

Everything is constantly emanating and absorbing Experiential Information, which information is due to Consciousness and is conveyed to and from all things, and stored in all things, by the bipartite Subquantum ether. Everything is and has Consciousness. The Formative World is the continuous potential of Creation which produces, sustains, and dissolves all physical forms. The physical experiential world rests on, and is the result of, activities in the Formative World. Whether or not the individual is cognizant of it, the Cosmic Harmony is always acting in, and as, Nature.

REFERENCES

[1] Boyd R. N. 2017. *Worlds within Worlds.* June 23. Accessed June 23, 2017. http://worlds-within-worlds.org/publications-and-research.php.

[2] Laughlin, Stormer and Tsui. 1998. *"The Nobel Prize in Physics 1998 - Advanced Information".* Nobelprize.org. https://www.nobelprize.org. Accessed 1998. https://www.nobelprize.org/nobel_prizes/physics/laureates/1998/advanced-physicsprize1998.pdf.

[3] Byckov, V., V. Krasnoholovets. 2000. "Real inertons against hypothetical gravitons. Experimental proof of the existence of inertons." *Indian Journal of Theoretical Physics,* Vol. 48, No. 1, 1-23.

[4] Krasnoholovets, V. 2014. "Submicroscopic viewpoint on gravitation, cosmology, dark energy and dark matter, and the first data of inerton astronomy." In *Recent Developments in Dark Matter Research*, Edited by: Nori Kinjo and Akira Nakajima with participation of Dr. Roldão da Rocha, 1-61. New York: Nova Science Publishers.

[5] Smith, Tony. 2008. "Segal's Conformal Physics and Conformal GraviPhotons." *Tony Smith's Home Page.* June 24. Accessed June 24, 2008. http://www.valdostamuseum.com/hamsmith/SegalConf.html#conformalEMG.

[6] Ali, Ahmed Farag, Saurya Das. 2014. *"Cosmology from quantum potential"*, Cornell University Library, https://arxiv.org. Dec 29. Accessed March 18, 2016. https://arxiv.org/abs/1404.3093v3.

[7] DePalma, B. 1991. "Magnetism as a Distortion of a Pre-Existent Primordial Energy Field and the Possibility of Extraction of Electrical Energy Directly from Space." *26th Intersociety Energy Conversion Engineering Conference, I.E.C.E.C., sponsored by the I.E.E.E.* Boston: I.E.E.E.

[8] Gariaev, P. P., M. J. Friedman, R.N. Boyd, and E.A. Leonova-Gariaeva. 2006. *"Resources."* Worlds-Within-Worlds.org. worlds-within-worlds.org/resources/GarFried06-01.doc.

[9] Bohm, D., B. Hiley, 1993. *The Undivided Universe: An Ontological Interpretation of Quantum Theory.* London and New York, Routledge Press.

[10] Cheng, David K. 1993. *Fundamentals of Engineering Electromagnetics.* New York: Addison-Wesley.

[11] Kozyrev, N. A. 1967. *Possibility of experimental study of properties of time*, Pulkovo. http://www.univer.omsk.su/omsk/Sci/Kozyrev/paper1a.txt.

[12] Wilcock, D. 2012. *The Source Field Investigations: The Hidden Science and Lost Civilizations behind the 2012 Prophecies.* New York: Penguin Publishing Group.

[13] Mishin, A. M. *The Ether Model as Result of the New Empirical Conception*, International Academy of MegaSciences, St. Petersburg, Russia. http://www.galactic-server.com/radio/mi-paper.htm.

[14] McTaggert, Lynne, 2008. *The Field Updated Ed: The Quest for the Secret* Force of the Universe. New York, HarperCollins.

[15] Kiehn, R. M. 2009. *The Category Theory of Topological Thermodynamics. url:* http://www22.pair.com/csdc/download/ParticlesWavesSH.pdf.

[16] Krasnoholovets, V. 2014. "Inerton field effects in nanosystems", in *Quantum Nanosystems: Structure, Properties and Interactions*, Toronto, Apple Academic Press, pp. 59-102. http://www.appleacademicpress.com/title.php?id=9781926895901.

[17] Sheldrake, R. 2009. *Morphic Resonance: The Nature of Formative Causation 4th Edition,* New York, Park Street Press. http://www.sheldrake.org/research/morphic-resonance/introduction.

[18] Mesmer, A. 1779. *Mémoire sur la dévouverte du magnétisme animal* [*Memory on the discovery of animal magnetism*], Geneva & Paris (For a short summary see: G. Vassilatos, 1999. *Lost Science*, Kempton, Adventures Unlimited Press, pp. 9-10).

[19] Vassilatos, G. 2001. *Secrets of Cold War Technology*, Kempton, Adventures Unlimited Press, pp 86-93.

[20] Haselhoff, E. H. 2001. *The Deepening Complexity of Crop Circles Scientific Research & Urban Legends,* Berkeley, Frog Ltd. Publishers. https://www.youtube.com/watch?v=9SJvVW8s_Oc.

[21] Hameroff, S., R. Penrose, 2014. Consciousness in the universe: A review of the 'Orch OR' theory. *Phys Life Rev*, 2014; Mar 11(1):39-78.

[22] Gasser, W., 2016. *"Superluminal information transfer confirmed by simple experiment".* http://www.pandualism.com/c/coulomb_experiment.pdf.

[23] Robnett, A., 2015. "Experimental evidence for the transfer of information at a speed greater than the speed of light." *General Science Journal,* Jan. 2015. http://gsjournal.net/Science-Journals/Research%20Papers/View/5873.

[24] de Sangro, R., G. Finocchiaro, P. Patteri, M. Piccolo, G. Pizzella, 2015. "Measuring propagation speed of Coulomb fields" *The European Physical Journal* C 75:137, Heidelberg. https://link.springer.com/content/pdf/10.1140%2Fepjc%2Fs10052-015-3355-3.pdf.

[25] Breit, G. (1932). "Dirac's Equation and the Spin-Spin Interactions of Two Electrons". *Phys. Rev.*, New York, 39: 616–624.

[26] Boyd, R. N. 2017. *Worlds Within Worlds*. June 23. Accessed June 23, 2017. http://worlds-within-worlds.org//doquarks.php.

[27] Penrose, R. 2007. *The Road to Reality*. New York, Vintage Books (Random House).

[28] Boyd, R. N. 2017. *Worlds Within Worlds.* June 29. Accessed June 29, 2017. http://worlds-within-worlds.org/obviousflawsinrelativity.php.

[29] Crothers, S. P. Robitaille, 2015. "The Theory of Heat Radiation Revisited: A Commentary on the Validity of Kirchhoff's Law of Thermal Emission and Max Planck's Claim of Universality", *Progress in Physics,* Vol. 11, Issue 2. http://vixra.org/pdf/1502.0007v1.pdf.

[30] Boyd, R. N. 2017. *Worlds Within Worlds.* June 29. Accessed June 29, 2017. http://worlds-within-worlds.org/refutationofheisenberg.php.

[31] Boyd, R. N. 2017. *Worlds Within Worlds.* June 29. Accessed June 29, 2017. http://worlds-within-worlds.org/propertiesofquantummatter.php.

[32] Fabricius, 2008. *Email correspondence.*

[33] Berndt Binder, 2007. *Personal discussions.*

[34] Dauxios, T., S. Ruffo, 2008. "Fermi-Pasta-Ulam nonlinear lattice oscillations", *Scholarpedia Encyclopedia Journal,* 5538. http://www.scholarpedia.org/article/Fermi-Pasta-Ulam_nonlinear_lattice_oscillations.

[35] Jadczyk, A., R. N. Boyd, 1998-2006. *Project Greenglow Forum Archive.*

[36] Evans, R. 2015. *Greenglow & the search for gravity control.* U.K., Troubador Publishing Ltd.

[37] Bohm, D. 2002. *Wholeness and the Implicate Order (Volume 135).* London, New York, Routledge Press.

[38] White, S. Krippner, 1977. *Future Science: Life Energies and the Physics of Paranormal Phenomena.* New York, Anchor Books.

[39] Fröhlich, H. 1986. *Modern Bioelectrochemistry* "Coherent Excitation in Active Biological Systems", pp. 241-261. New York, Plenum Press, Springer Link https://link.springer.com/chapter/10.1007%2F978-1-4613-2105-7_8.

[40] Grinberg-Zylberbaum, M. Delaflor, L. Attie, A. Goswami, 1994. "Einstein-Rosen-Podolsky Paradox in the Brain: The Transferred Potential". *Physics Essays,* Vol. 7, Number 4. http://www.deanradin.com/FOC2014/Grinberg1994.pdf.

[41] Thaheld, F. H. 1975. "Biological nonlocality and the mind-brain interaction problem". *Theoretical bases of cybernetics (the Information Field).* Lenigrad.

[42] Fels, D., M. Cifra, F. Scholkmann, 2015. *Fields of the Cell.* Kerala, India, Research Signpost.

[43] Becker, R. O., G. Selden, 1998. *The Body Electric: Electromagnetism and The Foundation of Life.* New York, William Morrow Paperbacks.

[44] Klein, A., R. N. Boyd, 2008. *"9. Information and the Ether: The "Subtle Energies".* Accessed July 2017. http://www.victorzammit.com/articles/Klein9.htm.

[45] Williams, R., R. Richards, S. Chin, P. Williams, E. Gloyne, 1992. *Clarus Transphase LLC,* http://www.holistichealthsecrets.com/aboutus.html.

[46] Payne, B. 1981. *"Discovery of the Biofield A Different Type of Magnetism?"* Accessed July 2017. https://nexusilluminati.blogspot.com/2011/08/discovery-of-biofield-different-type-of.html.

[47] Jones, R. L. 2010. *"Response of Growing Plants to Uniform Daily Rotation".* http://link.springer.com/content/pdf/10.1038/185775a0.pdf.

[48] Landahl, T., E. Mollo-Christensen, 1992. *Turbulence and Random Processes in Fluid Mechanics* (2nd ed.). Cambridge, U.K., Cambridge University Press. http://www.sjsu.edu/faculty/watkins/kolmo.htm.

[49] Whittaker, E. T. 2016. *A History of the Theories of Ether and Electricity: From the Age of Descartes to the Close of the Nineteenth Century* (Classic Reprint). London, U.K., Forgotten Books.

[50] Gurvich, A. G. 1944. *The Theory of Biological Field.* Moscow, Soviet Science.

[51] Savva, S. 1997. "A systems approach in biology and biophysics." *Monterey Institute for the Study of Alternative Healing Arts Newsletter* 18-19 (1997): 209.

[52] Van Wijk, 2012. *"Emission of Mitochondrial Biophotons and their Effect on Electrical Activity of Membrane via Microtubules".* https://arxiv.org/ftp/arxiv/papers/1012/1012.3371.pdf.

[53] Bockris, John O'M. 2005. *The New Paradigm: A Confrontation Between Physics and the Paranormal Phenomenon,* College Station, TX, D&M Enterprises.

[54] Xin Yan, Feng Lu, Hongjian Jiang, Xinqi Wu, Wei Cao, Hua Shen, Jun Wang, Ming Dao, Hui Lin, Runsheng Zhu, 2002. "Certain Physical Manifestation and Effects of External Qi of Yan Xin Life ScienceTechnology" *Journal of Scientific Exploration,* Vol. 16, No. 3, 381-411.

[55] Dobyns, Y. H., Y. H., B. J. Dunne, R. G. Jahn, R. D. Nelson, 2004. "The Mega REG Experiment: Replication and Interpretation". *Journal of Scientific Exploration,* Vol. 18, No. 3, 369–397. https://www.princeton.edu/~pear/pdfs/2004-megareg-replication-interpretation.pdf.

[56] Rauscher, E. A., B.A. Rubik, 1983. "Human Volitional Effects on a Model Bacterial System". *Psi Research* Vol. 2, No. 1, 38.

[57] Burr, H. S. 1972. *Blueprint for Immortality* U.K., C.W. Daniel.

[58] Ravitz, 1962. *History, Measurement, and Applicability of Periodic Changes in the Electromagnetic Field in Health and Disease.* Wiley Online Library, Annals of the New York Academy of Sciences, Vol. 98, No. 4, Version of Record online: 15 DEC 2006.

[59] Hunt, V. V., 1996. "Infinite Mind: Science of human vibrations of consciousness." *J.S.E.* Vol. 16, No. 3, 381-411.

[60] Togbey, K. 2004. "*Fractal Dimension and the 4/5 Allometric Scaling Law for the Human Brain*". https://www.researchgate.net/publication/251745742.

[61] Kurakin, P. 2014. "Hidden variables and hidden time in Quantum Theory". Modern Bioelectrochemistry, Plenum Press, pp. 241-261. *Phys Life Rev,* 2014; Mar 11(1):39-78. q-bio/0510039.

[62] Cramer, J. G. 1986. "The Transactional Interpretation of Quantum Mechanics". *Reviews of Modern Physics.* 58 (3): 647–688.

In: Old Problems and New Horizons in World Physics
Editors: Volodymyr Krasnoholovets et al.
ISBN: 978-1-53615-430-6

Chapter 5

NEUTROPSYCHOLOGY AND BEYOND: A SKETCH OF A RE-DESIGNED FREUDIAN MENTAL MODEL AND ITS IMPLICATIONS

Victor Christianto[*]
Satyabhakti Advanced School of Theology – Jakarta Chapter,
Jakarta, Indonesia

ABSTRACT

In this chapter, I give an outline of an ongoing study which is intended to go beyond the Freudian mental model. First, I review a recent book by our colleague: "*Neutropsychic personality*." Then, I discuss the possibility of reintroducing the spirit into human consciousness. While I am fully aware that much remains to be done, and I admit that this is only a sketch, I hope that this paper will start a fresh approach of research towards a more realistic nonlinear consciousness model.

INTRODUCTION

One of my colleagues, Professor Florentin Smarandache, recently published a new book, *Neutropsychic personality* [1]. In this book, Prof. Smarandache describes a possible extension of the Freudian mental model: Id-ego-superego, using his Neutrosophic Logic theory. He goes on to develop implications of this approach.

[*]Corresponding Author's Email: victorchristianto@gmail.com. URL: http://researchgate.net/profile/Victor_Christianto.

Later on, we thought that it would be necessary to push the boundary one step further, by considering a more realistic way to go beyond the classic Freudian mental model, i.e., by reintroducing the Spirit into the human consciousness model.

We are aware that many researchers have proposed such an extension, especially in Italian tradition, which was continually developed by students of Carl Jung, such as Assagioli and Piere Ferrucci, known as the *Psychosynthesis* movement (See for example [3]).

But here, we offer a different starting point of a mental model, based on Matthew 22, i.e., *The Great Commandments*. As far as we know, i.e., this is the simplest model of human consciousness, yet it is profoundly inspired by the Bible.

A Short Review of Neutropsychic Personality

In *Neutropsychic personality*, Smarandache described a possible extension of the Freudian mental model: *id-ego-superego*, using his Neutrosophic Logic theory. His definition of Neutropsychic is as follows:

> "Neutropsyche is the psychological theory that studies the soul or spirit using the neutrosophy and neutrosohic theories. It is based on triadic neutrosophic psychological concepts, procedures, ideas, and theories of the form (<A>, <neutA>, <antiA>), such as (positive, neutral, negative), (good behavior, ignorant behavior, bad behavior), (taking the decision to act, pending, taking the decision not to act), (sensitive, moderate, insensitive), (under-reacting, normally reacting, over-reacting), (under-thinking, normal thinking, over-thinking), and so on, and their refinements as (<Aj>, <neutAj>, <antiAj>)." [1, p. 29]

Among other things, he also refines the notion of human memory:

> "*Neutrosophic Memory*: Memory is thus divided into three main parts. It is a symmetric triad of the form (<A>, <neutA>, <antiA>) as in neutrosophy: 1) Conscious, meaning things that we are currently aware of. (It corresponds to <A>.) 2) Unconscious, which comprises things that we are not aware of; they are hard to access because they are deep inside our mind. It is the opposite of conscious. (It corresponds to <antiA>.)
>
> According to Webster Dictionary, unconscious means: "not endowed with consciousness; mindless; [...] (Psychoanalysis) the sum of all thoughts, memories, impulses, desires, feelings, etc. of which the individual is not conscious but which influence the emotions and behavior; that part of one's psyche which comprises repressed material of this nature" (p. 1453). We first cite a Webster definition since it is more objective by comparison with subjective definitions given by biased theorists. Freud [15, 16] has considered the unconscious as a repository for unacceptable desires and ideas by the

society, and traumatic memories. 3) Aconscious. We coin now for the first time the concept of "aconscious" (adj.), the "aconscious" (noun), and the derivatives "aconsciousness" (noun) and "aconsciously" (adv.), which etymologically means away from conscious and unconscious, or neither conscious nor unconscious, but in between, or a mixture of conscious and unconscious – a vague buffer zone between them.

Doing a search on American search engine Google (google.com), European search engine Bing (bing.com), and Chinese search engine (baidu.com) for the word "aconscious" on February 15th, 2018, we got no entry. Also, the concept "aconscious" does not exist in English, Latin, Greek languages – using Google's translation dictionaries. The consciousness, aconsciousness, and unconsciousness are the sources of positive, neutral (or blended), and negative emotions, thoughts, and behaviors throughout our lifespan." [1, p. 37-39]

But here we will not discuss such a conceptual extension of human memory, instead we will discuss a new conceptual model of human consciousness, beyond the Freudian mental model. First of all, let us discuss a basic problem with modern society.

PROBLEM WITH THIS MODERN SOCIETY

First of all, let us ask a question: how can we find the root cause of numerous problems in our modern society? Allow us to recall what Adam Grant emphasizes: the basic human motives are selfishness and altruism.

Figure 1. Adam Grant's model of human basic traits.

And also, we can recall from Genesis 3 that the first fall of our ancestors came from greediness. *How far have we fallen in this modern society, where greed has been hailed as highest virtue?* To emphasize this root problem in our modern society, allow me to quote Grekko's remark: *Greed is good.*

> "The point is, ladies and gentleman, that **greed**, for lack of a better word, is **good**. **Greed** is right, **greed** works. **Greed** clarifies, cuts through, and captures the essence of the evolutionary spirit. **Greed**, in all of its forms; **greed** for life, for money, for love, knowledge has marked the upward surge of mankind."[1]

We consider this the essence of our modern reality: all of us have been consumed and drowning in the ocean of greediness. The real irony is that greediness has eaten us alive, from our childhood, until we die. Most of us dream to be billionaires, right? Such a common dream has been captured in a popular song:

BILLIONAIRE

Travie McCoy

I wanna be a billionaire so freakin' bad
Buy all of the things I never had
Uh, I wanna be on the cover of Forbes magazine
Smiling next to Oprah and the Queen
Oh every time I close my eyes
I see my name in shining lights
Yeah, a different city every night oh right
I swear the world better prepare
For when I'm a billionaire
Yeah I would have a show like Oprah
I would be the host of, everyday Christmas
Give Travie a wish list
I'd probably pull an Angelina and Brad Pitt
And adopt a bunch of babies that ain't never had shit
Give away a few Mercedes like here lady have this
And last but not least grant somebody their last wish
It's been a couple months that I've been single so
You can call me Travie Claus minus the Ho Ho
Get it, hehe, I'd probably visit where Katrina hit
And damn sure do a lot more than FEMA did

[1] The Wall Street (1987). url: https://businessethicsblog.com/2010/10/12/wall-street-1987-greed-is-good/.

Yeah can't forget about me stupid
Everywhere I go Imma have my own theme music
Oh every time I close my eyes
I see my name in shining lights
A different city every night oh right
I swear the world better prepare
For when I'm a billionaire
Oh ooh oh ooh for when I'm a billionaire
Oh ooh oh ooh for when I'm a billionaire
I'll be playing basketball with the President
Dunking on his delegates
Then I'll compliment him on his political etiquette
Toss a couple milli in the air just for the heck of it
But keep the five, twenties tens and bens completely separate
And yeah I'll be in a whole new tax bracket
We in recession but let me take a crack at it
I'll probably take whatever's left and just split it up
So everybody that I love can have a couple bucks
And not a single tummy around me would know what hungry was
Eating good sleeping soundly
I know we all have a similar dream
Go in your pocket pull out your wallet
And put it in the air and sing[2]

While surely there is nothing so bad against the dream of being a billionaire, one should recall a famous saying of Jesus. The *New Testament* quotes Jesus as saying, "It is easier for a camel to go through the eye of a needle than for a rich man to enter the kingdom of God." Some interpreters say the term *"eye of a needle"* is used as a metaphor for a very narrow opening. It occurs several times throughout the *Talmud*. But some say the eye of a sewing needle is the part formed into a loop for pulling thread, located at the end opposite from the point.[3]

The truth is, this system makes profits from our bodies, our eyes, our jealousy, our heart, our mind, and our consciousness. Even if we die, there are greedy developers who sell a piece of cemetery with high price. Literally speaking, we are more or less like *walking zombies*. We are improved on the outside, but we are no more than rotten tomatoes deep inside. At this point, some may ask: How can we repair such a deep problem of our modern society?

---

[2] Songwriters: Philip Lawrence/Ari Levine/Bruno Mars/Travis Mccoy. Billionaire lyrics © Sony/ATV Music Publishing LLC, Warner/Chappell Music, Inc, Round Hill Music Big Loud Songs, BMG Rights Management

[3] https://en.wikipedia.org/wiki/Eye_of_a_needle.

Outline of Reasoning

Hebraic thought regarding the human being, is that people are integrated wholes, i.e., the wholeness of *body-mind-spirit.* But how can we come up with a model of human consciousness based on the Bible? As a starting point, we choose to begin with Jesus's sayings, rather than using other trivial sources. Let us begin by the *Greatest Commandments*:

Matthew 22:37-40 King James Version (KJV)

37 *Jesus said unto him, Thou shalt love the Lord thy God with all thy heart, and with all thy soul, and with all thy mind.*
38 *This is the first and great commandment.*
39 *And the second is like unto it, Thou shalt love thy neighbour as thyself.*
40 On these two commandments hang all the law and the prophets.

Our re-reading of the above commandments lead us to model a Trinitarian dialogue within human self: God, self, and others.[4]

When comparing this understanding with Adam Grant's "give and take" model of tensions inside our mind, let us consider parallels. "*Taking*" reflects a selfishness/greediness motive of ego, and "*giving*" reflects an altruistic motive of conscience.

Figure 2. Three directions of human love based on The Greatest Commandments in Matthew 22:37-40.

[4] For an alternative reading of Mat. 22, see Vern Poythress's article: https://frame-poythress.org/the-greatest-commandment-the-very-heart-of-the-matter/.

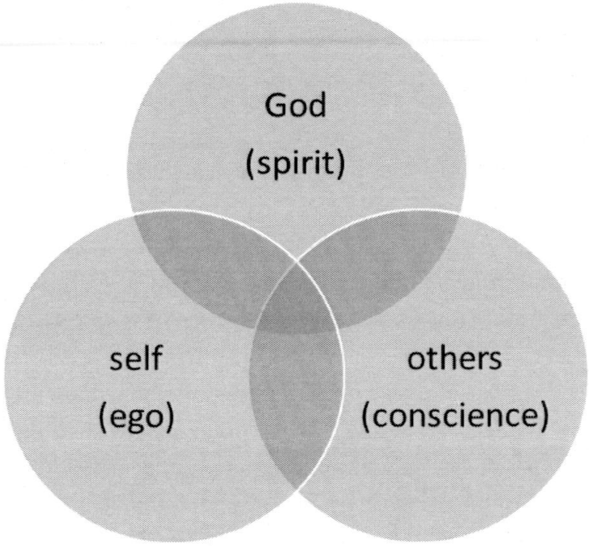

Figure 3. A model of human consciousness based on The Greatest Commandments in Matthew 22:37-40.[5]

In other words, now we have two parts of human consciousness: ego and conscience. There can be a deep tension between ego and consciousness, between selfishness and altruism. Along these two poles, we need a third entity which has purpose of easing and being an intermediary between these two motives. Approaching this problem with *Neutrosophic Logic*, allows us to submit wholeheartedly that the third entity, is actually none other than "the Spirit." (*pneuma* in Greek, *ruach* in Hebrew). The exact role of human spirit is to enlighten both ego and conscience.

At this point, some readers may ask question: what is new here? Doesn't it seem similar with Freudian id-ego-superego model?

No, it is really in contrast with Freud's model, which is purely materialistic in origin. The notion of spirit is rejected in Freud's model, which is why mankind is reduced to animals in his model, determined by his/her sexual instinct. And there is no way out of such animal instinct in his model.

Sometimes such an extended model of Freudian mental view is called transpersonal psychology:

> "Transpersonal psychology is a sub-field or "school" of psychology that integrates the spiritual and transcendent aspects of the human experience with the framework of modern psychology. It is also possible to define it as "spiritual psychology."

[5] This model may be compared to Jung's personality model, which includes individual unconscious and collective unconscious. (see Prof. Thee Houw Liong's added note in preface.)

An interesting argument of spiritual psychology has been discussed at the University of Santa Monica:

> "If you look up the word "psyche" in the dictionary, you will find "breath, principle of life, Soul." But if you look up "psychology," you will find "the science of mind and behavior." Somehow, in the translation from essence to practice, the most important aspect of "psyche" has been lost. At the University of Santa Monica, we recognize our task as reintegrating the spiritual dimension back into the essence of an authentic psychological inquiry. It is this reintegration that evokes the emergence of a Spiritual Psychology.
>
> Spiritual Psychology is the study and practice of the art and science of Conscious Awakening. To engage in this genre, we must begin by distinguishing the essence of human evolution—what does it mean to evolve? In short, it means learning how to identify, recognize, and navigate successfully within the Context of Spiritual Reality. Practically, it means learning how to surrender—or let go of—anything that disturbs one's peace. It also means sacrificing our illusions of separation. Essentially, this "surrendering" and "sacrificing" is work that can and has been called "healing," which includes healing on the physical, mental, and emotional levels in service to the deeper revelation of who we truly are as Loving, Peaceful, Compassionate, and Joyful beings. We refer to this level of awareness as the Authentic Self."

Then, how can we listen to God's voice?

We need to open up our heart and ask God to speak to us, like this song popularized by Robbie Williams:

> Not sure I understand
> This role I've been given
> I sit and talk to God
> And he just laughs at my plans
> My head speaks a language
> I don't understand ....

Maybe that voice does not really speak "*hetero lalein*" (speaking in foreign language), because most of the time the Holy Spirit speaks in our own mother language.[6] Of course, there are various approaches of spiritual psychology. But, what is really different in our simpler model?

---

[6] For interested readers, read a very good book by Cindy Jacobs: https://irp-cdn.multiscreensite.com/c21a6153/files/uploaded/TheVoiceofGod-Cindy-Jacobs.pdf.

A Socio-Economics Implication

In this time, we would only discuss the economic implications, based on modelling human identities into two opposites: (a) individualism (we call them: *fermions*), and (b) collectivism (we call them: *bosons*). In a recent paper, we discuss how to solve Mancur Olson's collective action problem [2].

Now, some of you may ask: By suggesting solution to Olson's collective action problem to save our humanity, where is the article heading? Are we advocating collective society as in the old day Marxism hammer? Or are we advocating how to escape from the curse of capitalism's social Darwinism?[7] Yes, normally you read numerous political-economics jargons, e.g., leftist, right wing, centrist left or centrist right and so on.

However, it is not our intention to submit another ideological parlance. In fact, these authors are scientists and mathematicians, so we are not so inclined to any parlance. In our opinion, our tendency to cooperate or compete is partly influenced by the culture that we inherit from our ancestors. One of us (VC) once lived for a while in Russia, and he found that many people there are rather cold and distant (of course not all of them, some are friendly). He learned that such a trait is quite common in many countries in Europe. They tend to be individual and keep a distance from each other. Using a physics term, they are like *fermions*.[8]

There is a developmental psychology hypothesis that suggests that perhaps such a trait correlates to the fact that many children in Europe lack nurturing or human touch from their parents, which makes them rather cold and individualistic. Of course, whether or not this is true should be studied.

On the contrary, most people in Asia are quite gregarious (except perhaps in big metropolitans). They tend to spend much time with family and friends, just like many Italians. They attend religious rituals regularly, and so on. Using a physics term, they are like *bosons*. Of course, this sweeping generalization may be oversimplifying.[9]

---

[7] See for example Richard Hofstadter: Social Darwinism in American thought. url: http://culturism.us/booksummaries/SocialDarwinismHofst.pdf.

[8] While our proposed simplifying analogy of human behaviour, i.e., individualism and collectivism sound not so common. Indeed such cultural psychology research has been reported since Harry C. Triandis et al. See for example: (a) The Self and Social Behavior in Differing Cultural Contexts, Psychological Review, vol. 96 no. 3; (b) Harry C. Triandis and Eunkook M. Suh, CULTURAL INFLUENCES ON PERSONALITY, Annu. Rev. Psychol. 2002. 53:133–60; (c) J. Allik & A. Realo, Individualism-collectivism and social capital, JOURNAL OF CROSS-CULTURAL PSYCHOLOGY, Vol. 35 No. 1, January 2004, 29-49. This last mentioned paper includes a quote from Emile Durkheim: "The question that has been the starting point for our study has been that of the connection between the individual personality and social solidarity. How does it come about that the individual, whilst becoming more autonomous, depends ever more closely upon society? How can he become at the same time more of an individual and yet more linked to society?"

[9] After writing up this article, we found that Sergey Rashkovskiy also wrote a quite similar theme, albeit with a statistical mechanics in mind. The title of his recent paper is: "'Bosons' and 'fermions' in social and economic systems." Here is abstract from his paper: "We analyze social and economic systems with a hierarchical structure and show that for such systems, it is possible to construct thermostatistics, based on the intermediate Gentile statistics. We show that in social and economic hierarchical systems there are elements that obey the Fermi-Dirac statistics and can be called fermions, as well as elements that are approximately subject to Bose-Einstein statistics and can be called bosons. We derive the first and second laws of thermodynamics for the considered

Therefore, it seems quite natural to us, why Adam Smith wrote a philosophy book suggesting that individual achievement is a key to national welfare (because he was a British which emphasized individualism).[10] It took more than a hundred years until mathematicians like John F. Nash, Jr. figured out that individual pursuit toward their own goals will not lead them to achieve a common goal as society.[11]

That is why we choose to work out Mancur Olson's theorem, because he is able to condense the complicated game theoretical reasoning (whether one should cooperate or not) into a matter of collective actions. So, which is better: *to be like fermions or bosons*? Our opinion is that just like in particle physics, both fermions and bosons are required. In the same way, fermion behaviour and boson behaviour are both needed to advance the quality of life. Fermion people tend to strive toward human progress, while boson people are those who make us alive. Just like the song "Ebony and Ivory"....they both make harmony in society.[12]

We hope this paper helps us to see that collective actions are what makes us a human society.[13] And it seems related to *social innovations* and also social *capital* too, in other words, a society with social capital and collective actions will ensure its sustainable future.[14] However, this is beyond the scope of this article, so let us leave such a discussion to economists.

By introducing this analogy from particle physics theories, we hope to resolve the classic clash between socialism and capitalism, which are no more than a cruel reformulation of the above basic human motives into political struggles, in attempt to put the entire mankind into eternal slavery.

---

economic system and show that such concepts as temperature, pressure and financial potential (which is an analogue of the chemical potential in thermodynamics) that characterize the state of the economic system as a whole, can be introduced for economic systems." Url: https://arxiv.org/ftp/arxiv/papers/1805/1805.05327.pdf.

[10] If only Adam Smith was born in Bangkok or Manila, probably he wrote his book in a different way.

[11] Imagine 10 players of a football team go simultaneously to make a goal to their opposite team, will they succeed? Of course no, they should arrange according to their coach's instruction: 1-4-4-2, or other type of arrangement.

[12] While we are fully aware that politically speaking, Marxism and capitalism are irreconcilable, recent studies indicate that there are possible intermediate states between these two, such as The Third way in Tony Blair's era, welfare states in most European countries, various cooperative systems in Scandinavian countries, state-driven capitalism in Vietnam and China, Japan inc. etc. In other words, from Neutrosophic perspective, dialogues between human tensions of individualism and collectivism remain open for future economics theorizing toward a much better and improved quality of life, while at the same time we ought to remove the dialectical clashes a la Huntington's NWO.

[13] In our country, there is a specific phrase for some people who work together to achieve a common goal: "*gotong royong*." (this phrase was popularized by Bung Karno.)

[14] Emily Ouma & Awudu Abdulai. Contributions of Social Capital Theory in Predicting Collective Action Behavior among Livestock Keeping Communities in Kenya. url: https://ageconsearch.umn.edu/bitstream/49994/2/Manuscript%20No.%20423_Social%20capital%20theory%20and%20collective%20action.pdf.

Too many decades have been wasted by numerous countries fighting on these ideologies, but the truth is these opposite ideological poles were crafted in order to trap mankind into eternal struggles.[15]

It needs to be stopped right now.

CONCLUSION

In this paper, we give an outline of an ongoing study which is intended to go beyond the Freudian mental model. First, we review a recent book by Smarandache, *Neutropsychic personality*. Then, we discuss the possibility of reintroducing the spirit into human consciousness. While we are aware that much remains to be done, and we admit that this is only a sketch, we hope that this paper will start a fresh approach of research towards spiritual psychology in a realistic nonlinear consciousness model.

Postscripts

A Short Note on the Plausibility and Experimental Vindication of the Proposed Model

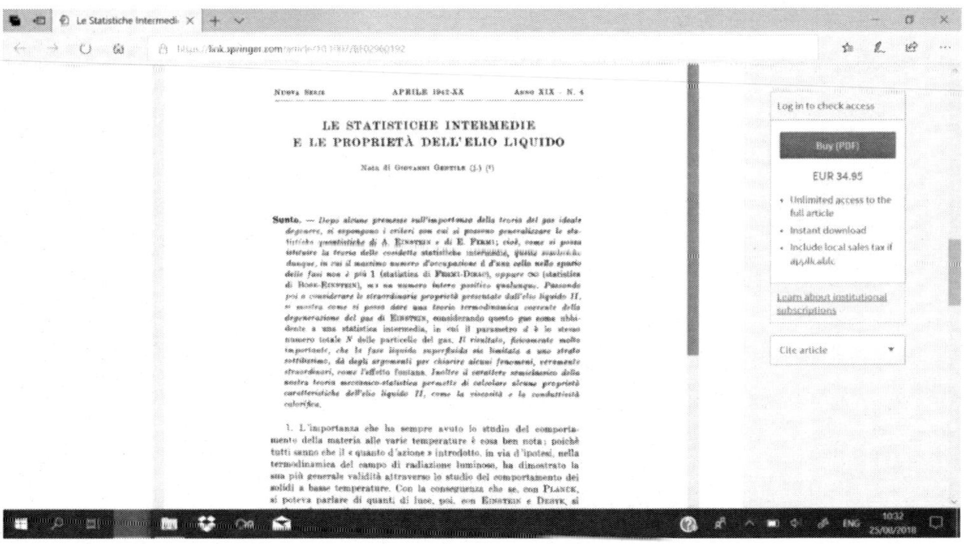

NUOVA SERIE APRILE 1942-XX ANNO XIX - N. 4

LE STATISTICHE INTERMEDIE
E LE PROPRIETÀ DELL'ELIO LIQUIDO

Nota di GIOVANNI GENTILE (J.) (*)

Sunto. — *Dopo alcune premesse sull'importanza della teoria del gas ideale degenere, si espongono i criteri con cui si possono generalizzare le statistiche quantistiche di* A. EINSTEIN *e di* E. FERMI; *cioè, come si possa istituire la teoria delle cosidette statistiche intermedie, quelle statistiche dunque, in cui il massimo numero d'occupazione d'una cella nello spazio delle fasi non è più* 1 (statistica di FERMI-DIRAC), *oppure* ∞ (statistica di BOSE-EINSTEIN), *ma un numero intero positivo qualunque. Passando poi a considerare le straordinarie proprietà presentate dall'elio liquido II, si mostra come si possa dare una teoria termodinamica corrente della degenerazione del gas di* EINSTEIN, *considerando questo gas come ubbidiente a una statistica intermedia, in cui il parametro d è lo stesso numero totale N delle particelle del gas. Il risultato, fisicamente molto importante, che la fase liquida superfluida sia limitata a uno strato sottilissimo, dà degli argomenti per chiarire alcuni fenomeni, veramente straordinari, come l'effetto fontana. Inoltre il carattere semiclassico della nostra teoria meccanico-statistica permette di calcolare alcune proprietà caratteristiche dell'elio liquido II, come la viscosità e la conduttività calorifica.*

1\. L'importanza che ha sempre avuto lo studio del comportamento della materia alle varie temperature è cosa ben nota; poichè tutti sanno che il « quanto d'azione » introdotto, in via d'ipotesi, nella termodinamica del campo di radiazione luminosa, ha dimostrato la sua più generale validità attraverso lo studio del comportamento dei solidi a basse temperature. Con la conseguenza che se, con PLANCK, si poteva parlare di quanti di luce, poi, con EINSTEIN e DEBYE, si

Picture 1. Screenshot of first page of G. Gentile's paper (1941). See [8].

[15] *Some authors wrote that such crafted political struggles only gave monetary advantages to those greedy financial institutions, by selling weapons to both sides of parties. See for instance a very good movie: The International.*

The author just thought of plausible vindication of the proposed intermediate state of fermion-boson, which may be called "*ferson.*" It may have a chance to get into real observation at CERN etc. It may be indeed interesting for particle physicists who wish to continue the service period of CERN's expensive facilities after the discovery of Higgs particle. As the readers may already know, they tried to extend the standard model to supersymmetry but it failed to come to detectors. Meanwhile, we have recently read that there are two possible theories which seem to correspond to an intermediate statistics we're looking for: (1) anyon fractional statistics by Franck Wilczek, which we have not studied, and (2) G. Gentile's statistics, which predict the existence of the "intermediate particle" between fermion and boson, but nobody has identified any experiments with such an intermediate particle so far. So, allow us to suggest to those interested readers to examine Giovani Gentile's original paper in *Nuovo Cimento* (1941).[16]

But that topic is reserved for another paper.

Toward Pneumatological Mind-Matter Interaction

Various models have been proposed to suggest the possibility of mind-matter interaction, but mostly fall within QM theory. Other experiments seem to suggest that the effect is quite real, albeit many aspects remain mystery.

There are vast amount of mind-matter interaction models of living systems, from Stuart Hameroff's model, to the *Semiotic Scaffolding* model of Jesper Hoffmeyer (which Brian Josephson suggests a new term: *Semiophysics*) etc. See for instance [9-11].

Jesper Hoffmeyer: "semiotic scaffolding of living systems"

Semiotic Scaffolding of living systems*

Jesper Hoffmeyer
Department of Biological Chemistry
University of Copenhagen, Denmark
jhoffmeyer@mail.dk

KEYWORDS: Biosemiotics Emergence, Scaffolding, Anticipation, Sign, Evolution

ABSTRACT

The apparently purposeful nature of living systems is obtained through a sophisticated network of semiotic controls whereby biochemical, physiological and behavioral processes become tuned to the needs of the system. The operation of these semiotic controls takes place and is enabled across a diversity of levels. Such semiotic controls may be distinguished from ordinary deterministic control mechanisms through an inbuilt anticipatory capacity based on a distinct kind of causation that I call here "semiotic causation" to denote the bringing about of changes under the guidance of interpretation in a local context. Anticipation through the skilled interpretation of indicators of temporal relations in the context of a particular survival project (or life strategy) guides organismic behavior towards local ends. This network of semiotic controls establishes an enormously complex semiotic scaffolding for living systems.

Picture 2. Hoffmeyer's first page of his paper: *Semiotic Scaffolding of living systems*. [9].

[16] See his short biography: https://www.luisabonolis.it/Giovanni_Gentile_Jr_files/GentileCronoFoto.pdf.

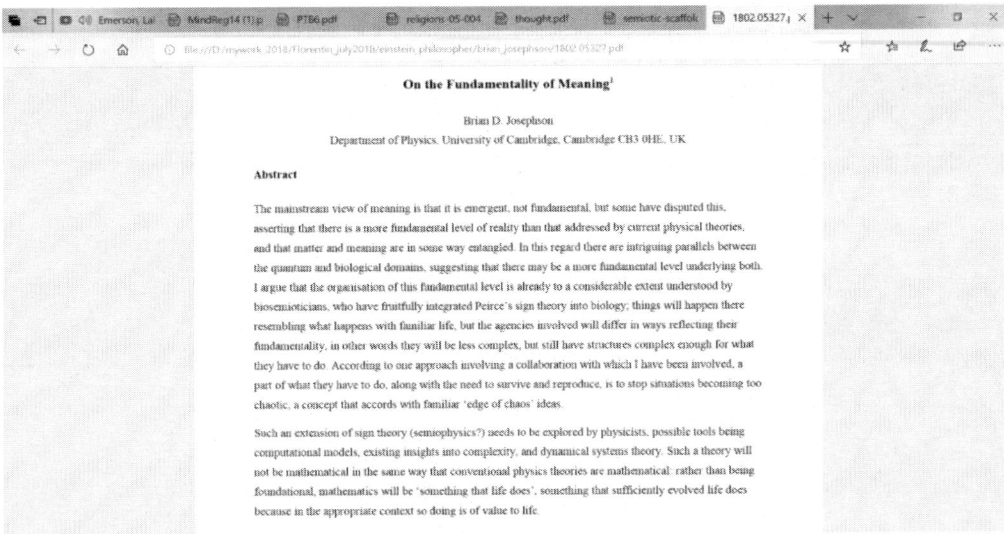

On the Fundamentality of Meaning[1]

Brian D. Josephson
Department of Physics, University of Cambridge, Cambridge CB3 0HE, UK

Abstract

The mainstream view of meaning is that it is emergent, not fundamental, but some have disputed this, asserting that there is a more fundamental level of reality than that addressed by current physical theories, and that matter and meaning are in some way entangled. In this regard there are intriguing parallels between the quantum and biological domains, suggesting that there may be a more fundamental level underlying both. I argue that the organisation of this fundamental level is already to a considerable extent understood by biosemioticians, who have fruitfully integrated Peirce's sign theory into biology; things will happen there resembling what happens with familiar life, but the agencies involved will differ in ways reflecting their fundamentality, in other words they will be less complex, but still have structures complex enough for what they have to do. According to one approach involving a collaboration with which I have been involved, a part of what they have to do, along with the need to survive and reproduce, is to stop situations becoming too chaotic, a concept that accords with familiar 'edge of chaos' ideas.

Such an extension of sign theory (semiophysics?) needs to be explored by physicists, possible tools being computational models, existing insights into complexity, and dynamical systems theory. Such a theory will not be mathematical in the same way that conventional physics theories are mathematical: rather than being foundational, mathematics will be 'something that life does', something that sufficiently evolved life does because in the appropriate context so doing is of value to life.

Picture 3. Prof. Brian Josephson's first page of his paper [10].

We prefer to suggest a simpler model based on the fact observed by Benvenistem and later by Maxim Trushin:[17] there is a kind of antenna or *sonic-mediated communication* between cells.[18] Therefore, we submit a model of mind-matter interaction by a new term: *Pneumatological cymatic mechanism*, i.e., by the human voice, soaked in the Holy Spirit, then it may affect the material/environment. Nonetheless, we admit that the exact mechanism of Pneumatological mind-matter interaction remains mystery, and this topic is reserved for future research. What we can say for now is that it seems the effect of mind-matter effect over long distance (more than 150km) has been reported, which suggests that this topic is very interesting for next research. [12].

Endnote

For a discussion on the implications of the proposed model of consciousness in integrative cancer therapy, see our paper: Victor Christianto, Florentin Smarandache (2018) "The World Within Us (Or A Sketch of Consciousness Space Beyond Freudian Mental Model and Implications to Socio-Economics Modeling and Integrative Cancer Therapy)." *BAOJ Cancer Res Ther* 4: 057. url: https://bioaccent.org/cancer-sciences/cancer-sciences57.php.

---

[17] Maxim Trushin. *J. Microbiol. Immunol. Effect*, 2003: http://nghl-ntge.s3.amazonaws.com/JMicrobiolImmunolInfect2003_36_153-160.pdf.

[18] Other authors such as Don Colbert or Paul Pearsall, also wrote on possible role of *neuropeptide* as transmitter from emotion in human brain into cells. But these assertions seem to require more study.

Acknowledgments

This chapter is inspired partly by Prof. Adam Grant's insightful book: *Give and Take*. The writer gratefully appreciates Elizabeth and Sujarwo in Jkt. The author dedicates this paper to Jesus Christ, *He is the True Savior of the Universe*. He is the Good Shepherd.

Version 1.0: 20th July 2018, pk. 19:30
Version 1.1: 20th July 2018, pk. 23:28
Version 1.2: 2nd November 2018, pk. 9:36
Version 1.3: 6th January 2019, pk. 19:43
VC

References

[1] F. Smarandache. *Neutropsychic personality. A mathematical approach to psychology.* Second ed. Belgium: Pons Publ., 2018.

[2] V. Christianto & F. Smarandache. *What we can do to save humanity in the coming era of global eavesdroppers.* Paper accepted in ICRIEECE, held in India. Presented online at 4th Aug. 2018. URL: http://www.rieece.org/.

[3] John Firman & Ann Gila. *Psychosynthesis: a psychology of the spirit.* Albany: STATE UNIVERSITY OF NEW YORK PRESS, 2002.

[4] University of Santa Monica. URL: https://www.universityofsantamonica.edu/why-usm/spiritual-psychology/.

[5] *Grekko's speech in The Wall Street* (1987). URL: https://businessethicsblog.com/2010/10/12/wall-street-1987-greed-is-good/.

[6] Richard Wurmbrand. *Marx and Satan.* Published by Crossway Books, a division of Good News Publishers, Westchester, Illinois, 1986.

[7] Richard Hofstadter: *Social Darwinism in American thought.* URL: http://culturism.us/booksummaries/SocialDarwinismHofst.pdf.

[8] G. Gentile. *"Le statistiche intermedie e le proprietà dell'elio liquido"*, Rendiconti del Seminario Matematico e fisico di Milano [*"Intermediate statistics and properties of liquid helium"*, Reports of the Mathematical and Physical Seminar of Milan] XV (1941-XIX), 1-19, 1941.

[9] Jesper Hoffmeyer. *Semiotic Scaffolding of living systems.* URL: https://link.springer.com/chapter/10.1007/1-4020-4814-9_6.

[10] Brian Josephson. *On the fundamentality of meaning.* arXiv: 1802.05327.

[11] Francisca Cho. Buddhist Mind and Matter. *Religions* 2014, 5, 422–434; doi: 10.3390/rel5020422.

[12] Patrizio Tressoldi et al. Mind-Matter Interaction at a Distance of 190 km: Effects on a Random Event Generator Using a Cutoff Method. *NeuroQuantology* September 2014|Volume 12|Issue 3|Page 337-343. URL: DOI Number: 10.14704/nq.2014.12.3.767.

In: Old Problems and New Horizons in World Physics
Editors: Volodymyr Krasnoholovets et al.
ISBN: 978-1-53615-430-6

Chapter 6

WHAT WE CAN DO WHEN THE WORLD IS RUNNING DOWN: TWELVE SMALL PIECES OF ADVICE FOR YOUNG SCIENTISTS

Victor Christianto*
Satyabhakti Advanced School of Theology – Jakarta Chapter,
Jakarta, Indonesia

ABSTRACT

In this chapter, we offer 12 impeccable pieces of advice for young scientists, based on our experiences throughout the years. We hope this advice will be found useful, albeit they may sound rather weird and unorthodox at first glance.

INTRODUCTION

Albert Einstein was notoriously famous not only for his theories, but also for his numerous witty remarks. At times, he also offered deep advice to young scholars. As per records in Princeton, he once said [1]:

"Why does this magnificent applied science, which saves work and makes life easier, bring us so *little happiness*? The simple answer is because we have not yet learned to make sensible use of it. In war, it serves that we may poison and mutilate each other. In peace, it has made our lives hurried and uncertain instead of freeing us in great measure from

* Author's Email: victorchristianto@gmail.com, URL: http://researchgate.net/profile/Victor_Christianto.

spiritually exhausting labor. It has made men into the slaves of machinery, who for the most part complete their monotonous long days' work with disgust, and must continually tremble for their poor rations. You will be thinking that the old man sings an ugly song. I do it, however, with a good purpose, in order to point out a consequence. It is not enough that you should understand about applied science in order that you may increase man's blessings.

Concern for man himself and his fate always forms the chief interest of all technical endeavors. Concern for the great unsolved problems of the organization of labor, for the distribution of goods, *in order that the creations of our minds shall be a blessing and not a curse. Never forget this in the midst of your diagrams and equations*" (italic emphasis by these authors).

Although that speech was translated from German, the essence remains relevant, even for today's life as a scientist, as Harry Gray once remarked:

"That was Albert Einstein on February 16, 1931, to the Caltech student body, translated by somebody and slightly retranslated by me. -- Obviously, what he said over 40 years ago has relevance to our situation today" [1].

In that article, Harry Gray offers his own receipt, called FEG formula, as follows:

But we understand that for thousands or hundreds of thousands of young scientists, such advice is too broad in scope.

Figure 1. The FEG Formula. (Source: Harry Gray [1]).

"*We need a set of simple axioms to start with*", says a mathematician. Probably, a large number of mathematicians among you are dreaming to be the next Gregori Perelman. In order to do that, you need to emulate him, to live in a way to reduce your life into a minimal number of axioms[1].

Therefore, in this paper, we offer 12 impeccable tips for younger scientists based on our experiences throughout the years. We hope these tips will be found useful, albeit they may sound rather weird and unorthodox at first glance.

Of course, if you don't find these words of advice stimulating, let alone useful, just dump this article in your PC's recycle bin.

12 Pieces of Advice for Younger Scientists

1. Have reality as your best friend. But the most important is, having God as your best ally will make a difference in your life journey.
2. Express your ideas as clearly as possible.
3. Do not torture geometry.
4. Respect and learn from Nature.
5. If you want mediocre improvements, refine your techniques. But, if you want a qualitative jump or breakthrough, double check your assumptions.
6. Always double check your facts.
7. While you really need them, do not rely too much on models.
8. Always be ready to unlearn old skills and relearn new skills.
9. Having pure mindfulness and consciousness always pays off. (Mindfulness in Sanskrit is known from ancient time as *sati*).
10. In times you may need to speak the truth to people in power…try to make your voice heard, otherwise it will be useless.
11. Having a big skull is one thing, but having a big heart will make you move ahead of your peers.
12. A healthy dose of common sense and humor will surely help you remain humble.
13. Corollary: "*if you do not know where to begin in your research, it may be useful to read some one-liner scientific jokes*".

Two examples:

a. Old joke: *topologists are those who cannot tell the difference between a coffee cup and a doughnut*. And guess what, the Nobel Prize in physics has been granted to

---

[1] www2.imperial.ac.uk/~mdavis/docs/MANIFOLD_DESTINY.pdf.

clarify such an old joke - for those who did groundbreaking work on topology in physics.[2]

b. Another joke commonly found by students majoring in physics: *Students belong to two classes: boson and fermion*. See our recent paper discussing how to model human society with such an analogy [4].

CONCLUDING REMARKS

Hopefully, this advice will be found useful by young scientists out there.

While we are fully aware that our world is running down, there are many things that need to be done and can be done. Read a lyrics of this old song by The Police:

When the World Is Running Down
The Police, Different Gear

Turn on my V.C.R.
Same one I've had for years
James Brown on the Tami Show
Same tape I've had for years
I sit in my old car
Same one I've had for years
Old battery's running down
It ran for years and years
Turn on the radio
The static hurts my ears
Tell me where would I go
I ain't been out in years
Turn on the stereo
It's played for years and years
An Otis Redding song
It's all I own
When the world is running down
You make the best of what's still around
When the world is running down
You make the best of what's still around
Plug in my M.C.I

---

[2] For those interested in applying topology in physics, see for instance: Nicolas Manton and Paul Sutcliffe, Topological soliton. *Cambridge monographs on mathematical physics*. Cambridge: Cambridge University Press, 2004.

To exercise my brain
Make records on my own
Can't go out in the rain
Pick up the telephone
I've listened here for years
No one to talk to me
I've listened here for years
When the world is running down
You make the best of what's still around
When the world is running down
You make the best of what's still around
When I feel lonely here
Don't waste my time with tears
I run 'Deep Throat' again
It ran for years and years
Don't like the food I eat
The cans are running out
Same food for years and years
I hate the food I eat
When the world is running down
You make the best of what's still around
When the world is running down
You make the best of what's still around
When the world is running down
You make the best of what's still around
When the world is running down
You make the best of what's still around
Turn on my V.C.R.
Same one I've had for years
James Brown on the Tami Show
Same tape I've had for years

Songwriters: Gordon Sumner

Version 1.0: 18th july 2018, pk. 20:16
Version 1.1: 19th july 2018, pk. 20:28
VC

Acknowledgment

VC is really grateful for very wise advice by a senior professor of physics, V.V. Kassandrov, back in his short period in Moscow: "*you ought to learn to have a big heart*". Thanks so much, Professor, for such deep advice! Many thanks also to Yunita Umniyati, for telling her jokes too. Special thanks go to Prof. Carlos Castro Perelman, Prof. Matti Pitkannen, Prof. Yu P. Rybakov, Prof. Alexander Yefremov, Prof. Diego Rapoport, Prof. Liek Wilardjo, Prof. Bambang Hidayat, Prof. Thee Houw Liong, Dr. George Shpenkov, Dr. Paskalis Nyoman Paska, Dr. Gatut Budiyono, Dr. Bambang Noorsena, Mr. Gani Wiyono, Dr. Robbie Chandra, Dr. Yonky Karman, Dr. Joas Adiprasetya and so many other physicists who have been corresponding with us in the past few years.

References

[1] Harry B. Gray. The Chemistry of Caltech: Caltech's 1975 Commencement Address - or How Commencement Speakers Are Chosen. *Engineering and Scien*ce, Oct.-Nov. 1975. url: https://core.ac.uk/download/pdf/46701190.pdf.

[2] F. Smarandache. *Neutropsychic personality. A mathematical approach to psychology*. Second ed. Belgium: Pons Publ., 2018.

[3] John Firman and Ann Gila. *Psychosynthesis: a psychology of the spirit*. Albany: State University of New York Press, 2002.

[4] V. Christianto and F. Smarandache. *What we can do to save humanity in the coming era of global eavesdroppers*. Paper accepted in RIEECE, to be held in India this month. July 2018.

[5] *Gekko's speech in Wall Street* (1987). url: https://businessethicsblog.com/2010/10/12/wall-street-1987-greed-is-good/.

[6] Alice Calaprice. *The ultimate quotable Einstein; with a foreword by Freeman Dyson*. Princeton: Princeton University Press and The Hebrew University of Jerusalem, 2011.

In: Old Problems and New Horizons in World Physics
Editors: Volodymyr Krasnoholovets et al.
ISBN: 978-1-53615-430-6

Chapter 7

THE GHOST OF MODALITY IN THEORETICAL PHYSICS

Akira Kanda[1,*], Renata Wong[2,†] and Mihai Prunescu[3,‡]
[1]Omega Mathematical Institute, Toronto, Ontario, Canada
[2]Nanjing University, Nanjing, Jiangsu, China
[3]University of Bucharest, Bucharest, Romania

ABSTRACT

Despite some fundamental issues, the concept of energy has taken over the world of physical research. The introduction of the concept was invalid at the very least, and threw the entire physics into confusion as a consequence. The fact of the matter is that modality has little place in physics. Once energy is defined via modality, the law of conservation of energy is violated at the very beginning of defining kinetic energy. This misconception has now spread all over physics. To make the matter worse, modern quantum mechanics has added probability (not statistics), which is yet another form of modality, into the theory of physics. We will study the consequence of introducing modality to all branches of theoretical physics, classical or modern.

---

* Corresponding Author's Email: kanda@cs.toronto.edu.
† Author's Email: renata.wong@protonmail.com.
‡ Author's Email: mihai.prunescu@gmail.com.

SECTION 1. FORCE FIELD: A PROJECTION OF ABSOLUTE SPACE TO RELATIVE SPACE

1.1. Force Field and Relativism: Refuting a Folklore

According to classical dynamics, a gravitational field is a spatial distribution of gravitational force per unit mass. A mass M in a space creates a gravitational force field

$$\mathbf{G} = \frac{GM}{|\mathbf{r}|^2}\frac{\mathbf{r}}{|\mathbf{r}|}$$

where G is Newton's gravitational constant and r is the location vector relative to the location of M.

Similarly, an electric field is a spatial distribution of electric force per unit electric charge, while a charge Q in a space creates an electric force field

$$\mathbf{E} = \frac{kQ}{|\mathbf{r}|^2}\frac{\mathbf{r}}{|\mathbf{r}|}$$

where k is Coulomb's constant.

If we place a mass m at location r, it will experience a gravitational force Gm upon it. Likewise if we place a charge q at location r, it will experience the Coulomb force Eq upon it. Here if the respective signs of Q and q agree, the force is a repelling force, otherwise it is an attracting force.

The physics community has failed to pay enough attention to the concept of virtuality and that of modality embedded in the said definition which is contingent upon the assumption that such a mass m or a charge q is placed at r. This is virtuality, not modality, as viewed by modal logicians of the Deleuze school. As the concept of modality is not, so far, properly fixed, we will use the term "modality" loosely when dealing with the issue at hand. The fact of the matter is that the concepts in question are not strictly factual.

Newton did not use the concept of Gravitational Force Field, the reason being that the concept violates the most important law, viz. the law of action-reaction, of his mechanics. M and Q are not getting any respective force from m and q within this conext. When the electromagnetic theory was developed, this issue, which is equally applicable to the electric field, was overlooked, and the entire theory was then developed using the force field concept.

Philosophically, all of this is very interesting. Newton claimed that the correct way to understand the absolute universe which God created and where we live in is to "observe" it by standing on the shoulder of God. His mechanics was an attempt not to observe from the planet Earth, whereas Galileo's perspective was to observe from the surface of planet earth. This may explain, at least philosophically, why Galileo could not reach any substantial result except the reclining tower experiment. In what follows, we shall exhibit a shocking comparison between the work by Newton and that by Galileo in the context of Galileo's reclining tower experiment.

Assume that m and m' are masses with $m > m'$. Let say m and m' are equidistant from the center of the Earth, and let r be that distance and M be the Earth's mass. The gravitational force between the Earth and m is

$$F = GMm/r^2.$$

Hence,

$$a_m = GM/r^2, a_M = Gm/r^2$$

where a_m is the absolute acceleration of m due to the gravitational pull by M. The absolute acceleration for M in absolute space is $-a_M$, the relative acceleration between M and m is then

$$a_M + a_m = GM/r^2 + Gm/r^2.$$

Thus, the time t_m needed for m to reach M is

$$(a_M + a_m)t_m^2 = 2r,$$

and so$t_m = \sqrt{\frac{2r}{GM/r^2+Gm/r^2}}$

For another mass m', we have

$$t_{m'} = \sqrt{\frac{2r}{GM/r^2 + Gm_0/r^2}}$$

The time required for m and m' to reach M is not the same unless $m = m'$. As $m > m'$, we have

$$t_{m'} > t_m.$$

It means "the heavier the faster it falls."

The interpretation adopted above is Newtonian and is based on an observation from outside the universe, whereas Galileo observed from the frame of M which is the frame of himself and the reclining tower. Galileo's observational perspective gives an entirely different conclusion. Within this perspective, M does not move and so the acceleration a_m between m and M is GM/r^2 . Similarly, the acceleration $a_{m'}$ between m' and M is GM/r^2. So both m and m' will reach the ground within the same time interval. This result is false, because the reasoning has ignored the law of action-reaction in between m and M.

Newton never used Galileo's result. Here is the historical fact: Newton derived the concept of gravitational force through Kepler's Law which mathematically represented his observational data of the orbits of other planets.

If we postulate an absolute universe, it seems clear that we will never observe motion, including our motion, in that absolute universe, since to be able to observe absolute motion it would require some kind of metaphysical observation which we have not. It is then more appropriate to say that Newton "induced" the law of gravity from Kepler's law.

1.2. Electromagnetic Field Theory

Under the influence of Heaviside and Hertz, the reluctant Maxwell had to accept the concept of force field and potential field in electromagnetic theory in order to attain mathematical nicety of vector algebra.

As discussed in the previous subsection, the electric field E created by a charge is defined as a spatial distribution of electric force $f_{\mathbf{E}}(r)$, which is "a force as action at a distance, exerted by the charge upon a unit charge when that unit charge is placed at r. This concept can be reformulated as " it is possible that the electric force $ef_{\mathbf{E}}(r)$ will be exerted on a charge e when e is placed at r." It is not necessarily that such a distribution of force is real. Philosophers call such a concept "modal concept". Magnetic field **H** is defined in a similar way.

The above definition assumes, wrongly, that the charge e in question will not affect the charge that creates the field **E**. This assumption seems to be equivalent to the assumption that no charge affects the other within a system of charges. It is a violation of the law of action-reaction. We are going to make an observation of some of the logical consequences of this assumption in the context of Maxwell's electromagnetic field theory.

Maxwell reduced the entire electromagnetic theory to the following field equations:

$$\nabla \times \mathbf{E} = -\frac{1}{c}\frac{\partial \mathbf{H}}{\partial t}, \nabla \cdot \mathbf{E} = 4\pi\rho, \nabla \cdot \mathbf{H} = 0, \mathbf{J} = \rho\mathbf{v}, \nabla \times \mathbf{H} = \frac{4\pi}{c}\mathbf{J} + \frac{1}{c}\frac{\partial}{\partial t}\mathbf{E}$$

where $c = \frac{1}{\sqrt{\epsilon_0 \mu_0}}$. Here ρ is the electric charge density and **J** is the current created by the charge density moving at a speed **v**. Maxwell obtained the last equation called the generalized Ampère's law under the assumption that the charges in **J** move at a constant speed. This law is to introduce the generalized current **J** that is not a conducting current but a displacement of electric field.

The introduction of **J** into the field theory is problematic. According to the definition of electric force field, charges placed in the field will not affect the field, which means that they will not affect other charges that create the electric field. Clearly the generalized Ampère's law allows charges in **J** to change **E** and thus the charges which have created **E**. Maxwell's field equations of electromagnetism are then logically inconsistent. It is a well know logical fact that an inconsistent theory is the most powerful. It proves anything. It gives us anything we want at the cost of giving us anything we do not want either.

There are also ontological issues. It is hard to understand **J** at such a basic level. How is it possible that mutually repelling electrons can form a coherent bundle? The answer is inside a conductor. Here is a vicious circle, though. Conductors are objects to be studied in materials science where we have to make use of electromagnetic theory. The problem somehow resembles the problem of tension rope in classical dynamics. The rope is supposed to be a massless entity that moves under acceleration. Then it must be the case that all such ropes will move with infinite speed. How does it connect two bodies?

From

$$\nabla \times \mathbf{E} = -\frac{1}{c}\frac{\partial \mathbf{H}}{\partial t}$$

which is called Maxwell-Faraday Law, the so-called Lorentz force

$$\mathbf{F} = q(\mathbf{E} + \mathbf{v} \times \mathbf{B})$$

follows. Here **v** is the speed of the charge q. This law has been empirically verified by many experiments. The problem is just that the dependence of the force upon a charge on the speed of the charge makes this law a violation of the second law of classical dynamics. Charge or not, the second law excludes any possibility for the force to be dependent on the speed. This means that the electromagnetic theory and the Newtonian dynamics contradict each other. This issue, which is a serious one, was never discussed in the past. It simply shows that the electromagnetic theory is invalid. That the Lorentz force's being verified by experiment does not make this theory consistent. It simply means that we cannot use this theory as the theory of electromagnetism.

Here is yet another consequence of the logical inconsistency discussed above. Maxwell obtained his electromagnetic wave equation

$$\nabla^2 \mathbf{E} = \frac{1}{c}\frac{\partial^2 \mathbf{E}}{\partial^2 t}$$

and

$$\nabla^2 \mathbf{H} = \frac{1}{c}\frac{\partial^2 \mathbf{H}}{\partial^2 t}$$

under the condition that $\mathbf{J} = 0$ (meaning no conducting current). From this wave equation, he calculated the speed of electromagnetic wave in a vacuum as $c = 1/\sqrt{\varepsilon_0 \mu_0}$.

This result of Maxwell's suggests that accelerating charges can create the so-called "EM wave" which affects any charge in the system. Indeed, if aether theory is correct, then charges under acceleration can affect the entire universe without getting reaction. After all, is it not the case that accelerating charges form a current?

A consequence of overlooking the aforesaid conceptual issue was a serious misunderstanding of the "EM wave". The EM wave is supposed to be a solution to the above wave equation which was discovered theoretically by Maxwell from his electromagnetic field algebra. His "algebra" however is in fact not really an algebra of electromagnetic fields since J is not a field. It is somehow ignored by physicists.

The so-called "EM wave" is not a wave that travels through a space, contrary to what we have been told for nearly two centuries. It is the transmission (at a distance) of the magnitude change of the EM force, generated at the emitter, from the emitter to the receiver of the "EM wave" located somewhere in the space. If there is no receiver, there is no emission of the magnitude change of the EM force.

Regarding the speed c of light, this quantity was theoretically obtained by Maxwell in his EM field theory and was defined as the speed at which the change of electric charge will "travel" through the vacuum between the emitter and the receiver. This original value had little to do with the "EM wave" of Maxwell. When Maxwell derived his "EM wave" as a solution of the EM wave equation in a vacuum, what he had obtained was that the speed of this mathematical (not ontological) wave is $c = 1/\sqrt{\varepsilon_0 \mu_0}$ as long as there is no current in the vacuum. As discussed above, the "EM waves" are just mathematical entities without any ontological status.

The problem here is a serious case of tension between empiricism and theory. Unless we could form a chain of receivers of EM magnitude change, we would not be able to observe the so-called EM wave. This simply means that we are not able to observe experimentally the "EM waves". In reality, what we have been doing is drawing a graph of the magnitude of EM force at a fixed receiving position with respect to time. This is not what we mean by the "EM wave" travelling through space.

One of the problems in making physics a mathematical theory is that ontology tends to be ignored and abstract mathematical argument takes over, yielding mistaken views of physics. It had become a serious issue in physics over the last century.

This problem of empiricism versus theory manifests itself more acutely in the case of relativizing classical electromagnetism. Historically, all the laws of EM theory were discovered within the laboratory frame of reference. Within this single reference frame, the theory, unlike Newton's classical dynamics, was developed upon many experiments; and in the end, it was the Faraday Induction which showed that it is impossible to relativize classical electromagnetic theory. This situation can be best explained as follows: When a charge is in acceleration in our frame, EM wave is emitted in our frame; nonetheless, in the frame of the charge, no EM wave is emitted. Unfortunately, we may not be able to verify or refute this experimentally.

Lorentz was aware of the problem that all axioms of electromagnetic field theory came from the experiments conducted within our frame, not in the absolute frame of aether. He applied the Galilean transformation between the aether frame and our frame to reflect the situation. He discovered that Maxwell's wave equation is not invariant under the Galilean transformation. This led him to consider a different transformation between the absolute frame if aether and non-aether frames which would preserve Maxwell's electromagnetic wave equation, i.e., the so-called "Lorentz transformation". Unlike Einstein who generalized Lorentz's work, Lorentz however did not consider transformation of non-absolute reference frames. Since then, nobody has noticed that wave functions with wave vector and frequency are invariant under the Galilean transformation. This gives rise to the fundamental question of whether wave equations are the right way to define waves. As is known, mathematically a wave equation has infinitely many different solutions. The same issue has dogged quantum mechanics with regard to the solutions of Schrödinger's wave equation. We shall revisit this issue in Section 6.3.5.

1.3. Aether Theory, Special Theory of Relativity and Lorentz Transformation

As it is now clear that the Michelson-Morley experiment was misinterpreted and there is no scientific ground to assume that the speed of light is constant in all reference frames, the aether theory of Maxwell/Lorentz should be reconsidered. Lorentz understood the problem of our observation, and what is really happening in the absolute frame which Maxwell considered to be filled with aether in order to explicate his axioms of electromagnetic theory.

So, the aether theory, for Lorentz and Maxwell, is in a sense the semantics of Maxwell's axiomatic theory of electromagnetism. Lorentz tried to derive "our observation" of electromagnetism through Lorentz transformation from aether to our reference frame. Einstein, a true relativist, pushed further and rejected the aether theory using a wrong interpretation of the Michelson-Morley experiment, hence liberalized humanity from the repressive absolute.

Neither Lorentz nor Einstein understood that the concept of relative frames violates the law of action-reaction,[1] and so there is no point in projecting the absolute to the relative, or in rejecting the absolute altogether and work on the relative frames alone. As stated above, the basic laws of Newton's dynamics and the early electromagnetism theory were written for absolute frame. The relativization attempts thus end up with some of the most fundamental laws of physics being violated.

However, these theories, classical electromagnetism theory in particular, have their own logical issues overlooked. In the case of Newton's classical dynamics, the concept of gravitational force was induced from Kepler's laws which came from studying the observed movement of planets from Earth. Newton's law was not derived, so to speak, from the shoulder of God. It was not a bird's-eye view of the planetary motions. In case of classical electromagnetism, all axioms of the theory came from lab experiments which were done within our frame of reference. It makes little sense to consider aether as the semantical counterpart of these laws of electromagnetism. In the end neither relativity theory nor aether theory are right on this point.

Notwithstanding, both Einsteinian relativity theory and Lorentzian projection theory share not only the Lorentz transformations but also the same shortcoming of measuring the speed v of moving reference frames which is a parameter in Lorentz transformation. In the case of Lorentzian projection theory, v is the absolute speed between the aether and the moving reference frame; and there is no known way to measure this speed. A correct interpretation of Michelson-Morley's method of using reflected light suggests that we cannot detect v in such a way. Moreover, Einstein's relativity theory has a serious problem on this issue.

The $v = d/t$ in the gamma factor $1/\sqrt{1-(v/c)^2}$ is the classically measured speed of a moving reference frame. Assume that an object Oj is stationary in a frame F_1, but moves a distance Δd in time Δt in frame F_2 (our frame). Let A and B be the respective positions of Oj at time 0 and time t in frame F_2. According to the perspective of Oj, A and B are moving points; and so, Oj will observe that the distance is $\Delta d'$, the time interval which took A to be in front of itself is $\Delta t'$, and the time interval which took B to be in front of itself is $\Delta t''$. So, Oj will observe that the speed of our frame is $\Delta d'/\Delta t'$. By time dilation

$$\Delta t' = \sqrt{1-(v/c)^2}\Delta t$$

and length contraction

$$\Delta d' = \frac{\Delta d}{\sqrt{1-(v/c)^2}}$$

[1] To be precise, the law of action-reaction does not apply to relative frames.

mathematically we have $v = \Delta d/\Delta t \neq \Delta d'/\Delta t' = v'$. This establishes that both frames F_1 and F_2 will not observe the other side moving with speed v. The $v = d/t$ in the gamma factor

$$\frac{1}{\sqrt{1-(v/c)^2}}$$

is the classically measured speed of a moving reference frame. Assume that an object Oj is stationary in a frame F_1, but moves a distance d in time t in our frame F_2. Let A and B be the respective positions of Oj at time 0 and time t in F_2 respectively. According to the perspective of Oj, A and B are moving points. So, Oj will observe that the distance is d', the time interval which took A to be in front of itself is t' and the time interval which took B to be in front of itself is t'. So, Oj will observe that the speed of our frame is d'/t' which unfortunately is not $v = d/t$. Now, the mutual relative speeds of F_1 and F_2 are not the same anymore. As v is measured classically, it must be shared as the mutual relative speed between F_1 and F_2.

Relativistic Lorentz transformation was derived from time dilation and length contraction combined under the principle of relativity. Roughly it goes as follows: Time dilation and length contraction yield a transformation from an inertial reference frame F to the other F' such that F' moves inside F with speed v in the direction of the x-axis as follows:

$$x' = \frac{1}{\sqrt{1-(v/c)^2}}(x - vt)$$

Similarly as F is an inertial frame which moves inside F' with speed v through time dilation and length contraction, we obtain

$$x = \frac{1}{\sqrt{1-(v/c)^2}}(x' + vt)$$

By solving these two equation for t', we get

$$t' = \frac{1}{\sqrt{1-(v/c)^2}}(t - vx/c^2)$$

Thus we arrive at the Lorentz Transformation

$$x' = \frac{1}{\sqrt{1-(v/c)^2}}(x - vt), y' = y, z' = z, t' = \frac{1}{\sqrt{1-(v/c)^2}}(t - vx/c^2)$$

This means that a point (x, y, z) in F moving with speed v will appear as (x', y, z) in F' and the time in F will appear as t' in F'. Now we have to consider the speed v' of this point in F in F'. The naive classical idea of speed will give us a possibility:

$$v' = \frac{x'}{t'} = \frac{x - vt}{t - vx/c^2} \neq v$$

Certainly this is not what we need. So, what went wrong here. Putting aside the fatal problem of time dilation and length contradiction that are based on the untenable assumption of relatively moving reference frames, we do have a serious problem of mathematical nature here. The attempt to replace the Lorentz transformation by a version of displacement such as

$$\Delta x' = \frac{1}{\sqrt{1-(v/c)^2}}(\Delta x - v\Delta t), y' = y, z' = z, \Delta t' = \frac{1}{\sqrt{1-(v/c)^2}}(\Delta t - v\Delta x/c^2)$$

will not do as

$$v' = \frac{\Delta x'}{\Delta t'} = \frac{\Delta x - v\Delta t}{\Delta t - v\Delta x/c^2} \neq \frac{\Delta x}{\Delta t} = v$$

Patently

$$v' = \lim_{t'\to 0} \frac{\Delta x'}{\Delta t'} \neq \lim_{t\to 0} \frac{\Delta x}{\Delta t} = v$$

This obviously does not solve the problem. The Lorentz transformation was obtained under the regulated condition that F_1 and F_2 share the same origin Oj at the same time 0. Classical calculus works fine as time and location are mutually independent coordinates. Once we allow mutual dependence of these two coordinates as we have done in the special theory of relativity as through time dilation, length contraction and the Lorentz transformation, we cannot define the concept of speed any more at the pain of contradiction. This is a totally expected consequence of the dimensional analysis of theoretical physics. [Time] is not defined in terms of [length] and [speed]. [Length] is not defined in terms of [time] and [speed].

We shall revisit this topic in subeection 6.1 The Special Theory of Relativity.

SECTION 2. ENERGY?

2.1. Energy Versus Work

In Newtonian mechanics, work W is defined as

$$\int \mathbf{F} \cdot d\mathbf{r}$$

where $\mathbf{F}$ is the force vector and $d\mathbf{r}$ is the infinitesimal displacement vector. Due to the nature of the dot product, work can be negative or positive. Energy is defined abstractly as the potential to do work. It is a standard argument that the work needed to accelerate from $m0$ to mv is

$$\frac{mv^2}{2}$$

This work is called the kinetic energy of mv. This standard argument, which a student of physics learns at the early stage of her physics studies, has overlooked one problem. The work needed to accelerate from $m0$ to mv depends on the way we accelerate m.

When the acceleration is constant, certainly the work needed is as above. But this is not the case if the acceleration is not constant, except in a few other cases. Let us demonstrate this point with an example. Assume that $a(t) = t$ instead of being constant. Then we can formulate the equation of the motion for the initial condition $v_0 = x_0 = 0$ as

$$v(t) = \int tdt = \frac{1}{2}t^2, x(t) = \int \frac{1}{2}t^2 dt = \frac{1}{6}t^3$$

Then, we have

$$W = \int (mt)x(t)dt = \int \frac{1}{6}t^3 mtdt = \frac{m}{24}t^4$$

However, as an identity of t,

$$\frac{m}{24}t^4 \neq \frac{mv^2}{2} = \frac{m}{2}\left(\frac{1}{2}t\right)^2 = \frac{m}{8}t^2$$

This invalidates a most important law of physics, which is the law of conservation of energy.

This mismatch between energy and work probably has yet another even more important consequence. Work can cancel each other as work can be negative; energy however cannot cancel each other as energy is always positive. Consider that with work W_1, we accelerate from position A with momentum $m0$ to position B with momentum mv, and from here with work $-W_1$ we decelerate from mv to the position C with momentum $m0$. The work done is 0. However, the energy needed to do this work is $2E$, where E is the energy needed to do the move m from A to B. This means that energy $2E$ is used to do no work. If such is the case, the law of conservation of energy is obviously false.

One of the main differences between the legitimate concept of momentum and the troubled concept of energy is that energy is defined as a "modality", what modal logicians call "possibility" in this case. It is an acknowledged fact that combining predicate reasoning and modal reasoning could lead to inconsistency. This could be a culprit of the inconsistencies that theoretical physics is facing in so many places. Certainly it is not quite plausible that energy, as a modal concept, can be used to formulate the most fundamental law of physics which is supposed to be the most precise physical science. This also invalidates the law of conservation of energy.

2.2. Potential Energy Field

A force field is a spatial distribution of force per unit mass. A potential energy field is a spatial distribution of energy. Considering that the concept of force field is a troubled one as discussed in Section 1, it becomes a shaky matter to assign the modal concept of energy to each location in a space. In case of a force field, such a force has violated the law of action-reaction. So what kind of force are we talking about? In case of the potential energy field, the concept of energy itself is not scientifically legitimate. It is a modality instead of a scientific fact. Moreover, it contradicts the conservation of energy claim without even involving "potential energy" which we will discuss in more details below.

In any case, here is how physicists create a "potential energy field" from a force field. Consider a "force field" in which the "force" (which disrespects the law of action-reaction) is allocated at each location of the space. Let $F(r)$ be the "force" a body experiences at a location r. When a body moves from a position r to r', the work done on this body is given as

$$W_{\mathbf{r}}^{\mathbf{r}'} = \int_{\mathbf{r}}^{\mathbf{r}'} \mathbf{F}(\mathbf{r}) d\mathbf{r}$$

In general, this integral value depends upon the path taken from r to r'. A force field in which this integral depends only upon r and r' is called a potential field.

It immediately follows that

$$W_{\mathbf{r}}^{\mathbf{r}'} = -W_{\mathbf{r}'}^{\mathbf{r}}$$

In a potential field, let

$$W_{\mathbf{r}}^{\mathbf{r}'} = -V_{\mathbf{r}}^{\mathbf{r}'} + e$$

where e is a constant. Then we say $V_{\mathbf{r}}^{\mathbf{r}'}$ is the potential energy of a body at location r' relative to the location r. Since

$$W_{\mathbf{r}}^{\mathbf{r}} = \boldsymbol{e}$$

$\boldsymbol{e}$ is the potential energy at the initial location r. It is clear that in the potential field we have

$$W_{\mathbf{r}}^{\mathbf{r}'} = \frac{m}{2}|\mathbf{v}'|^2 - \frac{m}{2}|\mathbf{v}|^2$$

where v and v' are the respective speeds of the body at r and r'. Therefore

$$\frac{m}{2}|\mathbf{v}|^2 = \frac{m}{2}|\mathbf{v}'|^2 - W_{\mathbf{r}}^{\mathbf{r}'}$$
$$\frac{m}{2}|\mathbf{v}'|^2 + V_{\mathbf{r}}^{\mathbf{r}'} = \frac{m}{2}|\mathbf{v}|^2 + e$$

From here we conclude that inside a potential field the sum of the kinetic energy and the potential energy of the body is constant. This is the law of conservation of energy.

All of this is mathematical nicety which is however marred by the conceptual and ontological obscurity of the whole argument. Well before the development of this potential field concept, the law of conservation of energy failed in purely simple dynamics. There is no such thing as conservation of energy as there is no such thing as kinetic energy. Adding the concept of force field will not remove this sad reality.

Fundamentally, the concept of force field, being the basis of the concept of potential field, is irrelevant to physics as it violates the law of action-reaction.

SECTION 3. COLLISION PROBLEM

Theoretical physicists developed the theory of collision via the concept of energy together with that of momentum. We shall examine a consequence of the concept of energy being not a physical quantity but a modality. The first half of this section deals with momentum.

According to Newton's second law, we have

$$\int_{t_1}^{t_2} \mathbf{F}(t)dt = \int_{t_1}^{t_2} m\frac{d\mathbf{v}(t)}{dt}dt = m\int_{t_1}^{t_2} \mathbf{a}(t)dt = \mathbf{P}(t_2) - \mathbf{P}(t_1)$$

It means that the impact $\int_{t_1}^{t_2} \mathbf{F}(t)dt$ is equivalent to the change of momentum $P(t_2) - P(t_1)$; and the law is called the law of impact. Conversely we can obtain the second law from the law of impact as follows:

$$\frac{d}{dt}\int_0^t \mathbf{F}(t)dt = \mathbf{F}(t) = \frac{d\mathbf{P}(t)}{dt} = m\frac{d\mathbf{v}(t)}{dt} = m\mathbf{a}(t)$$

Assume that two masses m_1 and m_2 collide at time t_0. The law of action-reaction can thus be formulated as

$$m_1\frac{d^2\mathbf{x}_1(t)}{dt^2} = -m_2\frac{d^2\mathbf{x}_2(t)}{dt^2}$$

Taking the integral on both sides of this equation, the conservation of momentum readily follows. Therefore we have

$$m_1\mathbf{v}_1(t') + m_2\mathbf{v}_2(t) = m_1\mathbf{v}_1(t) + m_2\mathbf{v}_2(t)$$

where

$$t' < t_0 < t.$$

After the collision ($t > t'$), due to the law of action-reaction, we have

$$m_1\frac{d^2\mathbf{x}_1(t)}{dt^2} = -m_2\frac{d^2\mathbf{x}_2(t)}{dt^2}$$

Notice that we have no mathematical description of the exact moment of the collision. This is because

$$\lim_{t\to t_0-0} \mathbf{v}_1(t) \neq \lim_{t\to t_0+0} \mathbf{v}_1(t)$$

and

$$\lim_{t \to t_0 - 0} \mathbf{v}_2(t) \neq \lim_{t \to t_0 + 0} \mathbf{v}_2(t)$$

which means that m_1 and m_2 have no speed at the moment of collision. This also means that there is no mathematical description of the motion of m_1 and m_2 after the collision as there is no initial condition for them.

Theoretical physicists claim that since the motion of m_1 and m_2 near the moment of collision could have an "approximately constant" speed. We can take some time t_{0-} and t_{0+} very close to t_0 and assume that m_1 and m_2 travel with constant speed $v_{t_{0-}}$ and $v_{t_{0+}}$ respectively immediately before and immediately after the collision. So the conservation of momentum yields

$$m_1\mathbf{v}_1(t_{0-}) + m_2\mathbf{v}_2(t_{0-}) = m_1\mathbf{v}_1(t_{0+}) + m_2\mathbf{v}_2(t_{0+})$$

Here $\mathbf{v}_1(t_{0-})$ and $\mathbf{v}_2(t_{0-})$can be analytically obtained but $\mathbf{v}_1(t_{0+})$ and $\mathbf{v}_2(t_{0+})$ cannot be obtained mathematically.

To resolve this mathematical problem, physicists introduce the following "empirical law".

Law of Collision

When A and B, moving with constant speeds v_A and v_B, collide and separate with constant speeds v'_A and v'_B, then for some $e = (e_1, e_2, e_3)$,

$$e_1 = -\frac{v'_{A1} - v'_{B1}}{v_{A1} - v_{B1}}, e_2 = -\frac{v'_{A2} - v'_{B2}}{v_{A2} - v_{B2}}, e_3 = -\frac{v'_{A3} - v'_{B3}}{v_{A3} - v_{B3}}$$

$e = (e_1, e_2, e_3)$ is called "collision vector". It can be readily shown that with e we can calculate $v_1(t_{0+})$ and $v_2(t_{0+})$.

If A and B are solid, then $e_1 = e_2 = e_3 = 1$. Only in this case, which we call "elastic collision", the law of conservation of energy holds. If $e_1 = e_2 = e_3 = 0$, then A and B move with exactly the same speed (move together as a single body) after the collision.

There are some confusions in this argument. Theoretically, we know nothing about (v'_A, v'_B) because we do not know the initial conditions for $x_1(t)$ and $x_2(t)$ after the collision, and (v_A, v_B) are just one side limit of $(v_1(t), v_2(t))$ from below. The e here does not really help. Why do physicists have to use one sided limit for $v_1(t_{0-})$ and $v_2(t_{0-})$ if they are to rely on empiricism to deal with the problem? Theoretically, the moment of collision is a singularity. This is where the gravitational force between m_1 and m_2 diverges. It is questionable if the treatment of collision above is acceptable. One of the issues involved here is the relation between theory and empiricism. For the philosophers of physics, theory predicts and empiricism either refutes or supports the theory. It becomes problematic when empiricism tries to cover up the failure of the theory in such a way. Of course, as the development in this field is rather technical and sometimes conceptually confusing, even the majority of professional physicists have failed to see the aforementioned (serious!) errors in reasoning. It then becomes the duty of the philosophers to correct errors of this type in a very tense situation of this kind.

The involvement of the empirical element e here is troublesome in other aspects too. First of all, conceptually, in classical dynamics, Newton made it very clear that all objects are point objects, without which as an assumption, there is no way to consider applying force on a body with geometric dimension since force is a vector which is an arrow with a pointed head. Under the influence of the pragmatism of engineering, this restriction is constantly compromised causing internal contradictions. As Popper said, it is not worth to even consider empirical refutation or verification of logically inconsistent theories; and as some mathematical physicists have warned, the bodies with geometric dimensions must be considered in terms of their internal structure as is done in statistical mechanics. None of this is being done in the theory of collision.

Let us revisit the issue of modality. We have shown that the concept of energy is a modality and treating it as a physical material reality has caused a lot of confusion. This is already enough ground to eliminate the concept of energy and the law of conservation of energy. We have presented many instances in theoretical physics where the violation of the law of conservation of energy has occurred. Most shockingly, we have shown that even the very definition of kinetic energy violates the law of conservation of energy. Here in the theory of collision, it has become very clear again that the law of conservation of energy fails.

Despite all of these problems with the modal concept of energy, as the reconstruction of the theory of collision above shows, the conservation of momentum holds without the collision vector e. This clearly indicates that the energy concept is foreign to mechanics.

When a mass m with speed v collides elastically with another mass with initial speed 0, we have the following equation for the conservation of energy:

$$v^2 = v_1^2 + v_2^2$$

where v_1 and v_2 are the speeds of the two masses after the collision. The law of conservation of momentum gives us

$$\mathbf{v} = \mathbf{v}_1 + \mathbf{v}_2.$$

From the latter equation we have

$$v^2 = v_1^2 + 2\mathbf{v}_1 \cdot \mathbf{v}_2 + v_2^2 \, .$$

Therefore we have $v_1 \cdot v_2 = 0$, which implies that the exiting speed vectors must be perpendicular to each other if neither of the exiting speed is zero.

Logicians may suggest that if energy is not a reality but a modality, one can-not combine this modal concept with the physical reality of momentum without causing conflicting issues like those mentioned above.

SECTION 4. THERMODYNAMICS

4.1. Internal Energy, Thermal Energy and Temperature

Thermodynamics is a theory of physical systems studied just in terms of energy. So, what has been developed above on the issue of energy may have certain material impact on this theory.

In classical dynamics, physical objects are reduced to point masses and their interaction takes place through forces such as the gravitational force. This theory provides a better understanding of the world of physics at the most fundamental level. In thermodynamics, the body, which has geometric dimensions, is treated as a "massive" collection of point objects which we call particles. In modern terms, these particles are either atoms or molecules. Contingent upon the validity of the theory of energy in classical mechanics, such a system possesses dynamic energy as the sum of the kinetic energy of all particles and the potential energy created by the system of such particles under inter-particle force. This energy is called the internal energy of the physical body, and is different from the energy with which this physical body interacts with the outside world. The sum of all kinetic energy of the particles of a physical body is called the thermal energy of the physical body. In case of gas, inter-particle force is negligible, and so the internal energy of a particle and its thermal energy are the same.

Within this framework, the temperature t of a gas is defined as the thermal energy "level" for one mole of the gas which is to satisfy the following formula

$$U = \frac{3}{2}k(t + c)\; k > 0 \qquad (1).$$

When we set $U = 0$, we have $t = -c$. This means that $-c$ is the temperature of the gas in which all particles stop. This is the theoretical minimum temperature. k is a constant which is determined by the choice of unit for U. Theoretically, the most convenient unit for temperature is called the absolute temperature and is determined as follows. We set the theoretical minimum temperature to be 0 $[K]$, then the temperature at which the oxygen gas changes to liquid is set to be 185 [*K*]. With this, 1 [*K*] is determined. With this unit, (1) is reduced to

$$U = \frac{3}{2}kt \ k > 0 \qquad (2).$$

When we take the unit of U to be [*J*], we will find a specific number $R > 0$ such that

$$U = \frac{3}{2}Rt \qquad (3).$$

This R is called the gas constant. With this R, we can define the absolute temperature t of a non-gas body as a solution to

$$U = \frac{3}{2}Rt \qquad (4).$$

This gives us a desirable property that, regardless of the "phase" of a system of particles, the system's absolute temperature is determined by just the heat energy of the system.

4.2. Transfer of Heat Energy and the Second Law of Thermodynamics

When we let a high-temperature body A touch a low-temperature body B, the more active particles of A collide with the less active particles of B near the boundary and, following the law of conservation of energy, loose their kinetic energy by passing it to the particles of B. This continues until A and B are in the same heat energy state. In terms of heat, thermodynamics researchers call this process "transfer of heat energy from A to B". From this, we immediately have the following fundamental law of thermodynamics:

The second law of thermodynamics: When a high-temperature body A touches a low-temperature body B, the temperature of A drops and that of B rises until they have the same temperature.

Due to the law of conservation of energy in dynamics, we also have the following fundamental law in thermodynamics:

The first law of thermodynamics: Assume that we have exerted heat energy Q and work W to a body with heat energy U_1, and the heat energy of this body have become U_2, then we have

$$U_2 - U_1 = Q + W \qquad (5)$$

In case $W > 0$, we apply work to this body from outside. In case $W < 0$, we let the body do work.

In case where the gas body does not exchange the heat energy with the external world and we compress or expand the gas, $Q = 0$ and so we have

$$U_2 - U_1 = Q.$$

Now when we compress the gas, the temperature of the gas increases and when we expand the gas, the temperature drops.

In case the gas temperature does not change, we have

$$U_2 - U_1 = W + Q = 0.$$

So we have $Q = -W$, and all the work the gas has received from outside is passed to the external world as heat energy.

In case we allow the gas to expand, ($W < 0$), as $Q = -W$, all the work done by the expanding gas comes from the external heat energy.

4.3. Heat Engine and Heat Efficiency

An instrument which converts heat energy into mechanical work is called a heat engine. A heat engine receives heat energy Q_1 from outside and converts a part of it to work W and emits the rest as Q_2. In association with this process, we call

$$e = \frac{W}{Q_1} = \frac{Q_1 - Q_2}{Q_1}$$

the thermo-efficiency of the heat engine.

The heat engine with 100% thermo-efficiency is called the "second perpetual engine". It can readily be shown that the second perpetual engine does not exist, because if it did exist then heat energy would be 100% converted into work. Then we can convert the heat energy of a lower temperature heat source completely into work and pass it to a higher temperature heat source. This violates the second law of thermodynamics.

An instrument which keeps doing work without getting heat energy from outside is called a first perpetual engine. This type of engine does not exist either as it violates the law of conservation of energy.

4.4. Entropy

According to the second law of thermodynamics, when we put hot gas and cold gas together it does not happen that the hot gas get hotter and the cold gas get colder, though this process respects the conservation of energy principle. This is yet another example of questioning the validity of the law of conservation of energy in addition to the example where the law of conservation of energy fails as is discussed in subsection 2.1. All in all, our discussion has shown that the concept of kinetic energy is ill-defined and in thermodynamics everything is about kinetic energy.

Putting aside this fundamental problem with kinetic energy, the second law of thermodynamics says that an isolated system which is free of external influence will transform a "state of relative order" to a "state of relative disorder" until it eventually reaches the state of maximum disorder.

What is missing here in this naive observation is that we do not know precisely what the state of orderis. This concept is represented by the quantity

$$dS = \frac{\partial Q}{T}$$

where T is the temperature and ∂Q is an infinitesimal amount of heat. Clausius showed that for a cyclic process

$$\oint dS \leq 0$$

The equality holds only for reversible processes. This result of Clausius shows that the line integral

$$\int_L dS$$

is path independent. Therefore there is a state function S such that

$$dS = \frac{\partial Q}{T}$$

This function is called "entropy".

Third law of thermodynamics: $S = 0$ at absolute zero temperature for a perfect crystal.

With this law, taking integral of dS as above, we can obtain the state function S. Now we have

$$-\Delta S + \int dS = -\Delta S + \int \frac{\partial Q}{T} = \oint dS \leq 0.$$

This means that

$$\Delta S \geq \int \frac{\partial Q}{T}$$

where equality holds only for reversible transformations.

This means that for an irreversible process its entropy keeps increasing. This result is called the "principle of increase of entropy".

4.5. A Question

Despite this spectacular list of achievements, there is one problem shedding a shadow over the entire theory of thermodynamics. All the results of thermodynamics depend solely on the concept of energy which, as we have learned, is a loose modal concept; in other words, a myth. The very definition of kinetic energy is also invalid. Now what does it mean to Shannon's claim on information entropy as the degree of order is an interesting question. The concept of information entropy seems to be legitimate.

SECTION 5. WAVE MECHANICS

5.1. Transversal Wave Equation

We cover only transversal waves. Consider a string whose equilibrium is the x-axis. The motion of the string is limited only to a plane which includes this x-axis. Let y be the axis which is perpendicular to the x-axis. Then the motion of the string can be expressed by a function of two variables

$$y = f(x, t).$$

Then

$$y = f(x, t_1)$$

represents the pattern of the wave at time t_1. Also

$$y = f(x_1, t)$$

is the history of the y-value of the string at x_1. Moreover

$$\frac{\partial y}{\partial t}(x_1,t)\,,\frac{\partial^2 y}{\partial t^2}(x_1,t)$$

are the y-directional speed and y-directional acceleration of the point x_1 at time t, respectively. We assume that the string at equilibrium has a uniform linear mass distribution σ and a uniform tension T. We further assume that all of this does not change under the wave motion of the string.

Under the above assumption, we carry out the transversal wave motion by applying the second law to the segment between x and $x + \Delta x$. The tension force under the wave motion at a point is constant in magnitude but the direction is determined as the tangent of the point of the wave pattern. Let $\theta = \theta(x,t)$ and $\theta' = \theta(x',t)$ be the direction of the tension vector at x and x', respectively.

Then with respect to the segment in between x and $x + \Delta x$, the tension to the y-direction is

$$\Delta(T\ sin\theta) \;=\; T\ sin\theta' \;-\; T\ sin\theta'.$$

We assume that this force is the only material force as the rest is negligible. The mass of this segment is $\sigma\Delta x$. Let $\bar{x}$ be the center of mass for this segment. Then the acceleration at the center of mass is

$$\frac{\partial^2 y}{\partial t^2}(\bar{x},t)$$

which gives the following equation as the second law

$$\sigma\Delta x\frac{\partial^2 y}{\partial t^2}(\bar{x},t) = \Delta T sin\theta.$$

So,

$$\sigma\frac{\partial^2 y}{\partial t^2}(\bar{x},t) = \frac{\Delta T sin\theta}{\Delta x}.$$

Taking $\Delta x \to 0$, we have $\bar{x} \to x$, and so

$$\sigma\frac{\partial^2 y}{\partial t^2}(\bar{x},t) = \frac{\partial T sin\theta}{\partial x}.$$

When θ is sufficiently small, we have $sin\theta + tan\theta$. As $tan\theta$ is the slope of the tangent line, we have

$$tan\theta = \frac{\partial y}{\partial x}.$$

So

$$\sigma \frac{\partial^2 y}{\partial t^2}(\bar{x}, t) = \frac{\partial}{\partial x}\left(T \frac{\partial y}{\partial x}\right) = T \frac{\partial^2 y}{\partial x^2}$$

By putting $c^2 = \frac{T}{\sigma}$, we have

$$\frac{1}{c^2} \frac{\partial^2 y}{\partial t^2}(\bar{x}, t) = \frac{\partial^2 y}{\partial x^2} \tag{6}$$

This equation is called the "one dimensional transversal wave equation". This is the wave equation for the transversal wave with small slope everywhere.

5.2. Solutions of Transversal Wave Equation

Let $f(u)$ be a function such that $f'(u)$ and $f''(u)$ exist. Define $y(x, t)$ as

$$y(x, t) = f(x - ct) \tag{7}$$

where $c = \sqrt{T/\sigma}$. It can readily be shown that this $y(x, t)$ is a solution to (6). It is clear that

$$y(x, 0) = f(x)$$

is the initial wave pattern of the wave $y(x, t)$ which is the translation ct to the right of this initial wave pattern. So, the wave pattern does not change its shape as the wave moves on. This is called the "propagation (to the right) of the wave", and this wave is called "advancing wave". c is called the speed of the propagation of the wave. Here c is not the speed of the point of the string. The point on the string moves vertically and its speed is given as

$$\frac{\partial y}{\partial t} = -cf'(x - ct) = -c\left(\frac{\partial y}{\partial x}\right).$$

As we have assumed that the slope of the wave is very small, $\partial y/\partial x \ll 1$ implies

$$\frac{\partial y}{\partial t} \ll c$$

which means that the speed of the wave is much faster than the speed of the particle in the string. Now let

$$y(x,t) = g(x + ct).$$

It can readily be seen that this function satisfies the wave equation (6). This wave function has the same wave pattern but it moves in negative direction on the x-axis. We call such a wave a "receding wave".

Now we have the following important theorem on the general solutions of the wave equation (6).

Theorem 1. *Necessary and sufficient condition for the wave equation (6) to have* $y(x,t)$ *as a solution is that there are twice differentiable functions* $f(u)$ *and* $g(u)$ *such that*

$$y(x,t) = f(x - ct) + g(x + ct).$$

Proof. Sufficiency can be readily shown as follows. Define

$$\Box^2 = \frac{\partial^2}{\partial x^2} - \frac{1}{c^2}\frac{\partial^2}{\partial t^2}.$$

Then y is a solution to (6) iff $\Box^2\, y = 0$. By the linearity of the differentiation, it can readily be shown that

$$\Box^2\,(y_1 + y_2) = \Box^2\, y_1 + \Box^2\, y_2.$$

If y_1, y_2 are solutions to (6), then $\Box^2\, y_1 + \Box^2\, y_2 = 0$. So, $y_1 + y_2$ also is a solution to (6). Necessity can be shown as follows: Let $u = x - ct$ and $v = x + ct$. Then for any function $y(x,t)$, there is a function Y such that

$$y(x,t) = Y(u(x,t), v(x,t)).$$

This leads to

$$\frac{\partial^2 y}{\partial x^2} = \frac{\partial^2 Y}{\partial u^2} + 2\frac{\partial^2 Y}{\partial u \partial v} + \frac{\partial^2 Y}{\partial v^2}.$$

Similarly we have

$$\frac{\partial^2 y}{\partial t^2} = \frac{\partial^2 Y}{\partial u^2} - 2\frac{\partial^2 Y}{\partial u \partial v} + \frac{\partial^2 Y}{\partial v^2}.$$

If $y(x,t)$ is a solution to (6), we have

$$\frac{\partial^2 Y}{\partial u^2} + 2\frac{\partial^2 Y}{\partial u \partial v} + \frac{\partial^2 Y}{\partial v^2} = \frac{\partial^2 y}{\partial t^2} = \frac{\partial^2 Y}{\partial u^2} - 2\frac{\partial^2 Y}{\partial u \partial v} + \frac{\partial^2 Y}{\partial v^2}$$

This means

$$\frac{\partial^2 Y}{\partial u \partial v} = \frac{\partial}{\partial u}\left(\frac{\partial Y}{\partial u \partial}\right) = 0.$$

So,

$$\frac{\partial Y}{\partial v} = h(v)$$

for some function h. This means for some function $f(u)$,

$$Y(u,v) = \int h(v)dv + f(u).$$

So, let $g(x) = \int h(v)dv$ have

$$y(x,t) = f(u) + g(v) = f(x - ct) + g(x + ct).$$

This means that if $y(x,t)$ is a solution to (1), then for some f and g,

$$y(x,t) = f(x - ct) + g(x + ct).$$

It can readily be seen that f and g are differentiable twice.

It is a fact that in wave mechanics what we need comes from the solutions to the wave equation rather than wave equations per se. This is a well understood fact in the mathematical theory of equations. Unless we have a unique solution to an equation, the equation cannot tell us much about a particular solution. So, Lorentz's discovery that wave equations cannot be mapped into wave equations under the Galilean transformation is nothing surprising. The fact is that wave functions are mapped into wave functions under the Galilean transformations. There is no need for the obscure Lorentz transformation.

More pointedly, there is an interesting result on finite waves which can-not be obtained without solutions of wave equations. If a wave function $y(x,t)$ is 0 except for one finite interval on the x-axis, we call it a "finite wave". Assume that a solution

$$y(x,t) = f(x - ct) + g(x + ct)$$

of the wave equation (6) is made of two finite waves $f(x - ct)$ and $g(x + ct)$ and so itself a finite wave. Then initially, f and g move towards each other without overlapping. $f(x)$ and $g(x)$ are the wave pattern of f and g. As the waves keep moving, f and g overlap at a certain interval only, the wave pattern changes in this interval due to the superposition. As the waves move on further and the overlapping disappears, f and g will come back to the original wave pattern. This means that in some cases, f and g can completely cancel each other in this overlapped interval.

5.3. Energy and Momentum of Waves

The energy associated with a string wave is expressed as the energy per wavelength. For the sinusoidal wave, it is given as

$$\frac{1}{2}\sigma\omega^2 A^2 v$$

where σ is the mass density of the string, ω is the angular frequency of the wave, A is the amplitude of the wave and v is the phase speed of the wave.

What the momentum of a wave is is a puzzling question. Conventionally, the momentum of a wave is given through energy as

$$momentum = \frac{energy}{phase\ speed}$$

The direction of this vector is the direction of the wave propagation. Momentum is a quantity associated with a moving point mass. Even though we have managed to define energy of a wave "successfully" as above, there are still some conceptual issues to be resolved. In particle-based dynamics, there is no connection between momentum and energy: momentum is conserved but contrary to popular belief, energy is not conserved as discussed in Sections 2 and 3 which makes this definition of momentum for wave highly questionable. Indeed, in particle-based mechanics, momentum is the product of mass and its speed. It depends on mass. However, the concept of energy does not depend on mass. It depends on force and distance. It is therefore not quite convincing to define momentum (of a wave) in terms of energy and phase speed. Peskin speculates that this relation between energy and momentum of waves should hint at wave-particle duality at the classical level. We disagree for the reason stated above. Specifically, in wave mechanics, no particle is moving towards the direction of the wave. This means that the momentum of a wave must be 0, if any. After all, as we have shown, the very definition of kinetic energy is faulty because it ignores the fact that the work needed to accelerate from $m0$ to mv is not $mv^2/2$ in general. The standard definition of kinetic energy violates the law of conservation of energy.

This definition of energy of waves directly contradicts de Broglie's pilot wave theory in which he defended that, in his theory, either group speed or phase speed can go faster than c, meaning that the energy of waves is not carried by phase speed but by group speed. The problem with this "opinion" is that a string wave as discussed above has no group speed, or equivalently the group speed and phase speed of the wave are the same. So, the energy of a wave can travel faster than c.

A more serious issue is that the speed of a wave and that of a particle are different things. When a transversal wave is moving, no particle in the wave medium is moving in the direction of the wave. They are moving slowly up and down. It is the transmission of kinetic energy of the particles, moving up and down to the next particles, that moves in the direction of the wave. This is a different kind of motion, and so a different kind of speed. No particle moves in the direction of the wave, only the energy moves in the direction of the wave. Since there is no convincing reason to think that Einstein's special theory of relativity applies to this kind of "speed" and "motion", there seems to be nothing wrong with de Broglie's wave that moves with phase speed faster than c. The problem with de Broglie's idea is that it has nothing to do with physics as there is no ontology associated with it. How valuable it is mathematically is a more serious question. The problem we have here is that due to the obsession with the superiority of electrodynamics and relativity theory, theoretical physicists have been misled to think that c is the speed limit for everything that "moves", and have failed to understand wave mechanics. A system of particles can readily transcend the primitive theory of particle (electro-) dynamics. We have already seen this in super conductivity. In a conductor, a chain of electrons moves much more smoothly than a single electron.

A main cause of this mix up is ascribable to the dominating influence of Göttingen mathematical physics lead by Hilbert - everything in advanced physics of the last century was reduced to "mathematics", dropping all ontological issues. This is how theoretical physics, which thrived in the "world" of cosmology led by the late Prof. Hawking, have become irrelevant to physics and the real world.

Moreover, according to the scenarios presented above, when we consider a system of waves that cancel each other, wave energy may not in general be conserved. What about momentum? When waves cancel each other due to the destructive superposition, what happens to the conservation of momentum? Just like the case of particle-based classical dynamics, energy conservation fails for waves. Momentum is a concept which is naturally defined for particles. The conventional definition of momentum for waves is still in question. It violates the dimensional independence of momentum and energy in particle-based dynamics. The claim that relativity theory "has established" the astounding energy-momentum relation is unthinkable when we consider the dimensional analysis of [momentum] and [energy]. In what follows, we shall show that this energy-momentum relation of Einstein's is false. He obtained it by violating the assumption that v in gamma factor must be constant at the pain of contradiction.

Notwithstanding, a fundamental flaw in this definition of momentum of waves is that energy is a modal concept whereas momentum is a predicated (existential) concept. Moreover the phase speed is also a predicated concept. Then how in the universe is it possible that energy and phase speed can define momentum? It is this lack of logical and philosophical understanding, which is inherent in the culture of theoretical physics, that has led the entire physics into chaos.

Finally, as we will discuss in the forthcoming discussion on quantum mechanics, what we have discussed here has some impact on the so-called wave-particle duality.

SECTION 6. MODERN PHYSICS?

6.1. Special Theory of Relativity

As it is the case that Einstein's $e = mc^2$ is the most important equation in the entire modern physics and the Hamiltonian energy equation is a foundation of Schrödinger's wave equation, then certainly the invalidity of the concept of energy and the law of conservation of energy will have serious consequences in modern physics too. We will explore the issue here.

From a logical point of view, what is so astounding about the special theory of relativity is that it completely violates the conceptual hierarchy of basic physical dimensions, long established as in classical dimensional analysis. The most fundamental physical concept is that of [distance] which is the degree of separation of two points in a 3D geometric space. By adding the dimension of [time] which is a 1D space with the flow of time, Newton built a mathematical theory of motion, viz. kinematics. Upon these fundamental dimensions, kinematics "defines" the dimensions of [motion], [speed] and [acceleration] so as to develop the theory of motion.

The respective definitions of [motion], [speed] and [acceleration] are:

- [motion] is a function from [time] to [distance].
- [speed] = [distance]/[time]
- [acceleration] = [speed]/[time] = [distance]/[time]^2.

This hierarchy disallows that [speed] will alter [time] and [distance]. Unfortunately this is precisely what the so-called "time dilation" and "length contraction" did.

The issue manifests itself in the clock scenario: If a clock slows down in a space where the length contracts, then there must be a coherent explanation of how the length contraction

$$\Delta d' = \frac{\Delta d}{\sqrt{1-(v/c)^2}}$$

on the clock's mechanism will cause the clock to slow down, thus yielding the time dilation

$$\Delta t' = \sqrt{1-(v/c)^2}\Delta t$$

Such a slowdown process is not explained. This poses a serious problem for the special theory of relativity, which is a positivistic theory and in which Einstein proposed that the time-is-clock reading.

Notwithstanding, in case of Lorentz's projection theory, the v in the gamma factor of Lorentz transformation is the absolute speed between the aether and the moving reference frame; but there is no known way to measure this speed. A correct interpretation of Michelson-Morley's method of using reflected light tells us that we cannot detect this speed v in such a way. Moreover, both Einstein's relativity theory and the Lorentzian transformation theory suffer from a serious inconsistency problem. Moving a reference frame in another reference frame immediately leads to a contradiction.

For example, assume that an arrow with length l moves with speed v towards a box with depth l. According to the perspective of the arrow, the box shrinks and the arrow cannot fit. According to the perspective of the box, the length of the arrow shrinks and the arrow can fit.

Remark 2. *A time version of the same logical problem has produced a more famous controversy of the twin paradox. Assume that one of the twins leaves and comes back with a constant speed v, then each twin will see that the other twin is younger. The usual response from mainstream physicists is that as the twin, who has left and come back, has gone through acceleration, it is not the twin who has come back that sees the other twin* *younger. First of all, the special theory of relativity assumes total symmetry under the principle of relativity between two inertial frames. This is why reference frames are restricted to only inertial frames. The response is not a legitimate one as it has broken the symmetry by allowing acceleration. In this sense, the statement of the twin paradox itself is meaningless as the process of leaving and coming back involves acceleration.*

All of this seems to suggest that the only reality of physics, as Newton said, lies in the absolute frame which is not observable by us from outside as we are inside the absolute frame. The only way to understand physics is to, so to speak, "stand on the shoulder of God" (i.e., from the outside of this universe) and observe metaphysically. Neither projection from aether nor pure relativity works.

All of this is a natural consequence of the following errors in the development of the special theory of relativity:

One can derive the time dilation and length contraction as follows: Assume that a train runs on a track. When the tip of the power pole of the train touches the power line at point P, a spark occurs at P. An observer A located in the train straight down the tip of the power pole P will observe that the spark of lightcomes straight down from the point P which is a point of the train. An observer B located on the embankment will observe that the light travels diagonally from P, which is a point of the power line being stationary in the frame of B, to the observer A in the moving train. Under the assumption of the constancy of the speed of light which says that the speed of light is constant c in all inertial reference frames, one can derive time dilation from which we can derive length contraction too.

This thought experiment yields more than Einstein had discovered, unfortunately. There is a contradiction in the frame of the observer A. For the observer A, the point P is also a point of the power line and it is moving backward. So, A will also observe that the spark of light travels diagonally from the point P which is a stationary point of the power line. Logicians will point out that this comes from the assumption of a reference frame that moves inside another reference frame. If we allow a geometric point to move inside a geometric space, it destroys the topological continuum of the geometric space. This is why we do not consider a geometric point moving in a geometric space. Then how is it possible to consider a geometric space which moves inside a geometric space? If a geometric space S1 moves inside another space S2, at each moment, a point P in S1 is also a point in S2, and P is also moving in S2. This is the situation from which Einstein derives dime dilation and length contraction.

Logically, this means that the derivation of time dilation and length contraction is invalid. The inconsistency has an interesting consequence which cripples down the Lorentz group as presented by Poincaré.

As already discussed in 1.3, from time dilation and length contraction Einstein derives a coordinate transformation which expresses the effect of the mutual speed v of two inertial frames:

$$x' = \frac{x - vt}{\sqrt{1-(v/c)^2}}, y' = y, z' = z, t' = \frac{t - \frac{vx}{c^2}}{\sqrt{1-(v/c)^2}}$$

This Lorentz transformation causes some problems:

$$\frac{x'}{t'} = \frac{x - vt}{t - vx/c^2} = \frac{c^2(x - vt)}{c^2 t - vx} \neq \frac{x}{t}.$$

The relative speed is not the same magnitude any more.

There are other problems with the Lorentz transformation as well. Relativistic addition of speed $v \oplus v'$ is obtained using the Lorentz transformations. So, one cannot use this speed to define or justify the Lorentz transformation on pain of committing a vicious circle. Moreover, v in the Lorentz transformation is a relative speed between two inertial frames and it is then classical speed by definition. So, not only mathematically but also conceptually, there is something uneasy about the concept of the Lorentz transformation. In what follows, we shall observe that this manifests itself as a serious setback both theoretically and empirically.

Many theoretical physicists discuss the Lorentz group - the group of Lorentz transformations and their composition that was discovered by Poincaré. Assume that there are three inertial frames $F1$, $F2$ and $F3$. Let v and v' be the mutual speeds between $F1$ and $F2$, and $F2$ and $F3$, respectively. As these speeds are used to define the gamma factor, on pain of committing a vicious circle, they must be pre-relativistic speeds, i.e., classical speeds. So, $v + v'$ is the mutual speed between $F1$ and $F3$. The Lorentz transformation L between $F1$ and $F3$ is defined as

$$x' = \frac{(x - (v + v')t)}{\sqrt{1-(v+v')^2/c^2}}, y' = y, z' = z, t' = \frac{(t - \frac{(v+v')x}{c^2})}{\sqrt{1-(v+v')^2/c^2}}.$$

Let L and L' be the Lorentz transformations from $F1$ to $F2$, and from $F2$ to $F3$

$$x' = \frac{(x - vt)}{\sqrt{1-v^2/c^2}}, y' = y, z' = z, t' = \frac{(t - \frac{vx}{c^2})}{\sqrt{1-v^2/c^2}}.$$

$$x'' = \frac{(x' - v't')}{\sqrt{1-(v')^2/c^2}}, y'' = y', z'' = z', t'' = \frac{(t' - \frac{v'x'}{c^2})}{\sqrt{1-(v')^2/c^2}}.$$

respectively. It is clear that $L \neq L' \circ L$ where $L' \circ L$ is the mathematical composition of L and L'.

Mainstream physicists contest that in the special theory of relativity, calculating $v + v'$ is the wrong thing to do. They claim that this should be replaced by the relativistic addition (transformation) of speed

$$v \oplus v' = \frac{v - v'}{1 - vv'/c^2}$$

This is to say that L should be

$$x'' = \frac{(x - (v \oplus v')t)}{\sqrt{1 - (v + v')^2/c^2}}, y'' = y, z'' = z, t'' = \frac{(t - \frac{(v \oplus v')x}{c^2})}{\sqrt{1 - (v + v')^2/c^2}}.$$

However, they remain silent when asked why isn't $v + v'$ replaced by $v \oplus v'$ in the gamma factor $(v \oplus v')$ in the Lorentz transformation is also problematic. As the Lorentz transformation is the agent to introduce relativity concepts, one cannot use the relativistic concept $v \oplus v'$ to define such a transformation. This is again conceptual vicious circularity. The fact that it works for composing Lorentz transformations, as claimed, doesn't make it valid. Unfortunately this version of L still "fails" $L = L' \circ L$. Mathematically, the Lorentz transformations are linear transformations from 4D space to itself. It is highly irregular that an algebraic composition of such transformations is not the desired transformation.

Conceptually, $(v + v')$ is classically measured speed. The reason why $(v \oplus v')$ was introduced in the first place in the numerator part is because classical addition $(v + v')$ does not work for the relativistic addition of speed. As the relativistic addition $(v \oplus v')$ is introduced by relativistic argument, it is viciously circular to use this relativistic version in the gamma factor.

Mathematically we have another problem. It is naturally expected that x''/t'' will serve as the observed speed $\oplus v'$. There is no way to prove that they are equivalent. Here is a simple calculation which leads to this conclusion.

$$\frac{x''}{t''} = \frac{(x - (v \oplus v')t)}{t - \frac{(v \oplus v')x}{c^2}} = \frac{c^2(x - (v \oplus v')t)}{c^2 t - (v \oplus v')x} = \frac{c^2 x - c^2(v \oplus v')t}{c^2 t - (v \oplus v')x}.$$

Note that even if we replace the classical $v + v'$ in the gamma factor with the relativistic $v \oplus v'$, the above calculation holds.

Now assume that

$$\frac{c^2 x - c^2(v \oplus v')t}{c^2 t - (v \oplus v')x} = v \oplus v'.$$

Then we have

$$\begin{aligned} c^2x - c^2(v \oplus v')t &= (v \oplus v')(c^2t - (v \oplus v')x) \\ c^2x &= 2c^2(v \oplus v')t - (v \oplus v')^2x \\ c^2x + (v \oplus v')^2x &= 2c^2(v \oplus v')t \\ \frac{x}{t} &= \frac{(v \oplus v')^2}{c^2 + (v \oplus v')^2}. \end{aligned}$$

There is no reason for this to be true.

Empirically speaking, even though it could be possible to measure v, what about v'? It is next to impossible to measure $v + v'$, is it not? If v is a speed of a star A moving away from us in distance and if v' is a speed of another star B moving away from A, then how can we measure the v' and so $v + v'$?

After all, the restriction to only inertial reference frames was not acceptable for Einstein, and so he moved into the special theory of relativity dynamics in which acceleration is involved. This was a must for Einstein as the dynamics involves the second law

$$\mathbf{f} = m\mathbf{a} = d\mathbf{p}/dt$$

where m is a mass, a is acceleration of the mass and f is the force exerted upon the mass. From here on, Einstein forgot about the restriction to just inertial frames, and through a thought experiment which involves collision of mass emitted from two inertial frames, he obtained the concept of relativistic mass

$$m = \frac{m_0}{\sqrt{1 - (v/c)^2}}$$

where m_0 is a rest mass. Here v is constant as it is the relative speed between the two inertial frames. With this, he defined the relativistic momentum as $p = mv$. Thus we have the following relativistic second law:

$$\mathbf{f} = \frac{d\mathbf{p}}{dt} = \frac{dm\mathbf{v}}{dt} = \frac{m_0\mathbf{a}}{\sqrt{1 - (v/c)^2}} \quad (second\ law)$$

With this, he obtained the famous $e = mc^2$ as follows:

$$\begin{aligned} de &= \mathbf{f} \cdot d\mathbf{r} = \frac{d(m\mathbf{v})}{dt} d\mathbf{r} \\ de &= d(m\mathbf{v}) \cdot \mathbf{v} = dm(\mathbf{v} \cdot \mathbf{v}) + m(d\mathbf{v} \cdot \mathbf{v}) \end{aligned}$$

$$de = dm(v^2) + \frac{1}{2} md(v^2)$$

$$dm = \frac{\frac{1}{2} m_0 d(v^2)}{c^2 \left(1 - \left(\frac{v}{c}\right)^2\right)^{\frac{3}{2}}}$$

$$de = m_0 c^2 \frac{1/2\, d(v^2/c^2)}{(1 - (v/c)^2)^{3/2}}$$

$$e = \frac{m_0 c^2}{\sqrt{1 - (v/c)^2}} = mc^2$$

Of course this proof is false as v is constant since it is in the gamma factor. So, the correct result is

$$\mathbf{F} = \frac{d\mathbf{P}}{dt} = \frac{dm\mathbf{v}}{dt} = 0$$

Using this famous equation $e = mc^2$, Einstein obtained yet another famous equation relating relativistic energy and relativistic momentum as follows:

$$E^2 = m^2 c^4 = \frac{m_0^2 c^4}{1 - (v/c)^2}$$

$$c^2 p^2 = \frac{m_0^2 v^2 c^2}{1 - (v/c)^2}$$

Therefore

$$E^2 - c^2 p^2 = m_0^2 c^2 \left(\frac{c^2 - v^2}{1 - (v/c)^2} \right) = m_0^2 c^4$$

The modal concept of energy is then smuggled back into the special relativity theory. As v in the second law is constant, what we get is $f = 0$ which in turn yields $e = 0$. So, in the special theory of relativity dynamics, we have

$$e = 0$$

instead!

In classical mechanics, as [energy] = [force] x [distance], [momentum] = [force] x [time], and [time] and [distance] are strictly independent, there is no connection between [energy] and [momentum]. One may wonder why Einstein got this phantom relation between [energy] and [momentum]. Simple dimensional analysis immediately tells us why.

In the special theory of relativity, through [speed] = [distance]/[time] in the "time dilation" and "length contraction", [time] and [distance] are related. Natural reasoning makes us wonder about the validity of the controversial concepts of time dilation and length contraction, since time dilation and length contraction both lead to ontological contradictions. This exposes where Einstein's theory of special relativity went wrong to begin with.

For this purpose, let us go back to the thought experiment which made Einstein come up with the ideas of time dilation and length contraction. On the basis of the same thought experiment mentioned earlier in this section, Einstein concluded that from the perspective of the embankment, the time in the frame of the train slows down; and also from the same perspective, the length in the train contracts. This apparently impeccable argument curiously caused the paradoxes of twins and flying arrows. A striking error in Einstein's argument is this: For the observer in the train, there is yet another story; as point A also is a stationary point of the power line which is a point moving backward in this observer's reference frame, the spark light can also travel diagonally to reach this observer. Hence, a contradiction occurs in the setting of the thought experiment.

In short, this problem came from Einstein's misunderstanding the way Newton set up his theory of motions. Newton knew that one cannot move even a single point in geometry as the motion is defined in terms of the mutually independent dimensions of [time] and [space] (or [distance]). It is impossible to let an entire geometric space move inside some other geometric space. The whole issue boils down to the fundamental conceptual errors of the dimensional analysis.

6.2. Doppler Effect?

6.2.1. Classical Doppler Effect

Despite treating blind acceptance as scientific consensus as Kuhn puts it, the Doppler effect is still a mystery, both in Galilean relativity and Einsteinian relativity. Assuming the Newtonian picture, A and B are moving towards each other with the absolute speeds v_1 and v_2 respectively. So the Newtonian relative speed between A and B is $v_1 - v_2$ which is the difference between the absolute speed of A and that of B respectively. From Newton's perspective, there is no mystery here at all. Assume that A emits light of frequency f which moves toward B. As there is no constancy of the speed of light, the light moves toward B with speed $c + v_1$ in the absolute frame. This however does not cause the change of frequency of the light that A observes. For B, A is approaching with speed v_2 and the light is coming with speed $c + v_2$. This makes B observe the frequency of the coming light to be not the same as the frequency of the light that A emitted. It is a higher frequency. So, B observes the light as having a higher energy than that emitted by A. If the principle of energy conservation is true, then this energy increase is caused by the motion of B. A similar story is the case that A is moving away from B in the absolute frame. Here we can say that when A moves away from B, the energy of light is used to push A further away. It all comes from the assumption that the principle of conservation of energy is correct; but the fact is that it is false. The bottom-line is, we do not know what energy is at all.

At this point, a serious question remains unanswered. What happens when an observer observes "on the shoulder of God". Does the light increase its frequency because B is approaching A? What about the effect of B moving on the wave? Does the frequency decrease? There are more questions than answers.

6.2.2. Einsteinian Doppler Effect

Now let us consider what happens in the special theory of relativity. Einstein applied time dilation and length contraction to study the effect relativistically. He assumed that two observers G and R, at their respective origins, measure the frequency of a light wave with plane wave fronts, and count the number of fronts that have passed in a specific time interval. Let θ and θ' be the observed directions of the propagation for G and R with respect to the positive x-axis and x'-axis respectively. Let f and f' be the respective observed frequencies of the wave. We further assume that $t = t' = 0$ when the origins coincide. Then, by simple geometry we have

$$t'f' = \left(1 - \frac{v}{c}\cos\theta\right)tf.$$

As

$$t' = t\sqrt{1 - \left(\frac{v}{c}\right)^2}$$

we have

$$f' = \frac{f(1 - \frac{\mathrm{v}}{\mathrm{c}}\cos\theta)}{\sqrt{1 - (v/c)^2}}.$$

Since $f'\lambda' = f\lambda = c$, we have

$$\lambda' = \frac{\lambda\sqrt{1 - (v/c)^2}}{1 - v/c\ \cos\theta}.$$

By simple calculation and elementary geometry, one can obtain

$$\sin\theta' = \frac{\sin\theta\sqrt{1 - (v/c)^2}}{1 - v/c\cos\theta}$$

$$\cos\theta' = \frac{\cos\theta - v/c}{1 - v/c\ \cos\theta}$$

Now assume that the direction of propagation is such that $\theta = 90°$. Then we have

$$f' = \frac{f}{1 - (v/c)^2}$$

This is the Doppler effect for transversal waves as postulated by Einstein. Theoretically, we can use this relativistic Doppler effect to empirically verify the claim of time dilation. Relativists say that it is difficult to arrange the light of precisely known frequency which comes in at an angle of precisely 90°.

The Doppler effect for longitudinal waves can be obtained by setting $\theta = \theta' = 0°$. This Doppler effect has been used to study the spectra of light from distant galaxies. As this version of the effect is relativistic, it would mean that, as Einstein claimed, the astronomy fails if the special theory of relativity fails.

What is strikingly distinct about this version of the story is that there is no consideration of energy. Everything used here is nothing but pure relativistic kinematics in which there isn't even acceleration. It is not clear how the energy issue as discussed above can come in here. There is nothing but elementary mathematical calculation.

What is so philosophically astounding in this regard is that in spite of our being told that $e = mc^2$ is the most important equation in the entire history of theoretical physics, energy is not considered in the Doppler effect. It may be the case that we have to use Planck's $e = hf$ to introduce energy. Still, all we get is an unexplainable energy loss (or gain) due to the relativistic Doppler effect. In the case of classical dynamics, we have a very clear explanation on this energy gain or loss under the principle of conservation of energy. The validity of the principle of conservation of energy aside, the classical version is admittedly more convincing.

6.2.3. Transformation of Energy and Momentum

In classical dynamics, as discussed above, there is no known relation between momentum p and energy e. The theory of collision in particular (Subsection 3.2) clearly shows that momentum, unlike energy, is conserved without the collision parameter experiment; meaning that there is no theoretical reason to think that energy must be conserved!

In contradistinction to this, Einstein presented the so-called "energy-momentum relation" in his special theory of relativity. Assume that an inertial reference frame F' moves in another frame F in x-direction only, then obviously we have

$$p_y' - p_y\ , p_z' = p_z\ .$$

According to the definition of momentum,

$$p'_x = \frac{m_0}{\sqrt{1-(v'/c')^2}}\frac{dx'}{dt'}$$

As between F and F' we have

$$dt'\sqrt{1-(v'/c')^2} = dt\sqrt{1-(v/c)^2} \ (I)$$

we have also

$$dx' = \frac{1}{\sqrt{1-(v'/c')^2}}(dx - vdt)$$

$$p'_x = \frac{m_0}{\sqrt{1-(v'/c')^2}}\left(\frac{dx}{\sqrt{1-(v/c)^2}} - v\frac{dx}{\sqrt{1-(v/c)^2}}\right)$$

Using the famous momentum-energy relation

$$E = m_0c^2\sqrt{1-v^2/c^2}$$

of Einstein's, we again have

$$p'_x = \frac{1}{\sqrt{1-(v/c)^2}}\left(p_x - \frac{vE}{c^2}\right)$$

We further use the Lorentz transformation for time and so

$$dt' = \frac{1}{\sqrt{1-(v/c)^2}}\left(dt - \frac{vdx}{c^2}\right)$$

Now from (I) and

$$E' = m_0c^2\sqrt{1-v'^2c^2}$$

we get

$$\frac{E'}{c^2} = \frac{1}{\sqrt{1-(v/c)^2}}\left(\frac{E}{c^2} - \frac{v}{c^2}p_x\right)$$

Putting them together, we have the following famous transformation of relativistic energy and momentum.

$$p'_x = \frac{1}{\sqrt{1-\left(\frac{v}{c}\right)^2}}\left(p_x - \frac{vE}{c^2}\right)$$
$$p'_y = p_y$$
$$p'_z = p_z$$
$$\frac{E'}{c^2} = \frac{1}{\sqrt{1-\left(\frac{v}{c}\right)^2}}\left(\frac{E}{c^2} - \frac{v}{c^2}p_x\right)$$

Going through this lengthy calculation, what becomes clear is that the entire development of this transformation is based on Einstein's famous relativistic energy-momentum relation. We have already addressed the following issues regarding this famous relation.

- Unlike momentum, which is a most fundamental ontological concept in physics, energy is a modal concept whose role in physics is not quite clear.
- The above immediately implies that there is no known relation between momentum and energy as we can see in classical dynamics.
- The concept of energy is corrupted as we can see that the concept of kinetic energy violates the principle of conservation of energy.
- In the theory of collision, momentum is always conserved but it is not the case with energy.
- As the famous $e = mc^2$ is false (its derivation is false), the so-called "energy-momentum relation" is also false.

In fact, the energy in a Lagrangian is not always conserved. It is not conserved if the kinetic energy is time-dependent. This is consistent with our result in subsection 2.1 that if v is time-dependent then the kinetic energy which is supposed to be the work needed to accelerate from m_0 to m_v is not $mv^2/2$. This is yet another piece of evidence that energy and momentum are not related, contrary to Einstein's claim that they are related through the energy-momentum equation. Ontologically, it is not acceptable that energy, which is a modal concept, can be related in any meaningful way to momentum, which is not modal.

Our next concern is the modality of probability in theoretical physics. Schrödinger-Born-Heisenberg made theoretical physics a glorified modal theory. We shall discuss this issue in what follows.

6.3. Quantum Mechanics

Quantum mechanics is yet another theory of physics which was developed upon the concept of energy of physical systems. In this regard, this theory is in need of re-evaluation as much as all other theories of physics that we have discussed, if not more. As the concept of probability is also a modal concept, quantum mechanics can be categorized as "the ultimate modal theory of physics".

There were three major reasons why physicists had to consider developing a new theory of physics:

- The blackbody radiation.
- The Rutherford model of atoms.
- The double slit experiment.

Among these three, the first two are directly related to the issue of energy. The blackbody problem was the discrepancy between theoretical prediction on the frequency-wise distribution of the electromagnetic energy emitted by the heated blackbody and the experimental results. The second problem was the problem of electrodynamics, according to which an electron orbiting about a positively charged nucleus of an atom must loose kinetic energy via electromagnetic wave emission and will fall quickly into the nucleus. This does not happen in reality. The last problem had to do with the so-called "double slit experiment" in which particles passing through a narrow slit on a wall behave like a wave. This problem in the end is also reduced to the problem of energy: the relation between energy for waves and that for particles appears to have some discrepancies. Contrary to the common belief, we do have some serious deficiency in our understanding of the concept of energy, none of the aforementioned problems are well formulated. At this point, the concept of energy does not belong to the exact sciences.

In the face of this fundamental conceptual issue of energy, physicists of the last century tried to resolve the difficulties by combining various different ideas involving the ill-fated concept of "energy" at the risk of making the theory inconsistent. The irony is that, as all mathematicians and logicians know well, inconsistent theories are most creative and productive as they can prove anything. But according to Shannon's information theory, such a theory is useless as it says "yes" to any proposition P including the negation of P.

6.3.1. Blackbody Radiation

This problem was resolved by Planck. As a mathematical convention, Planck proposed that the energy transfer of electromagnetic wave is to be formulated as nhf where n is a natural number, h is a constant called the Planck constant and f is the frequency of the electromagnetic wave. Planck called the minimum energy unit hf for an electromagnetic wave of frequency f "an energy quanta of frequency f".

This further led to the particle interpretation of hf as proposed by Einstein. Planck himself did not consider hf a particle carrying electromagnetic energy of frequency f. Einstein naturally considered this hf a particle of EM wave and called it a photon of frequency f. Naturally he assumed that the speed of a photon is the same as that of an electromagnetic wave, namely c. Einstein naturally assumed that this photon is a relativistic particle which obeys the laws of the special theory of relativity. This makes the relativistic energy

$$e = \frac{m_0}{\sqrt{1-(v/c)^2}}$$

of the photon diverge. To solve this mathematical difficulty, Einstein assumed that the rest mass of the photon is 0 because then the relativistic energy of the photon becomes $0/0$ which, Einstein and all theoretical physicists thought, could be any number. The reason behind this choice is because a related equation $0x = 0$ is indeterminate, meaning that any number can be its solution. Under this assumption, Einstein concluded that a photon's energy is

$$e = 0/0 = hf.$$

From this, we can derive the following contradiction:

$$e = \sqrt{(cp)^2 + (m_0)^2c^4} = cp = \frac{m_0vc}{\sqrt{1-\left(\frac{v}{c}\right)^2}} = \frac{0}{0}cv = c^2hf = hf\ .$$

We also can derive another contradiction:

$$e = \sqrt{(pv)^2} = \sqrt{c^2m_0/\sqrt{1-(v/c)^2}} = \sqrt{0/0} = \sqrt{hf} = hf$$

The reason why we have this problem is because it is mathematically wrong to assume that $0/0$ is any number. It is not a number. If it were, we would have the following contradiction:

$$\begin{aligned}\left(\frac{0}{0}\right) \times 3 &= 1 \times 3 \quad (law\ of\ fraction)\\ &= 3\\ &= \frac{(0\times 3)}{0} \quad (law\ of\ fraction)\end{aligned}$$

$$= \frac{0}{0}$$
$$= 1 \quad (law\ of\ fraction)$$

It is a mathematical principle that whenever we extend the concept of numbers, we have to assume that the newly introduced extended numbers ought to satisfy the existing laws of numbers. It is called the conservative extension principle. Indeed, when we extended rational numbers to real numbers, we made sure that the set of real numbers is closed under the arithmetic operations $+,-,\times$ and $\div$ because the set of rational numbers is closed under the same operations. The problem at hand is caused by the fact that $0x = 0$ and $0/0$ are entirely different things. The former does not involve division by 0 but the latter does.

Einstein's idea to consider Planck's energy quantum hf as a particle of rest mass 0 is a mathematical problem. Yet another problem has to do with this move Einstein made and this time it is not a mathematical issue but an issue of misunderstanding wave mechanics. As we know, when the tsunami hit north eastern Japan in 2011, the water did not move from the epicenters to the coast of Japan with a speed of more than 500km/h. It was the vibration of water molecules created by the earthquake which moved with this high speed. In any physical wave created in a wave medium, the medium's local vibration travels because of the intermolecular force of the medium. So, no particle moves in the direction of the wave. The easiest model for this is the string wave. No point mass located on the rope really moves in the direction of the wave. It is the vibration of the point mass on the rope which moves in the direction of the wave. All of this means that wave mechanics is yet another monster. Wave mechanics could well be the most sophisticated theory of physics ontologically and mathematically. Regrettably, theoretical physicists brush off wave mechanics as applied physics.

If Einstein and his followers, which means virtually all theoretical physicists of the last century, understood wave mechanics correctly, all of this confusion would have never happened. The point is that it is not a particle which is moving in the direction of waves. It is the vibration of the particles in the medium which is moving in the waves. Particles and vibration of particles are different things. It only means that we cannot consider Planck's energy quanta as physical particles like what Einstein did. Certainly the convention of particles with rest mass 0 has nothing to do with this ontological problem.

The real question now is what exactly is the vibration of particles in a wave medium. Is it just a form of energy? The answer is "no". Vibration is a physical reality but energy is just a modality. Again the modality of energy is causing us problems here. Is it then work? If so, then as work can be negative, the total work of a vibrating particle is 0. So, no work is passed in the direction of the wave. Either way, we lose.

It has been said that Einstein's photon as a particle that carries the energy of an electromagnetic wave has resolved the photoelectric effect for which Einstein received his

Nobel Physics Prizes. The authors of this paper suggest that just Planck's hypothesis is enough to resolve the problem, no particle concept is needed. Indeed, as we have shown above, photons are not particles. They are just vibrations.

6.3.2. Double Slit Experiment and de Broglie Relation

It was the double slit experiment which made Schrödinger use de Broglie's relativistic particle-wave duality to introduce a wave representation of particles. De Broglie obtained the following relativistic transformation of a plane wave:

$$k'_x = \frac{1}{\sqrt{1-(v/c)^2}}\left(k_x - v\frac{\omega}{c^2}\right), k'_y = k_y, k'_z = k_z, \omega' = \frac{1}{\sqrt{1-(v/c)^2}}(\omega - vk_x)$$

where $k = (k_x, k_y, k_z)$ is the wave vector and ω is the frequency. We denote the wave number $|\mathbf{k}|$ by k. So, $k = |\mathbf{k}|$. Comparing this and the momentum-energy transformation

$$p'_x = \frac{1}{\sqrt{1-(v/c)^2}}\left(p_x - v\frac{E}{c^2}\right), p'_y = p_y, p'_z = p_z, \omega' = \frac{1}{\sqrt{1-(v/c)^2}}(E - vp_x)$$

where $p = (p_x, p_y, p_z)$ is the momentum vector and E is the energy, he proposed the following association between a particle and a wave (called matter wave):

$$p = \frac{\hbar}{k} \quad E = \hbar\omega$$

where $\hbar$ is a constant. We call this de Broglie relation. Again as is discussed in Subsection 5.3, this allows to define momentum of a quantum particle as

$$momentum = \frac{energy}{phase\ speed}$$

In this case, we can be more specific on where things go wrong. It all came from $E = mc^2$ which makes the deduction of the energy-momentum relation

$$E^2 = m_0c^4 + c^2p^2$$

possible. But $E = mc^2$ is false as it was obtained by violating the restriction on the gamma factor $1/\sqrt{1-(v/c)^2}$ in which v must be constant.

As is established above, this momentum-energy transformation is based upon Einstein's ill-fated energy-momentum relation. We have pointed out again and again that energy is a modality and momentum is a physical reality. Moreover, the collision theory clearly showed that momentum conservation holds during the entire collision process while energy conservation fails in the same process. Lagrange also showed that the conversion from the Lagrangian to the Hamiltonian is possible only for constant rate acceleration. They all indicate that Einstein's energy-momentum relation fails. There is an error, as we have shown, in Einstein's proof for the relativistic energy equation $e = mc^2$, and his proof for the energy-momentum relation came from this false equation $e = mc^2$.

Without knowing any of this, de Broglie considered his particle-wave duality valid, and then he compared this to Einstein's particle-wave duality for light-photon

$$E = hf = pc \quad p = h/\lambda$$

which also is false because $E = 0/0 = fh$ for photons leads to a contradiction as we have shown.

Remark 3. *One may say that* $E = pc$ *is in agreement with the troubled definition of momentum in wave mechanics because* $p = E/c$. *The problem with this argument is twofold. First, if light is an EM wave, then it is not a wave in the sense of wave mechanics. There is no medium involved as Maxwell's aether theory is not successful. Moreover, as we have discussed many times, EM wave is not a wave at all. It is an action at a distance in which the change of electromagnetic force reaches a receiver of unit size at some location in the space. No wave travels to reach the receiver. There is no such thing as a wave whose magnitude is determined by the magnitude of the receiver either. This is a typical problem of the contemporary physicists who replace ontology with just mathematical calculations.*

De Broglie further assumed that associated with a particle with speed v was a wave having a "phase speed" $w = \omega/k$. He also assumed that the energy in the wave traveled along with a group speed $v_g = d\omega/dk$ which was identical to the particle's speed v. Here it is not clear what he meant by "energy" in the wave. The so-called "de Broglie relation" is nothing but a wild hypothesis based upon just the above mentioned "wave vector frequency" transformation and "momentum-energy" transformation analogies. Certainly this does not yield the concept of "energy in the wave". Moreover, as we can see in the theory of string waves, some waves have just a phase speed and no group speed, or the phase speed and the wave speed are equal.

The wave in wave mechanics carries no particles in the direction of the motion of waves (re subsection 5.3). It carries the vibration of particles in the direction of waves. So, what de Broglie did was something completely unrelated to wave mechanics.

Anyhow, it is easy to show that $c^2(\omega/c^2 - k^2)$ is invariant under the relativistic transformation as an analogy to the relativistic invariance of $c^2(E/c^2 - p^2)$.

We set

$$c^2(\omega/c^2 - k^2) = constant = C.$$

From $c^2(\omega/c^2 - k^2) = C$, we have

$$2\omega/c^2 \frac{d\omega}{dk} - 2k = 0$$

This leads to

$$v_g = \frac{d\omega}{dt} = \frac{c^2 k}{\omega}.$$

As the phase speed is $w = \omega/k$, we have

$$v_g = \frac{c^2}{w}$$

It follows that either the phase speed w or the group speed v_g could exceed c but not both. We do not know what this means to the special theory of relativity which asserts that nothing moves faster than the speed c.

All of this mathematical nicety is relative to the hypothesis that a particle with speed v has a wave dual called matter wave whose group speed is v_g and whose phase speed is $w = \omega/k$. A particle in motion carries energy and so it is expected that the wave dual of this particle also carries energy of the same amount if the law of energy conservation is to be respected. But for a wave to carry energy, it has to have wave medium. Wave according to wave mechanics is nothing but the motion of not particles but the vibration of particles in the wave medium. A concern we have here is that de Broglie's wave is a mathematical wave which appears to have no wave medium. One may say that the electromagnetic wave carries energy without having a wave medium. We have already pointed out that the electromagnetic field which carries "electromagnetic waves" is a fiction, a modality which plays no ontological role in physics. So, what happened to the energy issue of the matter wave? This question was answered by Schrödinger in a limited context.

6.3.3. Schrödinger's Wave Equation

Schrödinger used Hamilton's energy dynamics for the particle theory, and applied de Broglie's pilot wave theory to produce a satisfactory wave-particle duality which looks after the energy issue of de Broglie's relation. It goes as follows: All waves propagated along the x–axis obey the following wave equation

$$\frac{\partial^2 \Psi}{\partial x^2} = \frac{1}{\omega^2}\frac{\partial^2 \Psi}{\partial t^2}$$

where $\Psi(x,t)$ is the wave function and ω is the wave speed.

Here, we consider the wave function Ψ whose square yields the probability of locating a particle at any point in space. We consider only systems whose total energy E is constant, and whose particles move along the x–axis and are bound in space. Then the frequency associated, via the de Broglie relation which is hypothetical and relativistic, with the bound particle is also constant, and we can take the wave function $\Psi(x,t)$ to be

$$\Psi(x,t) = \psi(x)f(t).$$

As the frequency is assumed to be precisely defined,

$$f(t) = \cos 2\pi\nu t$$

then, we have

$$\frac{\partial^2 \Psi}{\partial x^2} = -\left(\frac{2\pi}{\lambda}\right)\Psi = -\left(\frac{p}{h}\right)^2 \Psi$$

where the wave length is $\lambda = \omega/\nu$ and the momentum of the particle is $p = h/\lambda$.

We take the particle of mass m to be interacting with its surroundings through a potential-energy function $V(x)$. The total energy of the system is given by

$$E = E_k + V = \frac{p^2}{2m} + V$$

where E_k is the kinetic energy of the particle. Then we have

$$p^2 = 2m(E - V)$$

and we have

$$\frac{\hbar^2}{2m}\frac{\partial^2 \Psi}{\partial x^2} + (E - V)\Psi = 0\,.$$

Therefore,

$$-\frac{\hbar^2}{2m}\frac{\partial^2\Psi}{\partial x^2}+V\Psi=E=i\hbar\frac{\partial\Psi}{\partial t}.$$

This derivation of the Schrödinger wave equation involves the de Broglie relation $p = h/\lambda$ which is relativistic. However, the resulting equation is not relativistic. Schrödinger was aware of the problem and tried to make his wave equation relativistic, but without success. The major difficulty here is that the Hamiltonian energy equation is classical and so the concept of energy here is modal, not ontological. Since it violates the law of energy conservation with respect to the definition of kinetic energy, the work needed to accelerate from m_0 to m_v is $mv^2/2$ only when the acceleration is constant. This energy in the Lagrangian is not always conserved. It is not conserved if the kinetic energy is time-dependent. The theory of collision shows that the momentum can be conserved through collision but it is not the case with energy. Only experiment can tell us if the energy is conserved or not. The bottom line is, "momentum" is an ontological concept whereas "energy" is a modal one. There is no energy-momentum relation as claimed by Einstein. To repeat ourselves, Einstein obtained the energy-momentum relation simply because he used the energy equation $e = mc^2$ which in turn was obtained by violating his own condition that v in the Lorentz transformation cannot be time-dependent on pain of violating the principle of relativity. In other words, he obtained $e = mc^2$ only by violating the principle of relativity. This is why we are not able to obtain a relativistic version of Schrödinger's equation.

Later, the frustrated Gordon-Klein made another attempt to relativize quantum mechanics by taking a shortcut. They replaced the energy variable and the momentum variable of Einstein's relativistic energy-momentum equation with an energy operator and a momentum operator, and claimed that they relativized quantum mechanics. Obviously they did not know that this equation of Einstein's is false.

6.3.4. The Rutherford Model

The failure of Rutherford's model is said to be yet another reason why physicists of the last century developed quantum mechanics. However, it appears that the problem is the failure of classical electrodynamics which cannot be resolved by jumping into some other paradigm.

Here is the issue: If an electron is orbiting around the nucleus of an atom, it must emit EM waves, thereby exhausting its energy and falling into the nucleus; but this is not the case. One of the problems is that we do not understand what we call an EM wave is. An EM wave is not a wave in any sense. It does not travel through space as a wave. It sends a vibrating EM force to a location in the space where there is a charge. The strength (amplitude) of the vibration relies on the magnitude of the charge. Otherwise, it does not send any time-varying EM force to anywhere. It is not a wave in the sense as it is used in wave mechanics. So, it is not likely that orbiting electrons will lose their mechanical energy and fall into the nucleus of the atom as quickly as many people think. Furthermore, the law of conservation of energy is false, and the Lorentz force also shows that the EM theory contradicts the second law. Therefore, electrodynamics is inconsistent. It appears that classical electrodynamics needs to have its problems fixed before launching the construction of a new world, that of quantum mechanics and relativity theory.

6.3.5. Von Neumann's Formalism and the Modality of Quantum Mechanics

As a consequence of the de Broglie relation, a particle with energy E has a wave function

$$\Psi(x,t) = Ae^{i(px-Et)/\hbar}$$

It follows that

$$i\hbar\frac{\partial}{\partial t}\Psi = E\Psi\,.$$

This together with the Schrödinger equation will give us

$$\frac{\hbar^2}{2m}\frac{\partial^2\Psi}{\partial x^2} + (\mathrm{E}-\mathrm{V})\Psi = 0\,,$$
$$\left(\frac{\hbar^2}{2m}\frac{\partial^2}{\partial x^2} - V\right)\Psi = E\Psi = i\hbar\frac{\partial}{\partial t}\Psi\,.$$

So, we can consider

$$E = \frac{\hbar^2}{2m}\frac{\partial^2}{\partial x^2} - V$$

as a representation of all energy aspects of this particle. It is called Hamiltonian. We denote it with H. So, Schrödinger's equation can be written as

$$i\hbar\frac{\partial}{\partial t}\Psi = H\Psi\,.$$

where the Hamiltonian can represent the energy situation of the particle more generally.

Heisenberg and Jordan presented an alternative formulation of early quantum mechanics using matrix and linear operators where operators are used to represent measurements. Though they were "shown" (?) to be equivalent, these two groups did not agree with each other.

It was von Neumann who unified the Heisenberg-Jordan theory with the Schrödinger theory, and added Born's view of probabilistic multiplicity in measurement. His idea was to represent measurement as Hermitian operators, and to yield the probabilistic distribution of the measured values as eigenvalues by way of eigenvalue analysis which is used in probability theory.

Von Neumann's Formalism

Axiom 1: In the absence of external influence, the quantum state Ψ changes smoothly in time t according to the Schrödinger equation

$$i\hbar\frac{\partial}{\partial t}\Psi = \hat{H}\Psi \,.$$

where $\hat{H}$ is the Hamiltonian operator over the complex Hilbert space.

Axiom 2: An observable is a self-adjoint linear operator over the complex Hilbert space.

Axiom 3: [Projection Postulate] When we apply an observable $\hat{A}$ to a quantum state Ψ (we call such application an observation $(\hat{A}, \Psi)$), then the quantum state jumps into one of its eigenvectors and one of the eigenvalues of the eigenstate is observed.

Axiom 4: The following represents the probability of one of the observed eigenvalues, a, of $\hat{A}$:

$$PR\left(\left(\hat{A}, \Psi\right) \approx \mathrm{a}\right) = \langle \Psi, \hat{P}\Psi \rangle$$

where $\hat{P}\Psi$ is an eigenstate of A with eigenvalue a

As is clear in this formulation, the standard quantum mechanics due to von Neumann yields the prediction on the "probabilistic distribution" of possible measurement values for each quantity to be measured. The uncertainty problem is therefore reduced to a problem of a probabilistic nature.

So, in this formulation of quantum mechanics, which is still the standard, there are two major modalities involved. First, the formalism is about energy described by a Hamiltonian; second, the probability. Regardless of whether the physicists are aware of it or not, this is the first serious scientific theory which is also a modal theory in its entirety. This new theory, as we have shown, was built upon contradictions resulting from combining classical reasoning and modal reasoning.

Moreover, what von Neumann called a quantum state Ψ is a wave function. It is not clear why a wave function is a state vector which changes in time. Is a state vector just a wave shape of that time in each moment? Why is a vector a wave shape in each moment? We shall revisit this problem in subsection 6.3.7-2.

6.3.6. The Uncertainty Principle

As we all know, nature around us is not modality, this modal theory of nature immediately encounters a mountain of contradictions. A deadliest contradiction consists in the trajectories of particles. According to the uncertainty principle, once we localize a particle, we lose information about the momentum of the particle. It is therefore next to impossible to localize a particle twice. This means that theoretically the possibility of observing a trajectory of any particle is next to impossible. How about the Wilson chamber? Particle trajectories appear whenever we carry out a particle experiment. That is to say, whenever we perform a particle experiment, quantum mechanics is refuted experimentally. This theory is the most empirically refuted theory in the history of physics. Without knowing this, theoretical physicists proudly claim that quantum mechanics is an empirically most verified theory in the history of science.

One of the issues regarding verification via experiment of a probabilistic theory like quantum mechanics is that we really do not know what it means to experimentally verify the theory. All we can do is to check the closeness of the experimental value with the expected value. To make any decisive conclusion on this matching is theoretically impossible as the relative frequency will converge only toward a limit of the calculated probability, viz. the Large Number Theorem of probability theory.

The uncertainty principle poses further problems. If we accept the Heisenberg/von Neumann theory of quantum mechanics, time is not measurable. There must be no time dilation as it appears just empirically. In conclusion, in quantum mechanics, we have the following obscurities emanating from the uncertainty principle:

1. If time is not measurable, how can we measure the speed of any particle or the speed of any wave? To utilize the de Broglie relation we must be able to measure time. If we have no time empirically, how can we measure the momentum which is supposed to be non-commuting with location, at least one of which is measurable as the uncertainty principle claims?
2. According to Dirac and Feynman, quantum electrodynamics is a combination of the special theory of relativity and quantum mechanics. But since time is not measurable in quantum mechanics, how can we mix the special theory of relativity and quantum mechanics? Time must be measurable in the former and it is not the case for von Neumann's and Heisenberg's quantum mechanics. All in all, there is serious confusion here as Schrödinger's wave equation, which was passed onto von Neumann's formalism, was obtained using the de Broglie relation which in turn came from the special theory of relativity in which time must be measurable.

6.3.7. Quantum Entanglement

6.3.7.1. Tensor Product of Hilbert Spaces

According to quantum mechanists, the "quantum mystery" stems from the particle-wave duality. Most of the mystery involving more than one particle is the manifestation of an interference between the involved particles as waves. Some of the interesting consequences of this "quantum entanglement" are the nonlocality and the loss of strict identity of particles in question.

In order to create a quantum system of two particles P_1 and P_2, each of which is formulated over the Hilbert spaces H_1 and H_2 with bases $\{e_1, e_2, \dots e_{n_1}\}$ and $\{f_1, f_2, \dots f_{n_2}\}$ respectively, we have to create the tensor product of H_1 and H_2 which expresses the interaction between P_1 and P_2. The standard direct product will not do.

Given two vectors $\phi_1 \in H_1$ and $\phi_2 \in H_2$, we write $\phi_1 \otimes \phi_2$ for (ϕ_1, ϕ_2) and call it the "tensor product of ϕ_1,ϕ_2". We do this to separate tensor product from direct product. With this we create a Hilbert space $H_1 \otimes H_2$ called the "tensor product of H_1 and H_2" which properly contains $\{\phi_1 \otimes \phi_2 : \phi_1 \in H_1 \; and \; \phi_2 \in H_2\}$ such that

1. $H_1 \otimes H_2$ has a base $\{e \otimes f : e \in \{e_1, e_2, \dots e_{n_1}\}, f \in \{f_1, f_2, \dots f_{n_2}\}\}$ such that for each ψ in $H_1 \otimes H_2$

$$\psi = \sum_{i=1}^{n_1} \sum_{j=1}^{n_2} \psi_{ij}(e_i \otimes f_j)$$

where $\psi_{ij} \in C$ and C is the set of all complex numbers.

2. For all $\alpha, \beta \in C$, $\psi, \phi \in H_1$ and $\chi \in H_2$,
$$(\alpha\psi + \beta\phi) \otimes \chi = (\alpha\psi) \otimes \chi + (\beta\varphi) \otimes \chi.$$
3. For all $\alpha, \beta \in C$, $\psi \in H_1$ and $\chi, \phi \in H_2$,
$$\chi \otimes (\alpha\psi + \beta\varphi) = \chi \otimes (\alpha\psi) + \chi \otimes (\beta\varphi).$$
4. For all $\alpha \in C$, $\psi \in H_1$ and $\phi \in H_2$,
$$\alpha(\psi + \varphi) = (\alpha\psi) \otimes \varphi = \psi \otimes (\alpha\psi).$$
5. For all $\psi_1, \phi_1 \in H_1$ and $\psi_2, \phi_2 \in H_2$,
$$< \psi_1 \otimes \phi_1, \psi_2 \otimes \phi_2 > = < \psi_1 \otimes \phi_1 > \times < \psi_2 \otimes \phi_2 >$$
6. For all $\psi_1, \phi_1, \phi_3 \in H_1$ and $\psi_2, \phi_2, \phi_4 \in H_2$,
$$< \psi_1 \otimes \psi_2, (\alpha\phi_1 \otimes \phi_2 + \beta\phi_3 \otimes \phi_4) > = \alpha < \psi_1, \phi_1 > \times < \psi_2, \phi_2 > + \beta < \psi_1, \phi_3 > \times < \psi_2, \phi_4 >.$$

For operators $\hat{A}_1$ and $\hat{A}_2$ on H_1 and H_2 respectively, we can define a tensor product operator $\hat{A} \otimes \hat{B}$ on $H_1 \otimes H_2$ as follows:

$$(\hat{A}_1 \otimes \hat{A}_2)\psi = \sum_{i=1}^{n_1}\sum_{j=1}^{n_2} \psi_{ij}(\hat{A}_1\mathbf{e}_i) \otimes (\hat{A}_2\mathbf{f}_j).$$

An observable A_1 for the system 1 can be represented by $\hat{A}_1 \otimes \hat{1}_2$ in the composite system and an observable A_2 can be represented by $\hat{1}_1 \otimes \hat{A}_2$ where $\hat{1}_1$ and $\hat{1}_2$ are identity observables for system 1 and system 2 respectively.

Here are some elementary results on the composite observables.

- Suppose a is an eigenvalue of $\hat{A}_1$ with an eigenvector v_a. For every vector ψ in H_2, we have

$$(\hat{A}_1 \otimes \hat{1}_2)\mathbf{v}_a \otimes \psi = \hat{A}_1\mathbf{v}_a \otimes \psi = a\mathbf{v}_a \otimes \psi$$

  Hence, a is an eigenvalue for $\hat{A}_1 \otimes \hat{1}_2$ with an eigenvector $v_a \otimes \psi$. This implies that even if a has only one eigenvector, with respect to $\hat{A}_1$, it has many eigenvectors with respect to $\hat{A}_1 \otimes \hat{1}_2$.

- Let $\hat{A}_1$ and $\hat{A}_2$ be observables of H_1 and H_2 respectively. We have

$$[\hat{A}_1 \otimes \hat{1}_2, \hat{A}_1 \otimes \hat{1}_2] = 0.$$

- So, $\hat{A}_1 \otimes \hat{1}_2$ and $\hat{A}_1 \otimes \hat{1}_2$ commute.

$$Pr(\hat{A}_1 = a_1, \hat{A}_2 = a_2; v_{a1} \otimes v_{a2}) = Pr(\hat{A}_1 = a_1; v_{a1})Pr(\hat{A}_2 = a_2; v_{a2}).$$

In the tensor product $H_1 \otimes H_2$, although $\psi \otimes \phi = (\psi, \phi)$ does not express any interference between ψ and ϕ, $H_1 \otimes H_2$ admits any superposition of its elements. So, if $\psi_1, \ldots, \psi_m \in H_1$ and $\phi_1, \ldots, \phi_m \in H_2$,

$$\sum_{i=1}^{m} \alpha_i(\psi_i \otimes \phi_i) = \sum_{i=1}^{m}(\alpha_i\psi_i) \otimes \phi_i = \sum_{i=1}^{m} \psi_i \otimes (\alpha_i\phi_i) \in H_1 \otimes H_2$$

$\sum_{i=1}^{m} \alpha_i(\psi_i \otimes \phi_i)$ represents the "interference" between $\psi_1, \ldots, \psi_m \in H_1$ and $\phi_1, \ldots, \phi_m \in H_2$.

In the context of this interference, particles expressed in $H_1 \otimes H_2$ are no longer particles in the classical sense. For example, some particles here are no longer local.

Proposition 4. *[Permutation postulate] Consider the n-fold tensor product*

$$H^{(n)} = \underbrace{H \otimes \ldots \otimes H}_{n}$$

and let

$$H^n = \underbrace{H \times \ldots \times H}_{n}$$

Let $P: H^n \rightarrow H^n$ be a permutation. Extend this into a self-adjoint operator $\hat{P}: H^{(n)} \rightarrow H^{(n)}$. We call this permutation operator. Then for all self-adjoint operators $\hat{Q}$ over $H^{(n)}$ and for all admissible state vectors ψ of $H^{(n)}$, we have

$$< \hat{P}\psi, \hat{Q}\hat{P}\psi > = < \psi, \hat{Q}\psi > .$$

This proposition states that empirically $\hat{P}\psi$ and ψ are indistinguishable. Remembering that $H^{(n)}$ is a space for n particles of the same kind, in any admissible state, each particle loses its own epistemological identity.

The ontological validity of this conclusion is impossible to verify as the same kind of particle can be located at any distance. This is the fate of relying upon purely mathematical formulation of physics. Remember, there is no ontology behind the infamous projection postulate of von Neumann's formalism. As Schwinger said, quantum mechanics is not to "understand" but to "calculate".

Let us now consider two particles 1 and 2 of the same kind, each of which has two quantum states, v_1 and v_2. According to the Permutation Postulate, the admissible states for these two particles of the same kind are as follows:

$$\mathbf{v}_1 \otimes \mathbf{v}_1$$
$$\mathbf{v}_2 \otimes \mathbf{v}_2$$
$$\frac{1}{\sqrt{2}}(\mathbf{v}_1 \otimes \mathbf{v}_2 + \mathbf{v}_2 \otimes \mathbf{v}_1)$$
$$\frac{1}{\sqrt{2}}(\mathbf{v}_1 \otimes \mathbf{v}_2 - \mathbf{v}_2 \otimes \mathbf{v}_1)$$

For bosons, the first three states are the only possible quantum states: while for fermions the last state is the only admissible state. Again, the results are obtained purely through theory and due to nonlocality, there is no way for us to verify this claim empirically.

6.3.7.2. Evaluation

To begin with, the theory of tensor product was not a part of quantum mechanics as per von Neumann. It was introduced only to consider the interaction of many particles in the setting of quantum mechanics. There are several important issues to be discussed.

Schrödinger and von Neumann did not consider a system of particles which are interacting with each other in their theory of quantum mechanics. Classically, this problem was considered to be impossible to handle mathematically (analytically) and so they went on to use potential energy field to discuss the influence of the rest of the particles on a specific particle as approximation. Luckily, potential energy field and force field superpose. Schrödinger produced a quantum theory of such a system to yield the wave equation which is supposed to describe the motion of the particle in question. This was the first materialization of the so-called "wave-particle duality". It was von Neumann who combined this idea of Schrödinger and the idea of Heisenberg-Jordan who expressed the measurement as a self-adjoint operator over infinite matrix space and presented the first unification of several models of quantum mechanics. He added Born's probabilistic interpretation to this, and proposed that the state of the quantum system is a vector of the complex Hilbert space.

In von Neumann's formalism, it is clear that the theory does not deal with the interaction of many particles because the effect of other particles is reduced to the potential energy field as in Hamiltonian mechanics. In short, field theory does not respect the law of action-reaction.

As a challenge to this limitation of von Neumann's formalism, the so-called "tensor product theory" of quantum mechanics was developed. The whole development of the theory has been outlined above. By theoretically proving non-locality, this theory detached quantum mechanics from empirical science. The nonlocal nature of the theorems of this theory makes it impossible to verify its claim empirically. We cannot carry out experiments of the same kind at a distance of 1 million light years away at the same moment, can we? To respond to this criticism, quantum mechanists introduced the concept of the Planck length to limit the space in which the quantum phenomena take place, yet they are claiming at the same time things like teleportation. On the one hand, they claim that the range of quantum mechanics is within the Planck distance and yet they claim that it can cover the distance of many light years too, in fact instantaneously. Even worse, no matter how they limit the scope of the theory, still the theory itself proves that nonlocality exists for any distance.

Classical mechanists discovered that it is impossible to solve the multi-body problem analytically due to its non-linear nature. Certainly, gravity is not linear. The potential energy field approach was a compromise (convention) to "overcome" this difficulty and it is not correct. This approach drops the law of action-reaction. It is astounding that quantum mechanists came up with the idea of tensor product of quantum states to facilitate the solution to the multi-body problem. It is a mathematical nicety but it clearly has no relevance to physics at all. As we said, the multi-body problem became a problem because of its nonlinear nature. Quantum mechanists found an alternative solution to this problem by considering the "linear interaction" of particles.

More fundamentally, multi-body problems in classical theory are reduced to the issue of a particle in a force field created by many other particles and from this Schrödinger's quantum wave equations stem. This view does not agree with the reality as it violates the law of action-reaction. So we have no reason to believe that this view is consistent with the wave view of the whole issue. Assume that a particle is interacting with many other particles in the classical sense, then this particle can be represented by a Schrödinger wave equation under the potential energy field created by the rest of the particles. In the same way, each particle will get a wave representing it under the potential energy field of the rest of the particles. Assume that we have a system of particles $p_1, p_2, \ldots, p_n$, Then for each i, we obtain the Schrödinger wave equation under the potential energy field created by the rest of the particles and make all of these waves interfere. This seems to be one way of describing a multi-body system. The problem with this is that the Hamiltonian equation used to describe the classical system is false as it ignores the law of action-reaction by representing it with the wrong idea of force field as we have argued at the beginning of the paper. Or we can consider some potential energy field for each p_i and get a Schrödinger wave equation for each p_i and make all waves interfere. Again it is not clear whether the interaction of all particles can be captured correctly. And in most cases, we cannot solve Schrödinger's wave equations analytically, which is a bigger problem.

So, what is happening with quantum entanglement? Is it the case that all of these problems can be resolved by quantum entanglement? Almost all theoretical physicists will say "yes". But we have doubts.

First, there appears to be a problem with amalgamating the tensor product which is used to describe multi-particle systems with the potential force field approach which in turn is essential in Schrödinger's wave mechanics which in turn is a most essential part of von Neumann's formalism. In the quantum entanglement approach, the state of the system, that is its wave function under the claimed wave-particle duality, is described by the tensor product of the wave functions of all particles:

$$\psi = \psi_1 \otimes \psi_2 \otimes \cdots \otimes \psi_n.$$

The total Hamiltonian of the n-particle system is given as:

$$H = (\hat{H}_1 \otimes 1 \otimes \cdots \otimes 1) + (1 \otimes \hat{H}_2 \otimes \cdots \otimes 1) + \cdots + (1 \otimes 1 \otimes \cdots \otimes \hat{H}_n).$$

Then,

$$H\psi = \hat{H}_1\psi_1 \otimes \hat{H}_2\psi_2 \otimes \cdots \otimes \hat{H}_n\psi_n.$$

This means that we need to separate each particle and all forces upon it, and specify a wave function for each particle which will then be included in the tensor product.

Consider the expanded Hamiltonian,

$$H = \sum_{i=1}^{n} K(p_i) + \sum_{i \neq j = 1}^{n} V(p_i, p_j)$$

where $K(p_i)$ is the kinetic energy of the particle p_i and $V(p_i, p_j)$ is the potential energy of the interactions between particles p_i,p_j (counted once only). For a two-particle system of the same mass m, for example,

$$H = -\frac{\hbar}{2m}(\nabla_1^2 + \nabla_2^2) + V(p_i, p_j).$$

This standard development in quantum mechanics does not allow separation of inter-particle interactions for each particle. Naturally, we should expect

$$H\psi = \hat{H}_1\psi_1 \otimes \hat{H}_2\psi_2$$

where

$$\hat{H}_1 = -\frac{\hbar}{2m}\nabla_1^2 + V(p_i, p_j) \;\; \hat{H}_2 = -\frac{\hbar}{2m}\nabla_2^2 + V(p_i, p_j)$$

This clearly indicates some difficulty in integrating quantum entanglement into von Neumann's quantum formalism. So, the issue of treating multiple particle systems in quantum mechanics is still up in the air.

In short, the interaction between particles expressed via a tensor product and that through potential energy do not commute.

Remark 5. *As usual, quantum mechanics is a typical product of amalgamating disjoint theories. This may not be a concern for theoretical physicists, but it is a serious concern for mathematicians. A similar situation also occurred in quantum electrodynamics. Dirac, justifiably frustrated by the fact that the Planck/Einstein photon did not come from the quantization of electromagnetic theory, introduced a new concept of photons from quantizing the vector potential of electromagnetic theory using the Fourier expansion of the vector potential. In this way, he obtained a lot more photons of different kinds associated with the creation and annihilation operators. This is called the "second quantization of photons". Nobody bothered to study the precise connection between these two concepts of photons which are vastly different, and so by mixing them up the Quantum Electrodynamics was created. This opportunism a la Dirac went further and he relativized Schrödinger's wave mechanics by replacing e and p in Einstein's energy-momentum relation with the energy operator and the momentum operator. In this way a quick* *quantization of relativity theory (which should be called quantization of relativity theory) was established.*

All of this seems to conflict with the fundamental assumption of the wave-particle duality in the quantum mechanics as per von Neumann, which is the standard. According to the von Neumann formalism, the quantum states are the vectors Ψ in the complex Hilbert space whose transition is specified by the Schrödinger equation.

$$i\hbar\frac{\partial}{\partial t}\Psi = \widehat{H}\Psi$$

As is discussed in subsection 6.3.5, we cannot use Schrödinger's wave equation to define the quantum state transition as the equation itself has little to do with quantum states. In Schrödinger's "wave equation", Ψ is not a state vector but a wave function. A vector in Hilbert space is not a wave; and a wave is not a state vector in the Hilbert space.

This simply means that the most fundamental claim of particle-wave duality is in question! If a quantum state is a (state) vector of the complex Hilbert space which is claimed by quantum mechanist, then there is no concept of particle involved. It is all about waves. We do not know what is happening in the quantum computing. Oddly, it appears that it is thought that the so called "quantum states" in the entanglement are the descriptions of particles.

Because time is not observable in von Neumann's formalism (in Heisenberg-Jordan's matrix formalism neither), it becomes highly questionable to claim that a wave function is a state vector which changes as time passes. Despite this kind of rather shocking obscurity in quantum mechanics, the current engineering project of quantum computing is to take advantage of quantum entanglement to dramatically speed up computation by executing parallel computing at quantum scale world. It has however started to appear that they are more interested in finding some other physical theory which would materialize this tensor product theory of interference. Indeed, the tensor product theory is an interesting formalism in the discussion of the interaction of multiple particle systems regardless of its relevance to von Neumann formalism.

We reckon that one of the deepest problems with quantum mechanics is that there is no such thing as the claimed wave-particle duality. To begin with, the Hamiltonian description of a particle system is wrong as it uses potential energy and ignore Newton's law of action-reaction; then it uses the special theory of relativity which is an inconsistent theory to convert particle systems into wave equations. If the wave-particle duality fails, we do not know what quantum mechanics is all about. This theory is experimentally refuted whenever accelerator experiment takes place. The theory says that there must be no trajectory and yet every experiment yields trajectories.

SECTION 7. MOMENTUM VERSUS ENERGY

In dimensional analysis, [impact] = [force] × [time] and [work] = [force] × [distance]. There is no clear conceptual connection between [impact] and [work]. So, [impact] and [work] are separated by the difference between [time] and [distance]. As, in classical theory, there is no "space-time symmetry", [time] and [distance] are independent dimensions. One cannot express [time] through [distance] and vice versa. This is why we have no explicit connection between [work] and [momentum].

To clear up the confusion on [work] and [energy], let us be reminded that [energy] is a modalization of [work]. They say that [energy] is a "potential" to do [work]. Going back to the original issue of separation of [time] and [distance], one mistake of Einstein's was allowing v in the gamma factor of his relativity theory to be time-dependent; and so it confuses the entire modern physics on the energy-momentum relation which "glued" these two independent concepts together. It was Einstein himself who at the same time stressed that this v must be constant so as to avoid contradiction [which comes from the violation of the law of action-reaction]. With this mistake, a revolution in theoretical physics occurred, and so he obtained the infamous $e = mc^2$ and the well-used energy-momentum relation which conceptually cannot exist for the reasons stated above.

All of this urges us to take the theory of dimensional analysis more seriously in order to avoid this kind of century-long diversion. Notwithstanding, dimensional analysis is at its very elementary level. So far this theory was used mostly to get the "physical unit" right for the purpose of empiricism and engineering.

A similar issue surfaced in mathematical logic and computer science half a century ago. Prof. John Peck tried to introduce the idea of physical dimensional analysis as a generalization of data types in ALGOL 68. This generalized concept of type was called "pouch". It was to consider the compatibility of programs as pouches and connections between different pouches.

Of course we are fully aware of the fact that the modality of probability has taken over the entire computer science in the form of Artificial Intelligence, producing "intelligent" machines which are more effective than human beings in many activities. For good or bad, this poses an interesting question about the modality against predicate reasoning. Theoretical physics is not alone. Finally, the issue of modality in the form of probability in theoretical physics can be a big challenge for empiricism. Due to the large number of theorems produced by probability theory, as Kuhn once pointed out, probabilistic prediction can-not be even refuted through experiments.

Though we did not make a special section to discuss statistical mechanics in our presentation, the modality in statistical mechanics and that in quantum mechanics are different things. In case of statistical mechanics, each particle contributing to the macro level behavior of a macro system is governed by determinism. There is no such thing as wave-particle duality at each particle level. It is just because there are way too many of these particles acting deterministically under classical mechanics, they have to take statistical average of the entire system to figure out the macro effect of these particles. On the contrary, when it comes to quantum mechanics, the behavior of each particle is nondeterministic. Not for a massive system of particles, but for each single particle, we predict probabilistically the expected value of measurement. This is what makes many physicists feel uneasy about quantum mechanics. When it comes to statistical quantum mechanics, they have to take the statistics of the probabilistic behavior of a massive number of quantum particles. ... We have come a long way. Is God probabilistic? Einstein said it differently, he asked "Does God throw dice?"

It is clear that momentum is a physical reality and it does exist. But as we have argued, the definition of kinetic energy for mv is false. It is not $mv^2/2$ in general. Moreover, energy is defined as possibility to do work. If so, then it is a modal concept and we do have to sort out some serious questions on the conflict between predication and modality. Does energy "exist"? As we know, as an intellectual common sense, existence is a concept which belongs to predicate logic. Why is it that advanced experts on modal logic never pay attention to this important instance of the conflict between modality and predication in the most important science, physics.

What makes the situation even more awkward is the concept of potential energy. Energy itself is already a modal concept. It is a potential to do work. It is not clear what then is the "potential to potential to do work"? This is a very clear indication that physicists have forgotten that energy is the "potential" to do work.

REFERENCES

Anderton R., *Private communication,* 2007.

Bacon F., *Novum Organum* [*New organ*], 1620.

Bennett J., *Locke, Berkeley, Hume: Central Themes*, Oxford University Press, 1971.

Berkeley, G. A., *Treatise Concerning the Principles of Human Knowledge*, Part I, 1710.

"Calligraphy: Ancient India and Japanese Sutras, Textiles: Banner and Banner Legs – The Gallery of Hōryū-ji Treasure". Tokyo National Museum. Retrieved 2017-11-03. "Calligraphy: Heart Sutra and Usnisa Vijaya Dharani Sutra in Sanskrit, Late Gupta dynasty, 7th–8th century (Important Cultural Property)".

De Broglie L., *"Recherches sur la théorie des quanta"* [*Researches on the quantum theory*], Thesis, Paris, 1924, Ann. de Physique (10), 1925.

Dingle H., *Science at the Crossroads*. London: Martin Brian & O'Keeffe. 1972.

Eilenberg S., MacLane S., *"General theory of natural equivalences",* Trans. Am Math Society, 1945.

Einstein A., "Zur Elektrodynamik bewegter Körper" [On the Electrodynamics of Moving Bodies], *Annalen der Physik,* 332 (10), 1905.

Engelhardt W. W., “Phase and Frequency Shift in a Michelson Interferometer”, *Physics Essays* 27 (4), 2014.

Heisenberg W., *The Physical Principles of the Quantum Theory*. Translators: Eckart, C., Hoyt, F. C., Dover, 1930.

Kanda A., Apostoli P., *“Inconsistency of Some Formal Theories”*, presented at the Meeting of Exact Philosphers, 2007.

Kanda A., Apostoli P., “Logicism v.s Empericism”, *International Seminar on the Structure in Cosmology and Theoretical Physics,* Finish Physics Foundation Society and Finish Society for Natural Philosophy, Helsinki, 2010.

Kuhn T. S., *The Structure of Scientific Revolutions*, The University of Chicago Press, Chicago, 1962.

Lawden D. F., *Elements of Relativity Theory*, John Wiley & Sons, 1985.

Lorentz H. A., “Electromagnetic phenomena in a system moving with any velocity smaller than that of light”, *Proceedings of the Royal Netherlands Academy of Arts and Sciences,* 6: 809–831, 1904.

MacLane S., *Categories for the Working Mathematician*, Graduate Texts in Maths. 5 (2nd Edition) Springer-Verlag.

Maxwell J. C., *“A Dynamical Theory of the Electromagnetic Field”*, Philosophical Transactions of the Royal Society of London 155, 459–512, 1865.

Miller N., Ashby S. C., *Principles of Modern Physics*, Holden Day, 1970.

Minkowski H., “Das Relativitätsprinzip” [The Relativity Principle]. *Annalen der Physik*. 352 (15): 927–938, 1907.

Robinson A., *Non-standard Analysis*, Princeton Landmarks in Mathematics, Princeton University Press, 1966.

Schrödinger E., “An Undulatory Theory of the Mechanics of Atoms and Molecules”. *Physical Review*. 28 (6) 1926.

Sells R. T., Weidner R. L., *Elementary Modern Physics*, Allyn and Bacon, Boston, 1973.

Suntola T., *Dynamic Universe*, Physics Foundations Society, 2011.

Woit P., *Not Even Wrong,* Basic Books, New York, 2006.

In: Old Problems and New Horizons in World Physics
Editors: Volodymyr Krasnoholovets et al.
ISBN: 978-1-53615-430-6

Chapter 8

THE ROLE OF EMPIRICISM IN PHYSICS

***Akira Kanda*[1,*], *Mihai Prunescu*[2,†] *and Renata Wong*[3,‡]**
[1]Omega Mathematical Institute, Toronto, Canada
[2]University of Bucharest, Bucharest, Romania
[3]Nanjing University, Nanjing, Jiangsu, China

ABSTRACT

It has been commonly understood that physics is an amalgamation of mathematization and empiricism. The founder of physics, Newton, made it clear that the only way to observe a physical system is to do it from ouside the system, as he said "I was standing on the shoulder of God." This view was seconded by Goethe who said that it is impossible to make a measurement inside the physical universe without changing the system to be observed. This problem became acute as we developed relativity theory and quantum mechanics. Due to the apparent success of modern physics, this important issue was pushed aside. Here we will reinvestigate this problem in a modern context.

"I know that most men, including those at ease with problems of the greatest complexity, can seldom accept even the simplest and most obvious truth if it be such as would oblige them to admit the falsity of conclusions which they have delighted in explaining to colleagues, which they have proudly taught to others, and which they have woven, thread by thread, into the fabric of their lives."– Leo Tolstoy

"If you don't read the newspaper, you are uninformed. If you read the newspaper, you're misinformed."– Mark Twain.

* Corresponding Author's Email: kanda@cs.toronto.edu.
† Author's Email: mihai.prunescu@gmail.com.
‡ Author's Email: renata.wong@protonmail.com.

SECTION 1: A BRIEF HISTORY

It was Aristotle who emphasized the importance of observing nature to understand the universe. Through abstracting on what was observed, he went on to present his cosmology. In this sense, we could call him the founder of empiricism. Curiously, he also ventured into an entirely different discipline, logic. He suggested using a formalization of logical reasoning upon what was observed so as to secure a correct understanding of the world.

His view was internally complimentary, and he exercised his empiricism selectively to justify his cosmology. Unfortunately, this criticism on him is fully applicable to the contemporary empirical sciences, physics in particular. In this sense, one may argue that little progress has been made since the time of Aristotle. Aristotle's "rationalist" philosophy was accepted as the backbone of medieval Catholic theology and cosmology, and was further developed by Augustine and others. While Vatican's theologians were preoccupied with their logical project of proving the existence of God, they neglected the issue of empiricism. Despite their rather astounding maturity in logical reasoning, their assumptions about nature were rudimentary. Naturally, they were confronted by many empirical circumstances which refuted their logical predictions.

All of this led to two consequences toward the end of the Middle Ages. First, empiricists dismissed logical reasoning as a fallacy which created wrong predictions. Second, they failed to understand that by obtaining wrong predictions about nature from wrong assumptions about nature, logic did what it was supposed to do perfectly. The fact to be reckoned with is that Vatican's theologians were intellectually honest by admitting that their attempt to prove the existence of God logically failed bacause all of their attempts to do so assumed the existence of God. For this reason they moved on to what we now call modal logic with which they tried to prove the necessity of God instead. This is in a sharp contrast to theoretical physicists who dismissed mountains of criticism with name-calling and blackmailing.

To further the reaction to this apparent shortcoming in Aristotelian "rationalism", empiricists adopted two theses regarding the scope and limits of human knowledge. The first thesis is called "genetic empiricism", which is a limitative psychological theory regarding the source and constituents of human knowledge. It denies that there is anything in human understanding which was not "previously given in sensory experience" such as sensory impression or ideas derived from sensory impressions. Berkeley and Locke's theory of ideas and Mill's "phenomenalism" are examples of psychological theories that uphold the first thesis. Humans acquire concepts as "inductive generalization" of particular past experiences. The second thesis, called "epistemic empiricism", is a normative thesis of epistemology regarding the justification of our knowledge. This is the claim that all human knowledge is directly or indirectly "justified" by our sense experience. For Bacon, Hume and Mach, such a justification is based on inductive generalizations from our sense experience of the past.

In short, empiricism is based on the belief system that our knowledge is experience-based, that is, it is a systematization through induction which uses our previously given sensory experiences as ingredients and this kind of knowledge must be somehow verified by sensory experiences. In the form of these two twin theses, empiricism has come to be the guiding philosophy of modern western civilization.

It is interesting to note that the second thesis contradicts precisely what Renaissance astronomy has taught us! For example, Copernicus showed that the actual structure of the solar system is not as it appears and it is not directly observable from Earth: to explain the observations regarding the planetary motion requires way more than an "induction" from what we observe; a proper conceptual model, in this case the heliocentric model of the solar system, and mathematical reasoning are needed. The second thesis is also contradicted by contemporary mathematics, as this pure form of human knowledge enables us to understand, manipulate and justify claims regarding non-empirical concepts such as actual infinity, which cannot be captured by our sensory systems. For example, transcendental numbers transcend our finitary sensory world. More recently, Fermat's last theorem showed that to understand empirical numbers, such as natural numbers, requires non-empirical numbers, such as real numbers.

The empiricist's first thesis was challenged in the last century by the legacy of the Soviet topologist Pontryagin. His unparalleled mastery of geometry is a counterexample to the first thesis, as he lost his sight in infancy. His abstract geometry which lead the 20th century mathematics may well not be a product of "induction" upon visual experience of 3D spaces.

Going back to physics, it was Newton, the founder of physics, who warned that he was observing while "standing on the shoulder of God" (Giant) outside the universe. This means that he was not observing in empirical sense. His "position" of being on the shoulder of God transcended empirical observations. Newton was not only a physicist or a mathematician, he was a theologian of the Orthodox faith as well as the founder of physics and mathematics of the periods of Renaissance and Reformation. He developed physics and calculus while in exile from the plague that had infested Cambridge. The Renaissance was a counter-thesis to Vatican's mediaeval theology and cosmology. Empiricism was an antithesis to the apparently "outdated" Vatican theology of the Middle Ages which reformed contemporary Catholics desperately rebuke. Throughout his life, Newton faced fierce political attacks by the reformers and empiricists, because of his absolutism. It is recorded in history that he viciously attacked empiricists also.

Though initially persecuted as the enemy of Catholicism, Galileo, the founder of relativism in physics, has been lifted in recent years to the position of "the Catholic Physicist" after the recognition of Einstein's relativity theory as a work of genius, the "ultimate truth". In Einstein's relativity theory, Newton's basic laws of dynamics were adopted as the most fundamental laws despite the fact that Newton's laws of physics are valid only for the absolute frame.

SECTION 2: GALILEO VERSUS NEWTON ON THE RECLINING TOWER EXPERIMENT

Assume m and m' are two masses with $m > m'$. Let r be the distance from the center of the Earth to both m and m'. Let M be the mass of the Earth. The gravitational force between the Earth and m is

$$F = GMm/r^2.$$

So,

$$a_m = GM/r^2, \quad a_M = Gm/r^2$$

where a_m is the absolute acceleration of m due to the gravitational pull by M. The absolute acceleration for M in the absolute space is $-a_M$. Therefore, the relative acceleration between M and m is

$$a_M + a_m = GM/r^2 + Gm/r^2.$$

Thus, the time t_m needed for m to reach M is such that

$$(a_M + a_m)t_m^2 = 2r.$$

So

$$t_m = \sqrt{\frac{2r}{GM/r^2 + Gm/r^2}}$$

For another mass m', we have

$$t_{m'} = \sqrt{\frac{2r}{GM/r^2 + Gm'/r^2}}$$

The time required for both m and m' to reach M is not the same unless $m = m'$. As $m > m'$, we have

$$t_{m'} > t_m.$$

This means that "the heavier a body is the faster it falls."

The Newtonian interpretation above is based upon the "observation" from outside of the universe. Galileo observed it from the frame of himself and the reclining tower which is the frame of M. This will give us an entirely different conclusion. According to Galileo, M does not move and so the acceleration a_m between m and M is GM/r^2. Similarly the acceleration $a_{m'}$ between m' and M is GM/r^2. So both m and m' will reach the ground within the same time interval. This result is false as the above reasoning ignores the law of action-reaction between m and M.

Newton never used Galileo's result. The historic fact is that Newton derived the concept of gravitational force through Kepler's laws which mathematically represented his observational data of the orbiting of other planets. One may argue that this view of Newton's on our planetary system is not absolute either. Newton's position on this is that the absolute frame is the frame of the Sun. This also is problematic as there is no convincing reason why this should be the case. It is more likely that our sun is also moving in the absolute space.

After all, only a metaphysical observer, which we are not, can observe the motion in the absolute universe. The best we can say is that from Kepler's laws, Newton "induced" the law of gravity. Philosophically there is nothing wrong with this limitation.

SECTION 3: NEWTONIAN RELATIVITY AND GALILEAN RELATIVITY

Newton, putting himself "on the shoulder of God", considered the absolute physical space in which physical events take place. In this way, he avoided the difficulty of the concept of relativity mentioned above. For him, there is no observer in the absolute frame. Upon this paradigm, he introduced the laws of dynamics, namely the the second law, the third law (the law of action- reaction) and the law of gravity. As the so-called first law is a special case of the second law, we will not discuss it here. These laws are written for the absolute frame. In this way, Newton defined "relative motion (speed or acceleration)" as the difference between two motion (speed or acceleration) vectors in the absolute space.

There is a problem of asymmetry among two reference frames that accelerate with respect to each other. It is rather disappointing that relativists have no comprehensible and fundamental explanation of why we can-not consider the accelerating frames in a relative setting. So, they simply ban accelerating frames and forces on the inertial frames.

Here is a fundamental analysis of the situation which leads to its full understanding. Assume that m and M are exerting force F on each other in opposite directions as specified by Newton's law of action-reaction. This is happening in the absolute frame. Deploying relativism leads to the following argument: For m, M is under the acceleration F/M and for M, m is under the acceleration F/m in the opposite direction. Assume that $m \neq M$, then M and m will both observe that the other is approaching with a different speed. Of course this is not what we experience in reality.

This situation would not pose a problem for Newton. As an observer standing outside the absolute frame, he "observes" that in the absolute frame both m and M are moving with the respective absolute accelerations F/m and $-F/M$. So, from Newton's point of view, both M and m are moving towards each other with the same relative acceleration $F/m + F/M$ inside the absolute frame. Thus they are moving towards each other with the same speed but in opposite directions.

The difficulty with the relativistic view of this event is due to the fact that by considering the perspective of M and that of m, i.e., putting observers on both M and m, we are violating the law of action-reaction in each frame. The introduction of the relativistic reference frames violates Newton's third law. Certainly we can get around this problem by limiting ourselves only to inertial frames. But then as discussed above, we are no longer doing dynamics.

Moreover, as will be discussed in section 5, we still have the problem with placing an observer on a matter which moves in the universe regardless of the speed, whether constant or accelerated, because the contradiction occurs regardless of whether the speed of the train is constant or accelerated. A "quick fix" of restricting the reference frames only to inertial frames is not a solution.

The violation of the third law of Newton's dynamics occurs also in the attempt to introduce the concept of a force field (which seduced many theoretical physicists) as the medicine to counter the anxiety toward what Newton himself referred to as a "spooky" action happening at a distance. Apparently the deployment of the concept of action at an infinitesimal distance, as in the concept of the force field, resolves this problem; but if force is an action at a distance, then it is impossible to replace it by an action at an infinitesimal distance. Moreover, due to Newton's third law, when m and M are present in the universe, both m and M have to comply with the third law and so are under the influence of the gravitational force. If we take into consideration the force field created by M, we will violate the third law because M is under no force in this version of understanding. Like the moving reference frame, the force field is a forbidden paradise into which everybody wants to enter.

Observation and experiment are the key elements of theoretical dynamics. We haven't taken this important issue seriously enough. Many issues await resolution, it seems.

An extremely important issue, which has not been attended to, may even force us to rewrite the entire physics. Historians say that Newton did not approve of the usage of the concept of energy while he used the concept of momentum. Here is a scientific reason why it was so. Elementary dynamics claims that the energy needed to accelerate from $m0$ to mv is $mv^2/2$. This is not true. It all depends on the way a body accelerates.

In the Galilean (and Einsteinian) relativity theory, each body A has its own reference frame. Except in some extremely rare and strange situations, all other bodies are moving in A's frame. Furthermore, to avoid the well known contradiction, the speed of the moving bodies must be constant. Already this scenario has little to do with any empiricism we know of. This is a totally unrealistic artificial situation. To make the matter worse, as far as dynamics is concerned, all bodies interact with each other and so it is absolutely impossible to keep each body isolated from all other bodies that are inertial to each other, if any. So, what are we doing here? Is this not a fantasy world way more ideological than Aristotle's cosmology? What does this theory have anything to do with empiricism? Once we have a grasp of this problem, how is it then even possible to measure the speed of bodies moving in our frame? Where are those bodies that are not moving in our frame? They are needed as reference points to measure anything moving in our frame. Are they everywhere in our frame so that we can use them to measure the speed of other moving objects in our frame?

With all these problems, how can we assume that the laws of physics must be the same in all inertial reference frames? Is theoretical physics supposed to be ruled by "common law"? It has to be reminded that common law and empiricism are anti-theses. Empiricism is based upon the basic assumption that there is only one truth.

SECTION 4: ELECTROMAGNETISM

It was Heaviside and Hertz who convinced the reluctant Maxwell to accept the concept of force field and that of potential field in electromagnetic theory in order to attain the mathematical nicety of vector algebra. The problem with the use of force fields has been highlighted in Section 3. In the theory of the electromagnetic theory post Weber-Gauss, the violation of the third law resulted in a serious conceptual confusion that affects theoretical physics to this day.

The electric field **E** created by a charge is defined as a spatial distribution of the electric force $\mathbf{f_E}(\mathbf{r})$, which is a force as an "action at a distance", exerted by the charge on a unit charge when it is placed at **r**. This concept can be rewritten as "It is possible that the electric force $e\mathbf{f}\mathbf{E}(\mathbf{r})$ will be exerted on a charge e when it is placed at **r**." It is not the case that such a distribution of force is real. Magnetic field **H** is defined in a similar fashion.

Following the advice of Heaviside and Hertz, Maxwell reduced the entire electromagnetic theory to the following field equations:

$$\nabla \times \mathbf{E} = -\frac{1}{c}\frac{\partial \mathbf{H}}{\partial t}, \nabla \cdot \mathbf{E} = 4\pi\rho, \nabla \cdot \mathbf{H} = 0, \mathbf{J} = \rho\mathbf{v}, \nabla \times \mathbf{H} = \frac{4\pi}{c}\mathbf{J} + \frac{1}{c}\frac{\partial}{\partial t}\mathbf{E}$$

where $c = \frac{1}{\sqrt{\epsilon_0\mu_0}}$. Here ρ is the electric charge density and $\mathbf{J}$ is the current, created by the charge density, moving with speed **v**. Maxwell obtained the last equation, viz. the generalized Ampère's law, under the assumption that the charges in $\mathbf{J}$ move with constant speed. This law is to introduce the generalized current $\mathbf{J}$ which is not a conducting current but the change of an electric field.

In fact, $\mathbf{J}$ is a bundle of electrons moving toward the same direction, which is hard to imagine considering that electrons repel each other. It is said that the effect of all charges but one isreduced to that of the electric field. So what is $\mathbf{J}$ doing here? Is $\mathbf{J}$ affecting $\mathbf{E}$ and $\mathbf{H}$? It is claimed that it exists only in conductors. What are conductors? They appear to be massive clusters of particles. Don't we need what we call material science, which is supposed to be based on basic electromagnetic theory, to understand such a system? We have come full circle. From

$$\nabla \times \mathbf{E} = -\frac{1}{c}\frac{\partial \mathbf{H}}{\partial t}$$

which is the Maxwell-Faraday law, follows the Lorentz force

$$\mathbf{F} = q(\mathbf{E} + \mathbf{v} \times \mathbf{B})$$

Here $\mathbf{v}$ is the speed of the charge q. This law has been empirically verified by many experiments. However, the force exerted on a charge is dependent on the speed of the charge, so that such dependency renders this law a violation of the second law of classical dynamics. Charge or not, the second law excludes any possibility for the force to be dependent on speed. This means that the electromagnetic theory and Newton's dynamics contradict each other. This problem, which has never been discussed in the past, shows that the electromagnetic theory is invalid. The experimental verification of the Lorentz force does not make the theory consistent. It simply means that we cannot use this theory as the theory of electromagnetism. The present theory of electromagnetism is doubly invalid as it violates both the law of action-reaction (because of the usage of electric field) and the second law.

Maxwell obtained the electromagnetic wave equation

$$\nabla^2 \mathbf{E} = \frac{1}{c}\frac{\partial^2 \mathbf{E}}{\partial^2 t}$$

And

$$\nabla^2 \mathbf{H} = \frac{1}{c}\frac{\partial^2 \mathbf{H}}{\partial^2 t}$$

under the condition that $\mathbf{J} = \mathbf{0}$ (meaning no conducting current is present). From this wave equation, he calculated the speed of electromagnetic wave in vacuum as $c = 1/\sqrt{\varepsilon_0 \mu_0}$.

A consequence of overlooking this conceptual issue was a misunderstanding of the so-called "EM wave" as a solution to the above wave equation which was discovered by Maxwell from his electromagnetic field algebra. The proper understanding of the so called EM wave is that it is not a wave which travels through a space contrary to what we have been told for nearly two centuries. It is the transmission (at a distance) of the magnitude change of the EM force, generated at the emitter, from the emitter to the receiver of the "EM wave" located somewhere in the space. If there is no receiver, there is no emission of the magnitude change of the EM force.

Regarding the speed c of light, this quantity was theoretically obtained by Maxwell in his EM field theory and was defined as the speed at which the change of electric charge will "travel" through the vacuum between the emitter and the receiver. This original value had little to do with the "EM waves" of Maxwell. When Maxwell derived his "EM wave" as a solution of the "EM wave equation in a vacuum, what he had obtained was that the speed of this mathematical (not ontological) wave is $c = 1/\sqrt{\varepsilon_0\mu_0}$ as long as there is no current in the vacuum. As discussed above, the "EM waves" are just mathematical entities without any ontological status.

The problem here is a serious case of tension between empiricism and theory. Unless we could form a chain of receivers of EM magnitude change, we would not be able to observe the so called "EM wave". This simply means that we are not able to observe experimentally the EM waves. In reality, what we have been doing is drawing a graph of the magnitude of EM force at a fixed receiving position. This is not what we mean by the "EM wave" travelling through space.

One of the problems in making physics a mathematical theory is that ontology tends to be ignored and abstract mathematical argument takes over, yielding mistaken views of physics. It had become a serious issue in physics over the last century.

This problem of empiricism versus theory manifests itself more acutely in the case of relativizing classical electromagnetism. Historically, all the laws of EM theory were discovered within the laboratory frame of reference. Within this single reference frame, the theory, unlike Newton's classical dynamics, was developed upon many experiments; and in the end, it was the Faraday Induction which showed that it is impossible to relativize classical electromagmetic theory. This situation can be best explained as follows: When a charge is in acceleration in our frame, EM wave is emitted in our frame; nonetheless, in the frame of the charge, no EM wave is emitted. Unfortunately, we may not be able to verify or refute this experimentally.

Lorentz was aware of the problem that all axioms of electromagnetic field theory came from the experiments conducted within our frame, not in the absolute frame of aether. He applied the Galilean transformation between the aether frame and our frame to reflect the situation. He discovered that the Maxwell's wave equation is not invariant under the Galilean transformation. This led him to consider a different transformation between the absolute frame of aether and non-aether frames which would preserve Maxwell's electromagnetic wave equation, i.e., the so-called Lorentz transformation. Unlike Einstein who generalized Lorentz's work, Lorentz however did not consider transformation of non-absolute reference frames. Since then, nobody has noticed that wave functions with wave vector and frequency are invariant under the Galilean transformation. This gives rise to the fundamental question of whether wave equations are the right way to define waves. As is known, mathematically a wave equation has infinitely many different solutions. The same issue has dogged quantum mechanics with regard to the solutions of Schrödinger's wave equation. We shall revisit this issue in Subsection 6.2.7.

SECTION 5. THE SPECIAL THEORY OF RELATIVITY (STR)

5.1. Einstein's STR

5.1.1. Time Dilation and Length Contraction

One can derive time dilation and length contraction as follows: Assume that a train runs on a track. When the tip of the power pole of the train touches the power line at point P, a spark occurs at P. An observer A located in the train straight down the tip of the power pole P will observe that the spark of light comes straight down from the point P which is a point of the train. An observer B located on the embankment will observe that the light travels diagonally from P, which is a point of the power line being stationary in the frame of B to the observer A in the moving train. Under the assumption of the constancy of the speed of light which says that the speed of light is constant c in all inertial reference frames one can derive time dilation from which we can derive length contraction too.

This thought experiment yields more than Einstein had discovered, unfortunately. There is a contradiction in the frame of the observer A. For the observer A, the point P is also a point of the power line and it is moving backward. So, A will also observe that the spark of light travels diagonally from the point P which is a stationary point of the power line. Logicians will point out that this came from the assumption of a reference frame that moves inside another reference frame. If we allow a geometric point to move inside a geometric space, it destroys the topological continuum of the geometric space. This is why we do not consider a geometric point moving in a geometric space. Then how is it possible to consider a geometric space which moves inside a geometric space? If a geometric space $S1$ moves inside another space $S2$, at each moment, a point P in $S1$ is also a point in $S2$, and P is also moving in $S2$. This is the situation from which Einstein derives time dilation and length contraction.

$$\Delta t' = \frac{1}{\sqrt{1-(v/c)^2}}\Delta t \qquad \Delta x' = \frac{1}{\sqrt{1-(v/c)^2}}\Delta x$$

It is unfortunate but a natural consequence of the contradiction arising from the moving reference frames is that we have

$$v' = \frac{\Delta x'}{\Delta t'} \neq \frac{\Delta x}{\Delta t} = v$$

Both v' and v should be the speed of the motion of $S2$ in $S1$.

This means that any extension of the Galilean relativity theory is inconsistent, as it allows a reference frame to move inside the other. The idea of relative motion of a reference frame is untenable. Unfortunately we are stuck with the old Newtonian absolute frame theory of dynamics. There seems to be no obvious way of getting around this difficulty. Next, we shall proceed to discuss other contradictions coming out of Einstein's special theory of relativity which is the Galilean relativity theory plus the constancy of the speed of light.

5.1.2. The Michelson-Morley Experiment

Recently, new theoretical results have been obtained proving that we cannot detect v in the composed speed $v + c$ in the way Michelson-Morley measured the speed of light. This result holds for both treating light as wave and light as particle. For the light as wave case, readers are referred to Engelhardt. We present an argument for the case of light as particle. The design of Michelson-Morley experiment came from the realization that there are technical difficulties in synchronizing two clocks at a distance, and such synchronization was needed to measure the speed of light. Michelson-Morley used an ingenious design of using reflected light thus reducing the number of clocks involved for the experiment to just one. No more synchronization problem. The apparatus has two ends, say A and B, at a fixed distance. At A there is an emitter of light and a stop watch. At B, there is a mirror facing A. When light is emitted towards B from A, the stop watch is set. When it comes back after being reflected at the mirror the stop watch is stopped. As we know the distance and the reading of the stop watch we know the speed of light. What happens when we move the whole apparatus with speed v? In the classical kinematics, light moves with speed $v + c$ in the absolute frame. But as the mirror at B is also moving away with speed v, the effect of v is cancelled. When the light is reflected at the mirror, the speed of the light reflected is $c - v$. But as the point A is moving with speed v, the effect of $-v$ cancels. In the frame of the apparatus, we will not measure the speed of light as $c + v$ but as c. So, there is no empirical reason to assume the constancy of the speed of light which claims that the speed of light is constant c in all reference frames. That the constancy of the speed of light is an empirical assumption is thus unfounded. This is an interesting case in which the empirical result received a wrong interpretation leading to wrong theories with material consequences.

5.1.3. The Obscurity of the Concept of Relative Reference Frames

As the fundamental concept of relative reference frames renders the theory incoherent (as it should be), experiments, including thought experiments, of the theory could be ill-designed or misinterpreted. We now consider the infamous Twin Paradox as an example. According to this paradox, when the twins separate from each other as one of them departs from the Earth and then comes back with a constant speed, due to time dilation, each twin will see the other twin younger. To defend the special theory of relativity, theoretical physicists came up with the following argument: As the twin, who has left and come back, has gone through acceleration, they can know that it is this particular twin who is younger. What if both of the twins accelerate symmetrically and come back with constant speed relative to each other? Length contraction will also create the following paradox: Assume that there is a box and an arrow with the same length, and that the arrow is flying towards the box. According to the arrow, the length of the box shrinks and so the arrow will not fit in the box. According to the box, the arrow gets shorter and so it will fit in the box! All of this seems to suggest that there is some kind of tension between theory and empiricism in the special theory of relativity.

A similar issue arises in a more instrumentalistic scenario as follows: If a clock slows down in a space where the length contracts, then there must be a coherent explanation of how the length contraction

$$\Delta d' = \frac{\Delta d}{\sqrt{1-(v/c)^2}}$$

on the clock's mechanism will cause the clock to slow down, thus yielding the time dilation $\Delta t' = \sqrt{1-(v/c)^2}\Delta t$. As the special theory of relativity is a positivistic theory and time is the reading of clocks, this is certainly a problem. When it comes to reading clocks as time, its empirical performance encounters even more difficulties than the issues of time dilation and length contraction. Clearly when we read a clock as it comes toward us, the reading of clock speeds up; and when we observe a clock as it goes away from us, the reading will slow down. This is the Doppler effect on time. As this effect is linear, it supersedes the Einsteinian time dilation effect, if any, in most of the speed spectrum. The problem is that on the one hand, relativity theory completely ignores this dominating effect, on the other, it is claimed that time dilation is experimentally verified..

5.1.4. Lorentz Transformations

From time dilation and length contraction, Einstein derived a coordinate transformation which expresses the effect of the mutual speed v of two inertial frames:

$$x' = \frac{1}{\sqrt{1-(v/c)^2}}(x - vt), y' = y, z' = z, t' = \frac{1}{\sqrt{1-(v/c)^2}}(t - vx/c^2)\,.$$

It goes as follows: time dilation and length contraction yield a transformation from an inertial reference frame F to another frame F' such that F' moves with speed v in the direction of the x-axis as follows:

$$x' = \frac{1}{\sqrt{1-(v/c)^2}}(x - vt)\,.$$

Similarly, as F is an inertial frame which moves inside F' with speed $-v$, through time dilation and length contraction, we obtain

$$x = \frac{1}{\sqrt{1-(v/c)^2}}(x' + vt)\,.$$

By solving these two equations for t' we get

$$t' = \frac{1}{\sqrt{1-(v/c)^2}}(t - vx/c^2)\,.$$

Thus we get the Lorentz transformation

$$x' = \frac{1}{\sqrt{1-(v/c)^2}}(x - vt), y' = y, z' = z, t' = \frac{1}{\sqrt{1-(v/c)^2}}(t - vx/c^2)\,.$$

Unfortunately, time dilation and length contraction came from the wrong assumption of a reference frame moving inside another, which is mathematically impossible because of the train-power line contradiction. The consequence is:

$$\frac{x'}{t'} = \frac{x - vt}{t - vx/c^2} = \frac{c^2(x - vt)}{c^2 t - vx} \neq \frac{x}{t}$$

So the relative speed is not the same magnitude. After all, the Lorentz transformation was obtained from time dilation and length contraction under the assumption that $F1$ and $F2$ share the same origin at the same time 0. This problem, an expected result following from the untenable assumption of a frame moving inside another, continues to haunt us in the Lorentz transformation.

There is yet another obscurity left on this issue here. The relativistic addition of speed $v \oplus v'$ is obtained by using the Lorentz transformations. So, one cannot use this speed to define or justify the Lorentz transformation on pain of committing a vicious circle. Moreover, v in the Lorentz transformation is a relative speed between two inertial frames, it is then a classical speed. So, not only mathematically but also conceptually, there is something uneasy about the concept of the Lorentz transformation. In what follows, we shall observe that this will manifest itself as a serious setback both theoretically and empirically.

Many theoretical physicists discuss the Lorentz group - the group of Lorentz transformations and composition that was discovered by Poincaré. Assume that there are three inertial frames $F1$, $F2$ and $F3$. Let v and v' be the mutual speeds between $F1$ and $F2$, and $F2$ and $F3$, respectively. As these speeds are used to define the gamma factor, on pain of committing a vicious circle, they must be pre-relativistic speeds, i.e., classical speeds. So, $v + v'$ is the mutual speed between $F1$ and $F3$.

The Lorentz transformation L between $F1$ and $F3$ is defined as

$$x' = \frac{(x - (v + v')t)}{\sqrt{1 - (v + v')^2/c^2}}, y' = y, z' = z, t' = \frac{\left(t - \frac{(v+v')x}{c^2}\right)}{\sqrt{1 - (v + v')^2/c^2}}$$

Let L and L' be Lorentz Transformations from $F1$ to $F2$, and from $F2$ to $F3$

$$x' = \frac{(x - vt)}{\sqrt{1 - v^2/c^2}}, y' = y, z' = z, t' = \frac{\left(t - \frac{vx}{c^2}\right)}{\sqrt{1 - v^2/c^2}}$$

$$x'' = \frac{(x' - v't')}{\sqrt{1 - (v')^2/c^2}}, y'' = y', z'' = z', t'' = \frac{\left(t' - \frac{v'x'}{c^2}\right)}{\sqrt{1 - (v')^2/c^2}}$$

respectively. It is clear that $L \neq L' \circ L$ where $L' \circ L$ is the mathematical composition of L and L'.

Mainstream physicists contest that in the special theory of relativity, calculating $v + v'$ is the wrong thing to do. They claim that this should be replaced by the relativistic addition (transformation) of speed

$$v \oplus v' = \frac{v - v'}{1 - vv'/c^2}$$

This is to say that L should be

$$x'' = \frac{(x - (v \oplus v')t')}{\sqrt{1 - (v + v')^2/c^2}}, y'' = y', z'' = z', t'' = \frac{\left(t' - \frac{(v \oplus v')x'}{c^2}\right)}{\sqrt{1 - (v + v')^2/c^2}}$$

However, they remain silent when asked why isn't $v + v'$ replaced by $v \oplus v'$ in the gamma factor. $(v \oplus v')$ in the Lorentz transformation is also problematic. As the Lorentz transformation is the agent to introduce relativity concepts, one cannot use the relativistic concept $v \oplus v'$ to define such a transformation. This is again a conceptual vicious circle, but the fact that it works for composing Lorentz transformations, as claimed, doesn't make it valid. Unfortunately this version of L still "fails" $L = L' \circ L$. Mathematically, Lorentz transformations are linear transformations from 4D space to itself. It is highly irregular that an algebraic composition of such transformations is not the desired transformation.

Conceptually, $(v + v')$ is classically measured speed. The reason why $(v \oplus v')$ was introduced in the first place in the numerator part is because classical addition $(v + v')$ does not work for the relativistic addition of speed. As the relativistic addition $(v \oplus v')$ is introduced by relativistic argument, it is viciously circular to use this relativistic version in the gamma factor. Mathematically we have another problem. It is naturally expected that x''/t'' will serve as the observed speed $v \oplus v'$. There is no way to prove that they are equivalent. Here is a simple calculation which leads to this conclusion.

$$\frac{x''}{t''} = \frac{(x - (v \oplus v')t)}{t - \frac{(v \oplus v')x}{c^2}} = \frac{c^2(x - (v \oplus v')t)}{c^2t - (v \oplus v')x} = \frac{c^2x - c^2(v \oplus v')t}{c^2t - (v \oplus v')x}.$$

Note that even if we replace the classical $v + v'$ in the gamma factor with the relativistic $v \oplus v'$, the above calculation holds.

Now assume that

$$\frac{c^2x - c^2(v \oplus v')t}{c^2t - (v \oplus v')x} = v \oplus v'.$$

Then we have

$$\begin{aligned} c^2x - c^2(v \oplus v')t &= (v \oplus v')(c^2t - (v \oplus v')x) \\ c^2x &= 2c^2(v \oplus v')t - (v \oplus v')^2x \\ c^2x + (v \oplus v')^2x &= 2c^2(v \oplus v')t \\ \frac{x}{t} &= \frac{(v \oplus v')^2}{c^2 + (v \oplus v')^2}. \end{aligned}$$

This is not true.

Empirically speaking, there is yet another problem. Even though it could be possible to measure v, what about v'? It is next to impossible to measure $v + v'$ is it not? If v is a speed of a star A moving away from us in distance and if v' is a speed of another star B moving away from A, then how can we measure the v' and so $v + v'$?

5.1.5. Positivism: Heisenberg versus Einstein

Curiously, this problem has not been given serious attention in the history of STR. Is it because the problem is too subtle to be treated articulately in the common practice of physics? For Newton, there is no observer in the universe and there is no problem. Once we bring in observers into the physical universe, problems get more complex even for kinematics, never mind dynamics. For dynamics, as is discussed above, bringing in observers into the universe immediately results in a violation of the basic laws of dynamics, such as the law of action-reaction. As the theory includes measurements as its core part, which may be regarded as an extreme form of empiricism (positivism), let's look into the intricate connection between theory and measurement. In physics, measurement is not just an additional factor anymore. We need more than the traditional analysis of error in calculation. Theory and measurement are interwoven and the latter cannot be taken for granted anymore. It does not mean that we need to spend more money for expensive equipment. It means that we need to acquire a better understanding of the theory behind measurement. The Michelson-Morley experiment is a good example. We did not fully understand the theory behind the Michelson-Morley experiment and rushed into concluding, incorrectly, the constancy of the speed of light.

Finally, it must be reminded that Einstein retracted his positivistic position. When Heisenberg met Einstein and tried to impress him by pointing out that his quantum mechanics also is based on positivism, Einstein replied, "It is a good joke which should not be repeated often." Einstein departed from positivism and moved on to the general theory of relativity. It is said that he could not tolerate the loss of acceleration in the special theory of relativity.

5.1.6. Measurement as Metaphysics

What is so striking here is that measurement (empiricism) is a metaphysics contrary to common belief. Einstein was correct when he said that we measure time and distance using clocks and rods with scales. Without something like them, we cannot measure time and distance as we are not (standing on the shoulder of) God. This means that measurement at the most fundamental level is metaphysics as rods and clocks used at this level are not physical entities. They are things which we have to assume in order to graduate from mathematics to physics.

Yes, it is a metaphysics which transcends our intuition and expectation. Indeed, we can measure the clock of a moving object O, no problem. This can be t. When we measure the distance that the object has moved, we can do that by measuring our positions in the object's frame. This can be d. Then within our frame we have a contradiction in the "measurement" of the speed of the object O as discussed previously. If v is the speed of a star A moving away from us and if v' is a speed of another star B moving away from A, then how can we measure v' and so $v + v'$?

It is also said that we do not need to measure the clock and the distance so as to measure the speed, we can use the Doppler effect. Some of the issues involved here are:

- How do we know the original frequency of the light wave which is emitted by a star whose speed we are trying to measure? Not all stars have the same surface temperature.
- The behavior of light is still a mystery. We just pretend we know about it. The theory says that its speed is constant c in vacuum without a conducting current. Theoretically this is meaningless because to emit a light wave, the presence of currents of accelerating charges is required. Optics says that the speed of light is dependent on the frequency of light. EM theory itself says that in the presence of a current, the speed of light is not constant; again, it depends on the frequency of the light. So, how does this affect the measurement of the speed, via the Doppler effect, of a star which emits light.
- It is not clear how the energy account of the Doppler effect is understood at all. When the frequency of a wave is increased, the energy it carries also increases. Where does that energy increase come from? Physicists do not have a clear answer.

Yes, it comes from the metaphysical assumption that we can read the clocks and scaled rods in any reference frame. The problem with physicists is that we rush into hasty conclusion without examining where we stand. The special theory of relativity is a most fundamental theory of contemporary physics and so it must inevitably be based upon some serious and intricate metaphysical assumptions. What we need to do here is to check the full details of the metaphysical assumptions we use. The concept of reference frames is way more complex than physicists think. The word "reference frame" automatically assumes metaphysics of measuring time and distance. This means that neither time nor distance can be subject to change which is defined in terms of time and space.

To make the matter even more confusing, reading time with a clock is not as simple as we think. A clock moving towards us will appear to speed up and a clock moving away from us will appear to slow down, which we call "Doppler Effect" on time. This factor is completely ignored in the theory of relativity. Furthermore, there is certain obscurity in the concept of reading time with clocks and measuring distance with rods. This type of measurement requires light which we do not know what it is. Is light physical or metaphysical? We define light using our physical theory, Maxwell's electromagnetic field theory in particular, and this logically means that we have a vicious circlee problem when we use this to define the most fundamental issue of measurement.

What is unique in the special theory of relativity is that the empirical reasoning manifests itself mostly through the so-called "thought experiments". As the relativistic effect manifests itself only when v is close to c, even to this day, it is next to impossible to carry out "material experiments" to verify the "thought experiments" and their theoretical consequences.

SECTION 6. QUANTUM MECHANICS

6.1. Early Quantization of Electromagnetism

6.1.1. Quantum Theory of Light

On the one hand, quantum mechanics is a product of the "experimental problem" of blackbody radiation which did not agree with the "theoretical prediction" made by the classical electromagnetic theory of Maxwell. To make the theory agree with the experimental results on blackbody radiation, Planck reluctantly proposed a "mathematical solution" that the energy of electromagnetic wave can be passed to the blackbody only as a natural number multiple of hf, where h is what we now call the Planck constant and f is the frequency of the EM wave. Beyond the hesitation of Planck, Einstein took the view that EM wave is a system of particles called photons each of which has energy hf. As Einstein furthered his view in the context of the special theory of relativity which was the ill-fated special theory of relativity dynamics, the so-called "relativistic theory of light (photon)" was born. This was a timely thing at that time. Experimental physicists discovered the puzzling fact that a particle, when localized by physical means (measurement), turns into a wave, which they called the particle-wave duality. Einstein's interpretation of Planck's idea discussed above was considered to be a good example of such a particle-wave duality.

To encourage the reluctant Planck, Einstein tried to derive this equation $e = hf$ (or $E = nhf$) using the relativistic energy-momentum relation. He could not prove it within the context of the special theory of relativity. We do not know exactly how he tried to prove $e = hf$ (or $E = nhf$). But as long as light is an EM wave, this is impossible classically. "Light wave" is a continuum wave. This problem will not go away even if we treat the issue relativistically. Relativistic energy is not quantum but continuum. Relativistic kinematics is not discrete mathematics, it is continuum mathematics. From a continuum theory, one cannot derive a discrete theory. But as the special theory of relativity dynamics is logically inconsistent, it is possible purely mathematically to prove any equation including $e = hf$. But this bears no relevance in theoretical physics at all.

Expectedly, Einstein's idea of making Planck's photon a relativistic particle immediately ran into a mathematical contradiction which was "denied" by mainstream theoretical physicists. Einstein assumed that as a photon is a particle dual of a light wave (EM wave), its speed must be constant c. This makes his energy equation

$$e = mc^2 = \frac{m_0 c^2}{\sqrt{1 - \left(\frac{v}{c}\right)^2}}$$

diverge. So, he decided to set the rest mass m_0 of the photon as 0. He hoped that this makes the relativistic energy of photon $0/0$ which should be any number. So, he chose this $0/0$ to be hf. His mathematics told him that as $0x = 0$ means that x can be any number as $0/0$ could be any number. The fundamentals of mathematics clearly show that $0x = 0$ and $x = 0/0$ are entirely different things. Solving $0x = 0$ does not involve division by 0. But $x = 0/0$ does. Simple proof by contradiction establishes that $0/0$ is not a number. If so, we would have

$$\left(\frac{0}{0}\right) \times 3 = 1 \times 3 = 3$$
$$\left(\frac{0}{0}\right) \times 3 = \frac{(0 \times 3)}{0} = \frac{0}{0} = 1,$$

and a contradiction: $3 = 1$. This problem does not end with just a mathematical metaphysical problem. It has a serious implication in physics too. For example, we have

$$e = \sqrt{(cp)^2 + (m_0)^2 c^4} = cp = \frac{m_0 vc}{\sqrt{1 - \left(\frac{v}{c}\right)^2}} = \frac{0}{0} c^2 = c^2 hf = hf$$

for a photon of frequency f which obviously contradicts Planck's assumption that $e = hf$.

Going back to the interpretation of the blackbody radiation problem which is purely empirical, recently it has been pointed out by Suntola [23] that it was not the issue of the EM wave theory. It was the problem of the mathematical modeling of the blackbody problem. Electromagnetic waves used in this experiment are purely monochromatic. This is yet another example of how a mis-interpretation of experimental results crippled theoretical physics.

The theory of photon as particle dual to EM waves is also empirically problematic. If $e = hf$ is the energy quanta of light with frequency f, as there are uncountably (continuumly) many frequencies for light waves, there must be continuumly many particles called "photons". But empirically, we know that there are at most countably many (as many as natural numbers) particles in the universe. Many would say that there are only finitely many particles in our universe.

6.1.2. What about Millikan's Electrons (and Positrons)?

Curiously, many physicists seem to fail to understand that Millikan's electrons obtained through legitimate experiments were real quantization of Maxwell's EM theory. Millikan "showed" for negative charge Q, $Q = ne$ where n is the number of unit charges called "electrons". Unlike Planck's quantization which is almost metaphysical, Millikan's quantization is purely physical, with strong ontological support. What is mysterious and confusing here is that this "electric quantum" does generate an EM wave, which is a continuum entity, under acceleration. Then this continuum EM wave generates a "particle" called photon hf. What is the relation between e and hf? It is clear that electron is a respectfully legitimate concept introduced to electromagnetic theory through experiments and photon is not. It was a convention to fill the gap created by a mis-interpretation of the blackbody radiation experiment.

Millikan's result challenged the basic assumption of classical electromagnetic theory that a charge is not a single physical point entity. It is a complex system of micro charges each of which has a charge e or $-e$. This demanded a departure from Newtonian classical mechanics in which a mass is a continuum, not a quantum. No one seems to have taken this fundamental issue seriously, and physicists have mixed up these two mutually contradicting paradigms of point mass and electric charge.

The reason behind this failure is that mathematically at macro level, as in electrical engineering, the internal structure of electric charges is irrelevant despite the fatal fact that we still do not know how the internal structure looks like. Still, Coulomb's law worked in practice.

6.2. Quantum Theory of Non-Light Particles

The so-called "quantum double slit experiment" gave us yet another push towards the idea of particle-wave duality which has become the foundation of quantum mechanics. Along this line, physicists at the time were forced to create wave functions in order to represent a particle in classical motion. It was Schrödinger who applied the de Broglie pilot wave theory to Hamilton's classical energy equation of particles to acquire a wave representation of the particle.

6.2.1. Hamilton's Energy Dynamics

Hamilton's energy equation is a classical equation which involves no relativity at all. De Broglie's pilot wave theory is a relativistic theory. Logical common sense tells us what to expect from this unique marriage. Over this warning, the physics community pushed along this line and now we have the so-called "ultimate theory" which is not supposed to be understood.

Let us dig a little deeper into the issue. The first issue which we have to be aware of is that the concept of energy, unlike its competitor momentum, is not a material one. It is what philosophical logicians call "modality" which may cause consistency problems when mixed with classical predicate calculus which is the standard logic used in mathematics. Moreover, on its own, the concept of energy violates the expected conservation property. It is not true that an acceleration from $m0$ to mv requires precisely $mv^2/2$. It all depends on how we accelerate. Unfortunately, the Hamiltonians do not describe the classical dynamics correctly.

Certainly combining these theories, which are highly questionable at best, into a new theory of "particle-wave duality" is not recommendable. It was Phippe who questioned if concatenating bad ideas will create any good idea.

6.2.2. De Broglie's Pilot Wave Theory

De Broglie obtained the following relativistic transformation of a plane wave:

$$k'_x = \frac{k_x - v\frac{\omega}{c^2}}{\sqrt{1-(v/c)^2}}, k'_y = k_y, k'_z = k_z, \omega' = \frac{\omega - v k_x}{\sqrt{1-(v/c)^2}}$$

where $k = (k_x, k_y, k_z)$ is the wave vector and ω is the frequency. We denote the wave number $|\mathbf{k}|$ by k. Using the analogy between this and the momentum-energy transformation

$$p'_x = \frac{p_x - v\frac{E}{c^2}}{\sqrt{1-(v/c)^2}}, p'_y = p_y, p'_z = p_z, \omega' = \frac{E - v p_x}{\sqrt{1-(v/c)^2}}$$

where $p = (p_x, p_y, p_z)$ is the momentum vector and E is the energy, he proposed the following association between a particle and a wave (called "matter wave"):

$$p = \frac{\hbar}{k} \qquad E = \hbar\omega$$

where $\hbar$ is a constant and $p = |\boldsymbol{p}|$. We call this the de Broglie relation. Although this resembles Einstein's particle-wave duality

$$E = h\nu = pc, \quad p = h/\lambda$$

where λ is the wavelength, there is a fundamental difference. Unlike the photon-light duality where the speed of photon and that of light are equal, the phase speed of matter wave and the speed of particle can be different. De Broglie further assumed that, associated with a particle with speed V was a wave having a "phase speed" $w = \omega/k$. He also assumed that the energy in the wave travels along with a group speed $v_g = d\omega/dk$ which was identical to the particle's speed V. Here it is not clear what he meant by "energy" in the wave. The de Broglie relation above is nothing but a wild hypothesis based upon just the above-mentioned "wave vector frequency" transformation and "momentum-energy" transformation analogies. Certainly this does not yield the concept of "energy in the wave".

Anyhow, it is easy to show that $c^2(\omega/c^2 - k^2)$ is invariant under the relativistic transformation as an analogy to the relativistic invariance of $c^2(E/c^2 - p^2)$. We set $c^2\ (\omega/c^2\ - k^2) = constant = C$. From $c^2(\omega/c^2\ - k^2) = C$, we have

$$2\omega/c^2 \frac{d\omega}{dk} - 2k = 0\,.$$

This implies

$$v_g = \frac{d\omega}{dt} = \frac{c^2 k}{\omega}\,.$$

As the phase speed is $w = \omega/k$, we have

$$v_g = \frac{c^2}{w}\,.$$

This means that either phase speed w or group speed v_g could exceed c but not both. We do not know what this means to STR.

All of this "mathematical nicety" is relative to the "hypothesis" that a particle with speed v has a wave dual called matter wave with a group speed v_g and a phase speed $w = \omega/k$. More fundamentally, as mentioned above, the de Broglie wave is a purely mathematical wave and we have little idea what energy means there. Wave mechanics is not a unified theory. Everything is relative to the nature of the wave medium, especially when we consider the energy. A serious concern we have is that de Broglie's wave is a mathematical wave which appears to have no wave medium. So, how is it possible to consider energy for such waves?

Moreover the question of why using the Lorentz transformation, instead of the Galilean transformation, resurfaces again. Here, de Broglie did not apply the Lorentz transformation to wave equations but (plain) wave functions. We have already stated that as long as we transform wave functions, the results are always wave functions regardless of whether we use the Galilean or the Lorentz transformation.

The de Broglie relation came straight from Einstein's special theory of relativity which does not involve mass, acceleration and force. This is because accounting for these in the special theory of relativity ends up with violating the most important axiom of any relativity theory, namely the principle of relativity. This means that the momentum-energy transformation is invalid, and so de Broglie's theory is invalid.

6.2.3. Schrödinger's Wave Equation

It is not clear historically if de Broglie proposed his pilot wave theory as a response to the issue of double slit experiment. However, it is clear that Schrödinger used Hamilton's energy dynamics for the particle theory and applied de Broglie's pilot wave theory to produce wave-particle duality which was badly needed a century ago.

All waves propagated along the x−axis obey the following wave equation

$$\frac{\partial^2 \Psi}{\partial x^2} = \frac{1}{\omega^2}\frac{\partial^2 \Psi}{\partial t^2}$$

where $\Psi(x, t)$ is the wave function and ω is the wave speed.

Here, we consider the wave function Ψ whose square yields the probability of locating a particle at any point in the space. We consider only systems whose total energy E is constant and whose particle move along the x−axis and are bound in space. Then the frequency associated, via the de Broglie relation which is hypothetical and relativistic, with the bound particle is also constant, and we can take the wave function $\Psi(x, t)$ to be

$$\Psi(x, t) = \psi(x)f(t).$$

As the frequency is assumed to be precisely defined,

$$f(t) = \cos 2\pi\nu t \;;$$

then, we have

$$\frac{\partial^2 \psi}{\partial x^2} = -\left(\frac{2\pi}{\lambda}\right)^2 \psi = -\left(\frac{p}{h}\right)^2 \psi$$

where the wavelength is $\lambda = \omega/\nu$ and the momentum of the particle is $p = h/\lambda$.

We take the particle of mass m to be interacting with its surroundings through a potential-energy function $V(x)$. The total energy of the system is given by

$$E = E_k + V = \frac{p^2}{2m} + V$$

where E_k is the kinetic energy of the particle. Then we have

$$p^2 = 2m(E - V)$$

and we have

$$\frac{\hbar^2}{2m}\frac{\partial^2\psi}{\partial x^2} + (E - V)\psi = 0\,.$$

Quantum mechanists call this equation (non-relativistic) "Schrödinger's wave equation" as the energy equation involves non-relativistic mass m, E and V. This created a myth that quantum mechnics is a non-relativistic theory. The derivation of Schrödinger's wave equation involved the de Broglie relation which is nothing but a relativistic theory. But E, V and m are all classical.

It is not the case that we can resolve the problem by replacing energy with momentum. In the Lagrangean theory, we still have m and v which are not relativistic. Moreover, no relativistic theory can involve momentum because the conservation of momentum, which is equivalent to the law of action-reaction, fails in the relativistic setting. This means that unfortunately Landau-Lifschts and Feynmann did not improve on quantum mechanics. Their momentum-based quantum mechanics is as faulty as Schrödinger's energy-based quantum mechanics.

6.2.4. The Uncertainty Principle

On the basis of Schrödinger's established particle-wave duality, in which a particle's energy E and momentum p are measured as matter wave frequency ν and wavelength λ, the uncertainty in conducting measurements on waves is brought in to the uncertainty of conducting measurements on particles. Intrinsically, measuring these quantities on waves succumbs to the uncertainty

$$\Delta\nu\Delta t \simeq 1.$$

Since $E = h\nu$, the uncertainty above brings in the following uncertainty:

$$\Delta E\Delta t \simeq h.$$

Similarly, upon the measuring of momentum/location, we have the following uncertainty:

$$\Delta p\Delta x \simeq \frac{\hbar}{2}$$

Here we must be reminded that the uncertainty principle is brought in through de Broglie's wave-particle duality which is nothing but a hypothesis based on the analogy between the "wave vector-frequency transformation" and the "momentum-energy transformation". Applied physicists and engineers know well that wave mechanics and particle mechanics are very different things, and identifying them will lead to problems; which means that we do have problems with measuring the momentum and the energy of particles through the hypothetical wave representation of them.

Unfortunately, this problem does not disappear in a von Neumann's more advanced (mathematically) formulation of quantum mechanics. This is because his Hilbert space formulation still is based on Schrödinger's wave equation which in turn is based on de Broglie's wave-particle duality which also is based on the (logically) inconsistent STR. Indeed, it appeared in a more complex form as von Neumann added a probability feature to quantum mechanics following the idea of Born. One may wonder if Landau-Lifshitz-Feynman formulation of quantum mechanics which uses the Lagrangian momentum equation will do better. The utilization of de Broglie's pilot wave theory still causes the same problem; but it certainly is an improvement as the concept of energy is invalid whereas that of momentum is valid.

Notwithstanding, it appears that the original purpose of explicating the double slit experiment, which presented the existence of "wave-particle duality" experimentally, was well served by the uncertainty principle which explicates the issue as such: According to the uncertainty principle, when a particle is localized, it loses its particle nature and turns into a wave. This is why we observe the double slit experiment phenomenon.

It is rather unfortunate that this nicety of an abstract theory, explicating a rather unexpected experimental result, encounters the following hurdle. According to the uncertainty principle, a particle localized by hitting an H_2O molecule in the Wilson chamber should turn into a wave and should not leave any trajectory by hitting other water molecules in the chamber successively to form a line. Furthermore, the resolution of localization by hitting a water molecule is much higher than that of being localized at a mechanically-made slit on a wall. This means that by hitting a water molecule in the Wilson chamber, the particle should lose way more particle nature than going through the slit on a wall. So, the particle-wave duality of de Broglie and Schrödinger is experimentally refuted. This means that quantum mechanics is experimentally refuted and physics is an empirical science. In the end, we can say that we have the problem of double slit experiment which still has no theoretical explanation. But this experiment is real and still waiting for explanation.

Moreover, according to the theory of wave-particle duality between EM waves and photons, there is no uncertainty principle applicable to the measurement of the (momentum, location) and (energy, time). This poses some serious questions on the claimed analogy between Einstein's light-photon duality and de Broglie's matter wave-particle duality.

6.2.5. The Probability Interpretation of the de Broglie Waves

Classically, the intensity of illumination caused by the EM radiation of frequency f perpendicular to a screen is given as

$$I = \varepsilon_0 E^2 c$$

where ε_0 is the electric permittivity of vacuum, and E is the magnitude of the electric field at the point on the screen.

According to the photon interpretation of light, we have

$$I = (hf)N$$

where N is photon flux. As

$N \propto$ probability of observing a photon,

we have

$E^2 \propto$ probability of observing a photon.

Therefore, the probability of observing a photon at any point in space is proportional to the square of the electric field strength at that point. Using this as an analogy, physicists considered the following interpretation of the de Broglie waves: The matter wave function Ψ is that quantity whose square Ψ^2 is proportional to the probability of observing the particle dual of the matter wave.

There are issues which have not been as well attended to as they should have been. First the relation between the energy-momentum and the "wave number" frequency is just a mathematical analogy. Moreover it has not been clear if this analogy, which leads to de Broglie's wave-particle duality, yields the Einstein-Planck relation for photons when we consider light? So, it is not clear if the above claim on Ψ and Ψ^2 is acceptable.

A simple examination yields the conclusion that this analogy is invalid. For photon, $v = c$ and so the transformed wavenumber

$$k'_x = \frac{1}{\sqrt{1-(v/c)^2}}\left(k_x - v\frac{\omega}{c^2}\right)$$

and energy

$$\omega' = \frac{1}{\sqrt{1-(v/c)^2}}(\omega - vk_x)$$

diverge. This has a far reaching consequence: The naive claim that a photon is a quantum particle for light is false. Also the claim that "Ψ^2 is proportional to the probability of observing the particle dual of the matter wave." is groundless.

6.2.6. The General Schrödinger Wave Equation

As a consequence of the de Broglie relation, a particle with energy E has a wave function

$$\Psi(x,t) = Ae^{i(px-Et)/\hbar}.$$

It then follows that

$$i\hbar\frac{\partial}{\partial t}\Psi = E\Psi.$$

This together with the Schrödinger equation above will give us

$$\frac{\hbar^2}{2m}\frac{\partial^2\psi f(t)}{\partial x^2} + (\mathrm{E}-\mathrm{V})\psi f(t) = 0,$$
$$\left(\frac{\hbar^2}{2m}\frac{\partial^2}{\partial x^2} - V\right)\Psi = E\Psi = i\hbar\frac{\partial}{\partial t}\Psi.$$

So, we can consider

$$E = \frac{\hbar^2}{2m}\frac{\partial^2}{\partial x^2} - V$$

as a representation of all energy issues of this particle. It is the Hamiltonian. We denote it with H. So, Schrödinger's equation can be written as

$$i\hbar\frac{\partial}{\partial t}\Psi = H\Psi$$

where the Hamiltonian can represent more general energy situation of the particle.

6.2.7. Von Neumann's Formalism

Heisenberg and Jordan presented an alternative formulation of early quantum mechanics where matrix and linear operators are used to represent measurements. This group and Schrödinger never agreed on using eigenvectors and eigenvalues of linear operators to model quantum mechanical measurements. It was von Neumann who unified the Heisenberg-Jordan theory and Schrödinger's theory and then added Born's view of probabilistic multiplicity into the measurement.

Axiom 1: In the absence of external influence, the quantum state Ψ changes smoothly in time t according to the Schrödinger equation

$$i\hbar\frac{\partial}{\partial t}\Psi = \widehat{H}\Psi$$

where $\widehat{H}$ is the Hamiltonian operator over the complex Hilbert space.

Axiom 2: An observable is a self-adjoint linear operator over the complex Hilbert space.

Axiom 3: [Projection Postulate] When we apply an observable $\hat{A}$ to a quantum state Ψ (we call such application an observation $(\hat{A}, \Psi)$), then the quantum state jumps into one of its eigenvectors and one of the eigenvalues of the eigenstate is observed.

Axiom 4: The following represents the probability of one of the observed eigenvalues, a, of $\hat{A}$:

$$PR\left(\left(\hat{A},\Psi\right) \approx \mathrm{a}\right) = \langle\Psi, \hat{P}\Psi\rangle$$

where $\hat{P}\Psi$ is an eigenstate of $\hat{A}$ with eigenvalue a.

As is clear in this formulation, the standard quantum mechanics due to von Neumann yields the prediction on the "probabilistic distribution" of possible measurement values for each quantity to be measured.

As in Axiom 1, Schrödinger's wave equation was used to specify the transition of quantum states. There is no mention of specific wave which is a solution to the Schrödinger wave equation at all. Mathematically, there are infinitely many solutions to each Schrödinger wave equation. Then whatever actual wave von Neumann is talking about, Axiom 4 specifies the probabilistic distribution of the possible eigenvalues to be "observed". A particle system of Hamilton is already compromised due to the use of the potential energy field, which violates the law of action-reaction. Further, we know that the de Broglie relation is based on the inconsistent special theory of relativity. Hence the question arises whether by assuming the de Broglie relation such a system of particles determines the probabilistic outcome of the measurement through all the solutions (possibly infinitely many of them) of the Schrödinger wave equation?

Moreover, von Neumann's quantum state Ψ is really a wave function. It is not clear why a wave function is a state vector that changes in time. At each moment, is a state vector just a wave shape of that time? Why is a vector a wave shape at each time?

6.2.8. Hydrogen Model

A major urge for developing quantum mechanics was the failure of the old Rutherford model to explicate the discrepancy between empirical reality and theoretical predictions of classical EM theory in which the orbiting electron in the hydrogen atom must fall into the proton due to the continuous loss of kinetic energy by emitting EM wave. The Hamiltonian energy equation for such a system of an electron orbiting around the proton does contradict this empirical result that the electron never falls into the proton. It makes little sense to claim that by considering relativistic pilot wave and the orbiting electron together, as Schrödinger did, we can resolve this problem at all. As explained before, the Hamiltonian dynamics and the pilot wave theory are mutually inconsistent as the former is classical and the latter is relativistic. Only dialectical scientists will think that by putting together two mutually inconsistent theories we will create a consistent theory which will resolve the problem of the classical theory.

Considering that the only problem with the Rutherford model is its violation of the law of conservation of energy, is it not more reasonable to put the blame on the concept of energy itself which clearly violates its own conservation as we discussed in Section 3. We now know that the energy needed to accelerate from $m0$ to mv is not necessarily $mv^2/2$. It depends upon the way we accelerate.

Indeed, this question extends into the connection between empiricism and energy. In physics, since the concept of energy was introduced by the followers of Newton, it was assumed that energy was an empirical concept. A careful analysis indicates otherwise. Is energy really measurable? It is defined as a "potential" to do work. It is not work!

6.2.9. Wave-Particle Duality, Really?

Here is an interesting issue regarding a difference between wave and particle. It is a well-known fact that the tsunami waves that hit Japan in 2011 travelled very fast. They reached to more than 500km/h. It is not the case that the water at the epicenter of the earth quake moved with this speed. Each water molecule basically moved up and down. It is this up and down motion which moved with this speed. It is the inter-molecular force among water molecules that had created this transmission of up and down motion.

Conventional physics explains this as the transmission of energy of water molecules moving up and down. This is not quite correct as we do not know what energy is. That it is a modality is a big problem and moreover, the kinetic energy of mv is not $mv^2/2$ in general. It depends upon how we accelerate from $m0$ to mv. This brings us back to the question of de Broglie's pilot wave theory. De Broglie used "energy" to resolve the problem of phase speed of his wave exceeding the speed c. He claimed that since the wave does not carry energy with phase speed, there is no problem.

But then again, we have another puzzling question. How fast can such a transmission move? We have no idea how to establish that the speed of such a transmission must be less than c. As Einstein made it clear, the special theory of relativity does not involve any force. This is because under the relativistic setting the law of action-reaction is violated. So, there seems to be no way to discuss the ontology of the speed of water waves in relativistic terms. The same argument can be applied to any form of physical waves which involve physical mediums.

The claim that the kinematics alone can limit the speed in physics seems to be invalid. Moreover, what is light? If it is an electromagnetic wave, then it is not physical reality as the electromagnetic fields are just a mathematical convention which violates the most fundamental law of physics, namely the law of conservation of momentum. This mathematical fantasy has speed which is not constant c. To be precise, a real c is a constant associated with a vacuum regardless of the existence of light.

$$c = 1/\sqrt{\epsilon_0 \mu_0}$$

It is shown theoretically that the speed of electromagnetic wave is c only in a vacuum with no conducting current. The problem is, if there is no current then there is no electromagnetic wave. Empirical scientists then say that in reality EM waves do exist. Unfortunately they are not waves because there is no such a thing as the electromagnetic field. The change of the electromagnetic value from the source to a location is transmitted as the action at a distance to where there is a receiving charge. Where there is no receiving charge, there is no EM force to be transmitted to the location. There is no wave moving in space at all.

The problem with this empirical science is that despite their skill to design experiments, in many cases, empiricists failed to understand what the experiments actually showed. The first example was the Michelson-Morley experiment, whose interpretation was faulty. The black-body radiation is yet another example. More shockingly, the entire quantum mechanics was to explicate the double slit experiment. It failed too. If the uncertainty principle is correct, we should not have seen the double slit experiment phenomenon.

6.2.10. Quantum Electrodynamics

A series of mis-interpretations of the Michelson-Morley experiment, the blackbody radiation and the double slit experiment leads to the relativized and quantized version of electrodynamics which is now called quantum electrodynamics.

Quantum mechanics is a relativistic theory as Schrödinger's wave equation assumes the special theory of relativity. However, Schrödinger's wave equation is not invariant under the Lorentz transformation. m in the wave equation is not relativistic. In short, his wave equation is not relativistic. This clearly indicates that there is a problem in the special theory of relativity since there seems to be nothing wrong with Hamilton's energy equation.

A relativistic description of quantum mechanical systems, which oddly is already relativistic, was sought for by many theoretical physicists from the late 1920s to the mid-1940s. The first quantum-relativistic basis

$$-\hbar^2 \frac{\partial^2 \psi}{\partial t^2} + (\hbar c)^2 \nabla^2 \psi = (m_0 c^2)^2 \psi$$

was obtained by Klein–Gordon by inserting the energy operator and momentum operator into the relativistic energy–momentum relation:

$$E^2 - (pc)^2 = (m_0 c^2)^2$$

This led to Dirac's relativistic equation for the electron

$$\left(\frac{E}{c} - \boldsymbol{\alpha} \cdot \mathbf{p} - \beta m_0 c\right)\left(\frac{E}{c} + \boldsymbol{\alpha} \cdot \mathbf{p} + \beta m_0 c\right)\psi = 0$$

which in turn led to the prediction of positron.

Putting the falsity of Einstein's energy-momentum relation aside, it is odd that quantum mechanics is a relativistic theory since it assumes the special theory of relativity to convert particle energy equations to Schrödinger's wave equations. So, it is not quite right that the Schrödinger wave equation is not relativistic. This fundamental problem will not go away by "quantizing" Einstein's energy-momentum equation by replacing energy and momentum variables with energy operators and momentum operators, as Klein-Gordon did. We still have no physical (ontological) meaning to substituting energy with an energy operator and momentum with a momentum operator of quantum mechanics in this equation.

The same problem exists with the issue of photons. The introduction of photons by Planck and Einstein as discussed above is corrupted at best. To make the matter worse, Dirac added one more definition of photon to quantum mechanics.

The Planck-Einstein photons were obtained by quantizing the energy of EM waves. Dirac obtained his photons via a Fourier expansion of Maxwell's electromagnetic field equations (vector potential in particular). There is no clear connection between Einstein-Planck's photons and Dirac's photons. The latter have more features such as the creation and the annihilation of photons. It appears that there is not enough attention paid to this connection. Mathematically speaking "quantization" as Fourier expansion is not quite desirable. Mathematical object and its finite approximation are not the same thing. The idea of identifying them leads to not only mathematical problems but also ontological problems. One should not identify two different ontological entities just because they have the same finite approximation. Moreover, the removal of diverging terms through renormalization lacks in physical ontology. All of this constitutes an alarming abuse of mathematics. Mathematics in other sciences is not to dominate but to serve. Dirac went on further to quantize Schrödinger's wave equations for quantum particles through Fourier expansion and described the interaction of original particles through the interaction of these newly created particles. The same criticism stated above applies here.

Despite all of this, it has been frequently said that quantum electrodynamics is the most empirically verified theory in the history of physics. It is rather puzzling that a theory based on quantum mechanics, whose most important theoretical prediction, the uncertainty principle, refutes the existence of trajectories of particles, is the most empirically verified through measuring on the trajectories left in the Wilson chamber. One possible explanation is that in particle experiment, we measure the quantities by applying the formulas of the quantum electrodynamics to trajectories recorded, and then calculate. Well we are dealing with extremely small objects and so what else can we do to make measurement.

6.2.11. The Uncertainty Principle Revisited

As we have pointed out, Schrödinger's uncertainty principle came from the uncertainty of measuring the number of crests of a wave, which carries energy, passing by.

But von Neumann's (or Heisenberg's) uncertainty principle came from the multiple eigenvalues of measuring (self-adjoint) operators. According to the claim of the equivalence between Schrödinger's quantum mechanics and Heisenberg's quantum mechanics, these two versions of the uncertainty principle "must" be the same. But are they really?

In Schrödinger's formalism, time is "classically measurable", in the same sense as all classical dynamics assume. Putting its invalidity aside, he assumed that time measurement, which in fact belongs to the world of Newton observing from the shoulder of god, is "available" to the human observers in the universe. After all, this was the only way to bring down the metaphysical Newtonian physics to the level of empiricism. Under this assumption, for Schrödinger, time became "classically measurable".

Contrary to this, for Heisenberg and von Neumann, time cannot be represented by a self-adjoint operator. Time does not change by measuring it. So, time is not measurable. This is why in their formalism, we do not have the uncertainty principle for (time, energy). All of this makes it impossible to unify the special theory of relativity and quantum mechanics. As we know, time in the special theory of relativity is measurable as this theory is empirical. All physical quantities in this theory including time and location must be measurable. Time dilation and length contraction occur as we perform the measurement. The problem does not stop here. Schrödinger used the de Broglie relation to establish the so-called "particle-wave duality". The relation is relativistic, it came from the Lorentz transformation of waves. So, is time measurable or not? This is the question. Using the assumption that time is measurable, von Neumann came to the conclusion that it is not measurable. It seems that for an excellent "applied mathematician" like von Neumann, this problem is not important.

This gives rise to a question of the validity of quantum electrodynamics. This is because relativity theory is a positivistic theory in which every physical quantity must be measurable.

6.2.12. Towards an Elementary Particle Theory

The study on the nucleus of atoms presented a new challenge which quantum electrodynamics could not handle. An obvious question is how it is possible for neutrons and protons to form the nucleus of an atom considering that there is no electric force pulling them together. It was Hideki Yukawa who proposed the first "solution" to the problem. He introduced a hypothetical particle called pi meson and using the action-reaction force between the proton and the neutron to pull the meson, in effect forming a bond between them. This theory was "empirically" verified in elementary particle physics experiments. The experiments used trajectories which the theory of particle physics rejects in lieu of the uncertainty principle. This already is alarmingly problematic. In addition to that, more fundamentally, the quantum field theory is a quantized theory of fields and any field theory rejects the law of action-reaction by violating it. Yukawa's theory is defective, i.e., it is self-contradictory.

SECTION 7: THE GENERAL THEORY OF RELATIVITY

7.1. "Relativity" of the Speed, Addition of Speeds and That of Accelerations

Einstein thought that if an accelerating reference frame can be reduced to an inertial frame in which acceleration induces a "gravitational field", then it is possible to treat accelerating frames as inertial frames within the framework of the theory of relativity which, for a legitimate reason, rejects reference frames that are under acceleration. Einstein called this the "principle of equivalence". He introduced it as follows:

Assume that a spaceship is in inertial motion in our reference frame. Then a force accelerates this spaceship with rate α. A body m in the spaceship experiences a force f, which is due to the acceleration of the spaceship, and the acceleration of the spaceship makes the body m move with an acceleration of rate a in the frame of the spaceship. Putting aside what the force f is, this means that $f = ma$. Then from our perspective, m in the spaceship experiences the acceleration with rate $\alpha + a$. So, m will experience $f = m(\alpha + a)$. Therefore,

$$\mathrm{f} - \mathrm{m}\alpha = \mathrm{m}a \qquad \text{(IF)}$$

This means that "from our perspective" the acceleration α on the spaceship induces an "additional" force $-m\alpha$ on m, which Einstein called "inertial force", upon the mass m and the equation (IF) yields the force that the mass m experiences in the accelerating spaceship. This, Einstein called "the second law" in the accelerating frame of the spaceship. According to him, upon the modification of f to $f - m\alpha$, the second law is conserved under the choice of accelerating reference frames.

There are several issues to be considered here.

(1) Putting aside the invalidity of the special theory of relativity as discussed previously, some issues have been overlooked here. According to the special theory of relativity, even the addition of speeds is not a classical addition. One has to use the so-called "relativistic addition" of speeds $v \oplus v'$. Then, how can the addition of accelerations be the same as classical addition, while hoping that acceleration, in the words of Einstein, is still the time derivative of speed? This confusion is closely bound up with the issue of the disjointedness between the acceleration and the relativistic reference frames. For that reason, we have been told, the special theory of relativity excludes acceleration. Disciplined coherent reasoning is expected here and no intellectual conventionalism appreciated.

(2) As an important example, the inertial force is also closely bound up with the issue of "fictitious force" acting on a mass inside an orbiting object. Fictitious force means that it is not real. The reason why we have a problem with the fictitious force for an orbiting spaceship is because such a spaceship is under centripetal acceleration. Since it is under acceleration, it is not an inertial frame. It is wrong to say that just because inside the orbiting spaceship a fictitious force called "centrifugal force" is created, the orbiting spaceship becomes an inertial reference frame, is it not?

(3) This fictitious force is the creation of the bidiectional interpretation of the second law of Newton. This law can be interpreted in two ways. First, when a force f is applied to a mass m, it accelerates the mass with the rate $a = f/m$. Second, when m accelerates with a rate a, a "fictitious force" f appears. Ontologically the second interpretation is invalid. It is always the case that a force applied to a mass causes the acceleration of the mass. This becomes clear when we consider the case of an accelerating train: Assume that a stationary train accelerates, then a passenger in the train will feel that he is being pushed back (against the direction of the train's acceleration). This is simply because he tries to stay where he is in the train due to the first law of Newton. If he could see the outside, which is not moving, he would see that it is not him but the train that is moving under the force.

(4) Now it is clear that the problem of inertial force (fictitious force) is caused by the bidirectional interpretation of the second law of Newton, which is a wrong thing

to do. This is to say that the second law is not "relativistic". This fact further confirms that the "ideology" of relativism as per Galileo and Einstein is an untenable fallacy. Aside from the third law, now also the second law is violated by the relativism. There is no room for relativism in theoretical physics.

Remark: Regarding the "relativity" of the addition of speeds as discussed above, this problem is not limited just to the addition. Indeed, the concept of relativistic speed itself is not relativistic, making the very foundation of the relativity theory invalid. This can be argued in two different ways with the same conclusion. First assume that d' is the contracted length and t' is the dilated time, both resulting from the speed v. Then it is expected that

$$v' = \left(\frac{d}{t}\right)' = \frac{d'}{t'}.$$

It however is not the case. This can be argued in a more ontological way as follows: Assume that an object O in the other frame moves a distance d in time t inside our frame. Let A and B be the respective positions of O at time 0 and t in our frame. According to the perspective of O, A and B are moving points. So, O will observe that the distance d' between A and B is less than d. Also O will observe that the time interval t' it took A to be in front of O and the time interval it took B to be in front of O is larger than t. So, O will observe that the speed of our frame is d'/t', which is not d/t. This failure means that according to the relativity theory an observer observes the speed of a reference frame as v and as v' at the same time, while $v \neq v'$. In the presence of this kind of facts, a typical response from theoretical physics community is that mathematics is just a language for physics. They even say that relativity theory is empirically verified beyond doubt. We wonder what empiricism means for physics.

Remark: Regarding point (1) above, the following calculation confirms our argument: Classically

$$f = m\alpha + m\alpha' = m(\alpha + \alpha') = m\left(\frac{d(v + v')}{dt}\right).$$

Relativistically, assuming the naturally expected $m\alpha + m\alpha' = m(\alpha + \alpha')$ we have

$$f = m\alpha + m\alpha' = m(\alpha + \alpha') = m\left(\frac{d(v \oplus v')}{dt}\right) = m\left(\frac{d}{dt}\right)\left(\frac{v - v'}{1 - \frac{vv'}{c^2}}\right)$$

$$= m\left(\frac{\frac{d^2x}{dt^2} - \frac{d^2x'}{dt^2}}{\left(1 - \frac{1}{c^2}\left(\frac{dxdx'}{dt^2}\right)\right)^2}\right) \neq m\left(\frac{d^2x}{dt^2}\right) + m\left(\frac{d^2x'}{dt^2}\right) = m\alpha + m\alpha'.$$

7.2. The Inertial Mass and the Gravitational Mass

Einstein thought that his theory of the equivalence principle is to explain the equivalence, vaguely hypothesized by Newton, between the inertial mass and the gravitational mass. Inertial mass means the physical entity which resists the acceleration by force. This mass is mathematically given through the second law

$$\alpha_m = \frac{f}{m}.$$

Gravitational masses m and M induce an attracting force called "gravitational force" between M and m

$$f = \frac{GmM}{r^2}$$

where G is the universal gravitational constant and r is the distance between m and M. Mathematically speaking, there is no reason to believe that the two concepts of mass are equivalent. In fact, we can readily show that the gravitational mass is a special case of the inertial mass by restating Newton's law of gravity as follows:

$$\alpha_m = \frac{GM}{r^2}.$$

However, the other way around, i.e., the implication of the second law from the law of gravitation is obviously false.

7.3. Red Shift

Assume a free-falling laboratory under gravitational force. Suppose that a light beam is emitted upward from the floor to the ceiling. Due to acceleration, by the time the light reaches the ceiling, the ceiling is moving faster than the source on the floor at the time when the light beam left it. In other words, the receiver at the ceiling is approaching the source's position at the time when the light left it. Therefore we should expect a blue shift due to the Doppler effect. The observer in the lab will then notice the blue shift. This will make the observer notice the downward acceleration. However, this contradicts the equivalence principle which says that the free-falling body will not notice its own free falling. So, Einstein postulated that there must be a red shift due to the light moving upwards against the gravitational force so as to compensate the blue shift.

Unfortunately, it is not the observer in the lab who sees the ceiling moving towards the floor; it is an observer outside the falling lab whowill see the ceiling falling towards where the floor used to be. So, the observer inside the lab will not observe the blue shift.

This is an instance of the confusion coming from the ambiguity of a rest point in a frame F1 which is also a moving point in a reference frame F2, whereas F2 is moving relative to F1. The problem which had haunted the special theory of relativity in the context of time dilation and length contraction came back to haunt the general theory of relativity here.

Without a proper frame analysis, we carry out mechanical calculation to get the wrong results. Doppler's effect is one thing and applying it to a wrong event analysis is the other. Moreover, this calculation ignores the fact that relativistic Doppler's effect does not apply to accelerating frames. There is no such thing as special theory of relativity dynamics on pain of creating a contradiction. Due to the faulty frame analysis as we discussed above, the equivalence principle is false.

Relativity theory is a theory of cross-reference-frame observations. Due to the difficulty of carrying out actual experiments which would require massive speed, virtually all "experiments" are thought experiments. This creates additional issues for the task of building up an ontological theory.

7.4. Center of Mass in the General Theory of Relativity

It is an elementary mathematical fact that when we represent a motion of constant speed in 3D Euclidean space as its graph, it becomes a straight line in the 4D spacetime. If the motion is accelerating, then the graph becomes a curved line in the 4D spacetime. Einstein represented the gravitational force field, which will cause accelerating motions, as the 4D spacetime manifold so that all motions caused by the gravitational field will appear as straight lines (geodesics) in the manifold. Applying this idea to gravitational fields created by a system of masses, Einstein obtained the so called gravitational field theory. In this theory the spatial distribution of masses determines the 4D spacetime manifold representing all possible "motions". According to the theory of gravitation, such masses move around due to the gravitational pull among them.

According to classical dynamics, when we have a system of masses in a space, they converge to the point called the "center of gravity". It would be interesting to learn how this process of convergence is represented in the 4D spacetime manifold.

7.5. General Coordinate System

According to Lawden, Einstein assumed that the whole cosmos is occupied by a fluid whose molecules are clocks of any variety. This fluid can flow in any manner except that it will be assumed that there is no turbulence, so that the neighboring molecules always have almost equal "speed" and the velocity of the flow is a continuous function.

Each clock is allocated three coordinates (x_1, x_2, x_3) in such a manner that (1) no two clocks will have the same coordinates, and (2) neighboring clocks have neighboring coordinates (thus, coordinates are also continuous with respect to spatial displacement). It is understood that the coordinates of each clock remain the same through time. As time elapses on each clock, its readings are assumed to increase but the rate of increase is not necessary uniform as compared with a local standard clock. Again almost no attempt is made to synchronize distant clocks, neighboring clocks are assumed to be sufficiently synchronized so that the clocks' readings are continuous with respect to spatial displacement.

These problems are of mathematical nature and physical nature. Considering that this kind of issues deal with the most fundamental concepts of theoretical physics, great care must be taken: one can build a castle only upon solid foundation. Here are some questions.

i. On the one hand it is assumed that the fluid of clocks can flow (in any manner), then it is said that the coordinates of each clock remain the same. There is a contradiction. If the coordinates of the clock do not move, the clock will not move, so the fluid of the clock will not flow.
ii. Upon what time will the fluid of clocks flow? Logically, this "flowing motion" of clocks takes place in some time external to the clocks. One cannot use the time of the clocks when one moves the clocks on pain of committing a vicious circle. Each clock is a physical system defined on some time.
iii. There are "at most" countably infinite (as many as natural numbers) clocks in the entire universe. Most empiricists will say that there are only finitely many clocks. So, no matter how closely we put clocks together, we cannot form a "continuum" of clocks. Continuum is much denser than countably infinite. We cannot fill the real number line with natural numbers.

SECTION 8: SOME OTHER FUNDAMENTAL ISSUES

8.1. Basic Logical Issues in Measurement

Aside from the impossibility of making measurements within a system as Newton and Goethe pointed out, a reason why the measurement of particle physics has been done in the foregoing manner is because particles are too small to be measured directly. We measure through calculation so that the formulas are verified. It is what logicians call a "vicious circle". The same thing happens in cosmology and astronomy too. Stars and planets are way too large to be measured directly; so, we use our formulas of dynamics to do the measurement. Vacuously circular formulas used to calculate the measurement satisfy the formulas.

In case of cosmology and astronomy, the distance between stars and planets is way too large for any conventional engineering method to measure. It is also impossible to measure their speed directly. So, they use the Doppler effect of light to measure the speed of stars. A problem thus arises. As discussed above, we do not fully understand light. The Doppler effect on light hence presents an additional problem. When the wavelength of the light wave changes, naturally the energy of the light wave changes. Then what is the buffer to maintain the conservation of energy in the Doppler effect phenomena?

It was Russell who pointed out that we use "common sense" to design an experiment in physics. So, for experimental verification of physical theories we use "common sense". The same story with experimental refutation. Russell called this the "virtue" of physics. It is quite clear that he meant something more serious. It must be the case that when we verify a theory through an experiment, we must use the theory we are to verify to design the experiment. This is circularity. A good example of this is what we do in astronomy. We assume the general theory of relativity which asserts that the mass distribution in space bends the space. So, to measure the distance between stars, they assume the general theory of relativity. Then they say, good, the general theory of relativity is empirically verified.

It also is the case that when we refute a theory, we use the theory to be refuted to design the experiment. This is a contradiction. Naturally, when we measure the mass of an object, we assume the law of gravity. This seems to be a fundamental question to empiricism in physics.

It was Newton, the father of physics, who questioned empiricism fundamentally and rejected it. He said the right thing that measurement is metaphysical! This is to say that for Newton, physics (which he called "natural philosophy") is not an empirical science. Newton was condemned by his contemporary empiricists as a reactionary authoritarian figure. So, where does contemporary physics stand regarding this issue of "empiricism versus metaphysics"? Curiously, it is either that we made no progress from the time of Aristotle or maybe it went even worse. What we have regarding this issue is what we call the "principle of opportunistic choice". In face of a logical contradiction in a theory, they just do some experiments and "verify" the theory by ignoring (dropping) the prediction which does not agree with the experiment. Aristotle opportunistically chose the experiment which fit into his "theory". Modern physics opportunistically chooses the theoretical prediction that agrees with the experiment from mutually contradicting predictions. They call it the empirical verification of their theory. What is more problematic is that they see nothing wrong with this. When challenged, physicists say that all of these questioning is nothing but "misunderstanding" by those who have no background.

Finally, a most disturbing abuse of experimental result to defend logically invalid theory appeared in the special theory of relativity. It is bread and butter of logic that by adding a new axiom, one cannot remove a prediction from the theory. This is a consequence of what logicians call the "monotonicity" of logic. Einstein thought otherwise. Before the Michelson-Morley experiment, everybody thought that the Galilean addition of speed was okay. This addition came from the basic axioms of the Galilean relativity theory. When the Michelson-Morley result came in, everybody interpreted it incorrectly (see previous discussion). So, Einstein added the constancy of the speed of light axiom to the Galilean relativity theory to produce the special theory of relativity. He clearly had little understanding of the monotonicity of logic. By adding the constancy of the speed of light to the Galilean relativity theory, he did not suppress the derivation of the Galilean addition of speed which "apparently" was refuted by Michelson-Morley experience. Instead, he made the special theory of relativity inconsistent as the constancy of the speed of light contradicts the Galilean addition of speed prediction. More acutely, as a logically inconsistent theory, the special theory of relativity proves the so-called "relativistic addition of speed" which Einstein thought was a victory of his special theory of relativity. In fact, this result contradicts the Galilean addition of speed which is a logical consequence of the special theory of relativity.

All of this is simply the consequence of ignoring the basic principles of logic upon which our precise reasoning is based. Without proper reasoning, we cannot carry out precise science. Physics cannot be above logic!

8.2. Classical Limit and Measurement

Originally the issue of classical limit appeared as the mathematical connection between the classical theories and modern theories of physics. For example under the limit $v \to 0$, the special theory of relativity is supposed to be reduced to the classical Newtonian theory. However, at this limit, nothing is moving, except the so-called "light". This is not kinematics. This commonly used classical limit is false. An alternative version is $c \to \infty$. As far as we know, this classical limit was first considered by Anderton [1] of Natural Philosophy Alliance. We are not quite sure if this is ontologically acceptable though it works mathematically. c, after all, is an electromagnetic "constant". Moreover, we must emphasize that the special theory of relativity is kinematics on pain of creating a contradiction. So, under these classical limits, the special theory of relativity is expected to be reduced to Newton's kinematics.

In case of quantum mechanics, it was Dirac who said that the classical limit for quantum mechanics is $h \to 0$. The validity of this claim is still debatable. Indeed, it was shown that when we take Dirac's classical limit, Schrödinger's wave equation is reduced to two equations, one is classical and the other is not.

Essentially, these classical limits are ontologically invalid as both c and h are physical constants. Mathematically a constant converges only to itself.

Notwithstanding, there are some more serious issues to be considered. (1) If the issue of the classical limit is that of connection between modern theories and classical theories, how about the measurement problems? If there is a material difference between classical theory and modern theory, why we are using classical apparatus for measurement in modern physics? (2) As quantum mechanics apparently discusses measurement in its formalism, it is natural to expect that the von Neumann formalism, the final theory, will provide predictions of the actual values to be observed. In reality, it is not the case as the von Neumann model predicts not actually measured values. It is compared not with predicted specific values but with predicted probabilistically expected values to be measured. In short, there is no connection between the actually measured value and the theoretical prediction. In essence, quantum mechanics does not discuss actually measured values. There is no formal system in quantum mechanics that can be used to calculate the value actually to be observed. Probability theorists know that there is a theorem which has a shocking empirical implication. It is called the "large number theorem". It says that the relative frequency sequence converges only at the limit. This means that unless we repeat the same experiment infinitely many times, we cannot confirm a probabilistic prediction. This simply means that in a probabilistic theory, there is no possibility of experimental verification or refutation of the theory.

This problem with the validity of empirical verification or refutation is not limited only to quantum mechanics. In a different sense, in Newton's classical dynamics, measurement cannot be done inside the absolute frame on pain of creating a contradiction. The issue was seconded by Goethe later. Unfortunately, the issue has not been attended to properly.

8.3. Problem with Empirically Refuting Probabilistic Theories

Popper pointed out that theorems (predictions) are universally quantified sentences and so to verify them experimentally requires infinitely many experiments. We need one experiment for each specific value of the variable. On this ground, he rightly concluded that there is no such thing as empirical verification of physical theories (predictions). But he concluded, purely logically, that by presenting a counter-example, we can refute a physical theory experimentally. So, according to him a theory cannot be experimentally verified but can be experimentally refuted. Kuhn pointed out that Popper's refutation argument is problematic because to present an experimental counter-example requires experiments of "infinitary nature".

It appears that Kuhn [15] was correct in the context of quantum mechanics. Quantum mechanics predicts the probability of an outcome of an experiment. To (verify or) refute such a prediction requires infinitely many repetitions of the same experiment as the large number theorem states. So, in the end, there is no such a thing as an experimental refutation of a theory either. For example when we toss a coin, the relative frequency of getting heads up will not converge to $1/2$. This is because of the large number theorem and the impossibility of repeating the "same" experiment.

Unlike quantum mechanics, statistics involved in the statistical mechanics does not support that the motion of particles is "under-defined". It is needed to deal with the situation where there are way too many particles moving in the limited space. So, the statistics is used to give expected value of the measurement of the particle system which will appear on the macro scale measurement instrument. There is a solid connection between the macro-measurement and the particle world. Each particle involved is interacting with the macro-measuring device. In case of quantum mechanics, we assume that the motion of each particle is nondeterministic and there is no known mechanism of interaction between the particle and the measuring device which is of macro nature. The possible values to be measured are given just mathematically, ignoring any physical reality, as eigenvalues of self-adjoint operators.

In the end, by the large number theorem again, statistical mechanics "should" fail to offer predictions that are empirically verifiable or refutable. In reality, in case of statistical mechanics, the number of experiments repeated in probabilistic experiments works as the number of experiments, which is massive, for each particle and so we get quite a reasonable match between the theoretical prediction and the actual measurement.

8.4. Mathematics, Physics and Philosophy

8.4.1. Copernicus, Galileo, Kepler and Newton on the Heliocentric Model.

It was Copernicus who proposed a hypothesis that planets are all orbiting around the sun, i.e., the heliocentric model of the planetary system. It however is questionable to call it an empirical process. It is more like an intelligent speculation. Considering that ancient Greeks had a totally different mathematical model of our planetary system, which predicts exactly the same thing as the Copernican model, does make us still wonder how our planetary system looks like when observed from outside our planetary system. Galileo apparently made no scientific contribution to this, he openly supported the Copernican view.

Kepler, a mathematician, presented an elliptic model instead of the circular model of Copernicus. For this model, he obtained the Kepler's law. In this law, he proposed that

- Every planet's orbit around the sun is an ellipse at one of whose focal points the sun is located. The sun's position is a focal point for all orbiting planets.
- For the same time interval, the area which each planet covers in the ellipse is the same.
- The total orbiting time for planet $\mathrm{p1}$ and planet $\mathrm{p2}$ have the ratio $a1^{3/2} : a2^{3/2}$ where $a1$ and $a2$ are the distance between two focal points for $\mathrm{p1}$ and $\mathrm{p2}$ respectively.

Newton, studying Kepler's law for a simpler case of circular orbiting, came up with his mechanics which consists in the first law, the second law, the third law and the law of gravitation. He attributed the centripetal force for circular motion to the gravitational force between the sun and the planet.

All of this strongly suggests that the development of Newton's mechanics is not an empirical process but a metaphysical process. How Newton's Orthodox theology contributed to the development of his theory of dynamics is yet to be studied.

8.4.2. Wave Functions and Wave Equations

The reason why Lorentz developed the Lorentz transformation is because the Galilean transformation applied to the electromagnetic wave equations did not yield an EM wave equation. He managed to show that the Lorentz transformation of an EM wave equation is an EM wave equation.

All of this is a mathematical fact and what it means to physics is a different issue. What is interesting is that the most basic mathematical representation of physical waves are not wave equations but sinusoidal functions over time and location. It is rather surprising that Lorentz did not check the consequence of applying the Galilean transformations to wave functions instead. The mathematical fact shows us that the Galilean transformations of wave functions are indeed wave functions.

What this means is yet to be studied. Is it not the case that wave equations are incomplete means of specifying physical waves? If so, then Lorentz's discovery of the variance of wave equations under the Galilean transformation means the inadequacy of using wave equations to describe waves, does it not?

Moreover, the alleged invariance of wave equations under the Lorentz transformation needs more explanation. Is the Lorentz transformation of a wave function a wave function? Apparently yes, as we have

$$k'_x = \frac{k_x - v\frac{\omega}{c^2}}{\sqrt{1-(v/c)^2}}, k'_y = k_y, k'_z = k_z, \omega' = \frac{\omega - \nu k_x}{\sqrt{1-(v/c)^2}}.$$

However, it is not clear at all why we have to choose Lorentz's transformed version over Galileo's transformed version. At best, that the Galilean transformation of wave functions are wave functions seems to suggest that the theory of the Lorentz transformation of wave functions is rather self-serving. That Lorentz transformation, which came from time dilation and length contraction, causes paradoxes (contradictions) seems to suggest that there are more fundamental things that have to be re-examined in regard to the theory of the Lorentz transformation.

8.4.3. Classical Limit in Physics and Mathematics

We have discussed the issues of classical limit in modern physics. The large number theorem in probability theory can be considered as the classical limit in mathematics that connects the finite world of relative frequency in statistics with the infinitary world of probability. It is an interesting thought.

8.4.4. Points and Continuum in Mathematics and Physics

In theoretical physics, following the first step taken by Newton, a physical body is reduced to a point mass whose size is that of a geometric point. This is a mathematical convention without which we cannot formulate motion as a mathematical function. When Newton developed the mechanics and calculus together, he was not fully aware of the complexity and depth of the calculus he developed for building mechanics. Late 19th century and early 20th century mathematicians revealed the complex world of real continuum as topology in their attempt to understand the world of calculus (real analysis). In the mid-20th century, using mathematical logic (model theory), Robinson made a replacement of topology by formulating his enigmatic concept of infinitesimals, which was originally developed by Newton. Now we have two mathematical solutions to the enigma of real continuum. The problem we still have is that the mathematization of physics in theoretical physics is still at the level of Newton's time, and theoretical physicists are not willing to upgrade their mathematical formulation of physics. We have already discussed this issue in Sections 6.1.1 and 7.5.

8.4.5. Are Mathematicians Innocent?

As discussed in Section 5.1.4, the composition of Lorentz transformations, which are linear transformations from 4D spacetime onto itself as per Minkowski, does not yield physically (ontologically) desirable transformations. The fact that a desirable composition has to be defined in terms of the Lorentz transformation of speed poses a serious concern on the mathematical integrity of the special theory of relativity. It appears that we need different kind of mathematics to develop a satisfactory mathematical theory of physics. This is the issue of the conflict between empirical reasoning and mathematics. This kind of conflicts split mathematicians working in theoretical physics and pure theoretical physicists.

Theoretical physicists did not acknowledge that there are serious issues to be resolved and went on to develop highly questionable theories which were to be defended not scientifically but "politically". In the end, they produced theories such as quantum mechanics which is totally refuted experimentally, or such as the relativity theory for which it is impossible to carry out any direct experiment. The only empirical "verifications" for this theory came from quantum electrodynamics which is empirically false. As we know, quantum mechanics predicts that there must be no trajectories and yet trajectories in quantum mechanical experiments are supposed to provide empirical verification of the special theory of relativity. After all, both of them are logically and mathematically inconsistent.

Here is one plausible explanation of why it appears that the "experiments" of quantum electrodynamics verify this theory so well. It indeed is curious that a theory that is fundamentally defeated empirically "verifies" that theory. The most fundamental issue here is that all experimental data produced have nothing to do with quantum electrodynamics as the theory predicts that there must be no such data which are all trajectories. But when we apply a formula of this theory, it fits well with the trajectories produced. A little bit of logical reasoning will give us the following explanation: When we apply a formula of this theory to experimental data, we are making the "measurement" using this formula through calculation. No wonder that electrodynamics is the most experimentally verified theory in the history of physics.

It is tragic that many mathematicians who worked in mathematical physics chose to leave the discipline instead of correcting the errors. They went back to mathematics as Woit wrote in his book "Not even Wrong!". To be fair to physicists, it must be stressed that mathematicians such as Minkowski, Hilbert, Noether went for the basic mathematical errors such as $0/0 = h\nu$. They went further to "justify" these basic mathematical errors by building "mathematical", "advanced" theories which are inevitably inconsistent. Minkowski's 4D spacetime theory, in particular, helped to "cover up" the fundamental ontological inconsistency of the special theory of relativity. This is a regretful systematic abuse of mathematics which never took place before.

Specifically, the blame should be heavily placed in the fallacies of "formalism" in mathematics as a "solution" to the inconsistency problem of the most important theory of mathematics, namely set theory. As Cantor's set theory, which Hilbert described as the "paradise" that Cantor gave to mathematics, turned out to be inconsistent, and nobody with a concern for the foundations of mathematics knew what to do, they chose the most irresponsible way of formalism to define set theory, hoping that it is consistent. The ZF axiomatic set theory was produced by this process of listing up whatever people wanted from set theory.

So, how is it relevant to the issue of 20th century theoretical physics? Those theories in 20th century theoretical physics also were produced through the same process of axiomatization. As it became clear, basically none of the axioms of relativity theory and quantum mechanics reflects any ontology. In fact they are in serious conflict with ontology. In physics our hands are more tied than in mathematics where we have more freedom to consider mathematical models of axiomatic theories. In physics, models of formal theories must be ontologically secured. We have seen that basically none of the 20th century theoretical physics has anything to do with ontology. Science and mathematics are logical disciplines and there is no room for speculation and creativity.

8.4.6. What about Philosophers?

It has been said that philosophers must be "wise". Considering the "achievement" of modern day philosophers of physics, it is unfortunate that we have had very few philosophers who did what they were supposed to do. If they had done their job, theoretical physics would have been in a much better position now. Where there is no critical discussion, there is no improvement.

History unfortunately shows that modern day philosophers of physics have been preoccupied with cashing in the "glory of theoretical physics".

All of this may well be connected to the last century's trend toward the politicization and big science orientation of academia through funding policy which is now recognized to be the cause of the current problems from which Western academia suffers. Another factor may well be the trend of excessive mathematization of scientific disciplines which started at Göttingen University under David Hilbert. Theoretical physics became applied modern mathematics promoted by the mathematicians who did not fully understand physics, which is an empirical science, and physicists who did not fully understand mathematics which, used as a fashion, is metaphysics. This confusion somehow got into the works of Dirac, Heisenberg and von Neumann in particular. With this excessive mathematization of physics, real philosophers of physics were phased out. All of this created the religion of "genius worshipping" in which philosophers of physics found their active role in its promotion. Some outstanding philosophers such as Russell, Popper and Kuhn gave material warnings which were ignored. Certainly, we have shown that there are many serious conceptual issues in theoretical physics. The mathematization initiated by the Hilbert school at Göttingen diverted people's attention away from the more important issues to which philosophers could have made some serious contribution. Kuhn's observation that science is a political process of consensus building certainly drove physics into the "witch hunt" of those who questioned the established theories. In this, interference from politics and business played major roles. Very few philosophers were able to counter the pressure. Many however played the role of consensus builders and witch hunters to advance their career.

To end this monograph of philosophical nature, let us sum up the role of some dedicated philosophers who tried to make a difference in the development of physics and yet have been unfortunately ignored. It was Karl Popper who first pointed out that any physical theory that is logically inconsistent must be thrown out categorically. This was the first warning from serious logicians to the community of theoretical physicists who grew up in an environment of Hegelian dialectic in which the pure nonsense that contradiction is the mother of new truth was taught. They were taught that by adding to inconsistent theories they could resolve the inconsistency and build a better theory. This we have witnessed precisely in Einstein's special theory of relativity in which Einstein added the constancy of the speed of light axiom to the Galilean relativity theory to resolve the conflict between the Galilean relativity theory and the assumption of the constancy of the speed of light. This is logical nonsense but it was a victory of Hegelian dialectic according to Einstein and his followers. The monotonicity of logic dictates that by adding more axioms, one cannot retract the consequences of the original logic. Popper made more contributions to physics. He pointed out that it is impossible to experimentally verify any theory because this would require infinitely many experiments. He also pointed out that the only thing we could do was to refute a theory experimentally by presenting an experimental result which contradicts the theory's prediction modulo "repeatability". It was Kuhn who pointed out that we cannot even refute a theory of physics when the theory is probabilistic because verifying a probabilistic prediction requires infinitely many repetitions of the same experiment. This argument came from the large number theorem of probability theory.

At a fundamental level, Russell nailed physics completely. He pointed out that the experimental verification of a physical theory is viciously circular because the experiment requires the theory to design the experiment. Moreover, he incisively pointed out that one cannot refute a physical theory by experiment as the experiment requires the theory, which is to be refuted, to design it. He said that this is a contradiction. His warning was completely ignored by the community of physicists who already reached their consensus.

Unfortunately by the time these totally legitimate warnings were issued, the process of building consensus in physics already took over.

8.4.7. The Process of Physics as an Empirical Science under Hegelian Dialectic

Logically and theoretically, we have shown that relativity theory is inconsistent. The concept of moving reference frames is untenable. This is because physical objects (mass, charge etc.) and physical space are entirely different categories. Physical objects can take up space and move in the absolute frame. But no subspace of a physical space, including a single geometric point, can move in any geometric space. We can locate ourselves in the absolute space and we can move inside the absolute space. However as Newton made it clear, we cannot observe the dynamics which takes place in the absolute space. Only when we "stand on the shoulder of God" outside the absolute space, can we observe the dynamics taking place in the absolute space. But as we are "out of the universe", we must use metaphysical means to observe from outside. This prevents the problem, as per Goethe, of our measurements interfering in the dynamics taking place in the absolute space.

All of this presents a challenge to yet another major element of physics, which is empiricism. A natural consequence of the discussion above is that we cannot measure the reality happening in the absolute space from our "fatally distorted view". This problem has deeper implication than the yet another political "criticism" of relativity theory among the communities of dissidents such as the Natural Philosophy Alliance (NPA), which is based upon the disagreement on the empirical results they have observed. Nevertheless, both of them are empiricists. After all, we have shown that our "measurement" is distorted and therefore leads us to a wrong picture. This distortion is not just numerical as is the case with the "uncertainty principle" in quantum physics. This is a structural distortion from which we cannot escape as long as we are human beings living in this universe. This has been a perpetual criticism on the so-called "physical science" from the view of metaphysics which, under the dominance of Masonic belief system, has been politically brushed away as "reactionary" nonsense.

Historically, this issue of the material versus the metaphysical has been a theme for perpetual discussions. The unbiased reality against empiricism is twofold: An empirical conclusion is always approximate. No matter how many times it has been "verified" through experiments, it never means that it is true. This was pointed out for the first time by Karl Popper. He categorically stated that there is no such a thing as an empirical verification of a theory. The Galilean addition of speed was common sense up until the ill-fated Michelson-Morley experiment came in. To be fair to Newton, this classical addition of speed came from Newton. It should apply correctly only in the absolute frame metaphysically observed "from the shoulder of God". Why? Because, as we have discussed above many times, the Galilean theory of relativity is false. It leads to contradiction.

Anyhow, this one incident turned the entire world of physics upside down. As we and Engelhardt have shown, it turned out to be that this Michelson-Morley experiment does not imply the constancy of the speed of light which became the empirical foundation for the development of Einstein's relativity theory.

Philosophically, it is more interesting. It was not the error in the experiment of Michelson-Morley, but the error in understanding their experiment that is philosophically significant. This is a severe warning to the dogmatic process conducted within the physics community which places empiricism as the most important factor for developing physics. We have shown that the experiment was correct but the interpretation was wrong. We want to stress, at this point, that mathematics is nothing but metaphysics upon which physics is built. When mathematicians say that the theory of physics which used mathematics is mathematically false, it is false. Definitely, the Einstein-Planck equation

$$e = \sqrt{(cp)^2 + (m_0)^2c^4} = cp = \frac{m_0 vc}{\sqrt{1 - \left(\frac{v}{c}\right)^2}} = \frac{0}{0}c^2 = c^2hf$$

is false. Ironically, this criticism goes beyond mainstream physics. Many prominent NPA members, such as Professor H. Thim of Linz University, claim that $e = mc^2$ is derivable from classical electromagnetic theory and object to the usage of Einstein's relativity theory. So, for them, it is all but an anti-Einstein campaign. As many of them also accept Planck's $e = hf$ for photon, in the end their theory also will accept the above equation which is false. The judgement that we have arrived at is now simple, the entire modern physics is false. In the world of mathematics, there is zero tolerance to any contradiction. What makes theoretical physicists think that they are above that?

Regarding the NPA, one point is interesting. Their argument avoids Einstein's relativity theory. They derive the same conclusion $e = mc^2$ using purely classical electromagnetic theory. Neither of them understands the second equality. This may imply that the source of all of this confusion came from the electromagnetic theory. What is most urgently needed is a serious investigation into the classical electromagnetic theory. It is unfortunate that because of the trend set by modern physics, instead of understanding the source of the problems we encounter, we tend to jump hastily into fanciful new "theories". We have already had more than enough thoughtless alternative theory proposals from the critics. Mainstream physicists ignore the fatal problems of their own theories. The critics, without really nailing the problems, jump into alternative theories. There seems to be no serious attempt to correct the errors we have made in the last 100 years or so.

As we have discussed previously, the above equation also has some alarming issues. e is energy which is a modal concept and p is momentum which is a physical concept, thus not modal. Philosophers have some serious opposition to this equation which connects modality and non-modality in this way. Einstein/Planck's "genius" will not make this violation of reasoning work. Only under the political process over science, this kind of irregularity has been forced upon us.

Finally, why all of this questionable activity is going on in physics? We find the roots in the so-called "Hegelian dialectic". This philosophy came from the ancient Buddhist teaching, Hannya Shingyo, which taught that beyond apparent contradiction, we will see the ultimate truth. This is a very deep teaching which was mistaken by Western physicists. The true meaning of this one of the deepest teachings of ancient Buddhism is that when we project the higher truth we always end up with contradiction.

Interestingly, this deep philosophy has been confirmed totally mathematically by the "adjoint functor theorem" of category theory which was denounced as "general abstract nonsense". This theorem in the most general setting shows that when we project we lose some vital information at the most general level. It is considered to be one of the deepest results in mathematics done in the most general setting. This is a deadly serious and precise argument that has nothing to do with the fancy "philosophy" of mixing the East and West from the Western perspective. One must understand that fantasy upon contradictions will not do. We have seen a catastrophic failure of this Western philosophy in the development of modern physics. It is quite clear that for some prominent mathematicians of the last century, most abstract mathematics such as category theory is nonsense. Certainly some theories like ZF which is an abstract formal theory of sets are nonsense. Even after 100 years, we still can-not show what it means. There is no known model for it as we do not know if the theory is consistent or not. However, just like abstract algebra, category theory is an abstraction of existing mathematical structures. It didn't come from "creative imagination". It is the extraction of the most fundamental and common features shared by many mathematical structures. Strikingly, the mathematics is now synonymous to solving some open problems in number theory as theoretical computer science is reduced to just concrete complexity theory.

REFERENCES

Anderton R., *Private Communication*, 2007.

Miller N., Ashby S. C., *Principles of Modern Physics*, Holden Day, 1970.

Bacon F., *Novum Organum* [*New organ*], 1620.

Berkeley, G. A., *Treatise Concerning the Principles of Human Knowledge*, Part I, 1710.

Dingle H, *Science at the Crossroads*. London: Martin Brian & O'Keeffe. 1972.

De Broglie L., "*Recherches sur la théorie des quanta*" [*Researches on the quantum theory*], Thesis, Paris, 1924, Ann. de Physique (10), 1925.

Eilenberg S., MacLane S., "*General theory of natural equivalences*", Trans. Am Math Society, 1945.

Einstein A., "Zur Elektrodynamik bewegter Körper " [Electrodynamics of moving bodies], *Annalen der Physik*, 332 (10), 1905.

Engelhardt W. W., "Phase and Frequency Shift in a Michelson Interferometer", *Physics Essays* 27 (4), 2014.

Heisenberg W., *The Physical Principles of the Quantum Theory*. Translators: Eckart, C., Hoyt, F.C., Dover, 1930.

Bennett J., *Locke, Berkeley, Hume: Central Themes*, Oxford University Press, 1971.

"*Calligraphy: Ancient India and Japanese Sutras, Textiles: Banner and Banner Legs – The Gallery of Hōryū-ji Treasure*". Tokyo National Museum. Retrieved 2017-11-03. "*Calligraphy: Heart Sutra and Usnisa Vijaya Dharani Sutra in Sanskrit, Late Gupta dynasty, 7th–8th century* (Important Cultural Property)."

Kanda A, Apostoli P, "Inconsistency of Some Formal Theories", presented at *the Meeting of Exact Philosphers,* 2007.

Kanda A, Apostoli P, "Logicism v.s Empericism", *International Seminar on the Structure in Cosmology and Theoretical Physics,* Finish Physics Foundation Society and Finish Society for Natural Philosophy, Helsinki, 2010.

Kuhn T. S., *The Structure of Scientific Revolutions*, The University of Chicago Press, Chicago, 1962.

Lawden D. F., *Elements of Relativity Theory*, John Wiley & Sons, 1985.

Lorentz H. A., "Electromagnetic phenomena in a system moving with any velocity smaller than that of light", *Proceedings of the Royal Netherlands Academy of Arts and Sciences,* 6: 809–831, 1904.

MacLane S., *Categories for the Working Mathematician, Graduate Texts in Maths*. 5 (2nd Edition) Springer-Verlag.

Maxwell J. C., "A Dynamical Theory of the Electromagnetic Field", *Philosophical Transactions of the Royal Society of London* 155, 459–512, 1865.

Minkowski H., "Das Relativitätsprinzip" [The relativity principle]. *Annalen der Physik.* 352 (15): 927–938, 1907.

Robinson A., *Non-standard Analysis*, Princeton Landmarks in Mathematics, Princeton University Press, 1966.

Schrödinger E., "An Undulatory Theory of the Mechanics of Atoms and Molecules". *Physical Review.* 28 (6) 1926.

Suntola T., *Dynamic Universe*, Physics Foundations Society, 2011.

Sells R. T., Weidner R. L., *Elementary Modern Physics*, Allyn and Bacon, Boston, 1973.

Woit P., *Not Even Wrong*, Basic Books, New York, 2006.

In: Old Problems and New Horizons in World Physics ISBN: 978-1-53615-430-6
Editors: Volodymyr Krasnoholovets et al.

Chapter 9

KARL POPPER AND EMPIRICISM: RESPONSE TO AKIRA KANDA ET AL.'S PAPER

Robert Neil Boyd*

Princeton Biotechnology Corporation, Barrington, NJ, US

ABSTRACT

In response to Kanda's criticism toward empiricism, Boyd argues that Popper did not assault empiricism. He supported and codified it in his Criteria for Reliability in the Sciences. Empirical science is based on reproducibly observable facts, as opposed to theoretical "science" which is based on sets of believable beliefs. Popper's Criteria are about the falsifiability of a given theory.

INTRODUCTION

Sir Karl R. Popper's philosophy of science illustrates the methodological view espoused by General Semantics as a discipline, a very useful discipline, which I have previously attempted to bring to your attention.

We now examine some of the key features of Popper's [1, 2] philosophy and show how major paradigm shifts in scientific beliefs over the millennia, conform to the system described by Popper and this will demonstrate the response to Kanda's (see, e.g., Refs. [3-5] and other works) criticism toward empiricism.

* Author's Email: rnboydphd@comcast.net.

METHODS OF REASONING

Almost everyone is familiar with the classical method of reasoning know as modus ponens. The well known example goes as follows:

> If Socrates is a man, then Socrates is mortal.
> Socrates is a man.
> Therefore, Socrates is mortal.

Few know that the progress of science no longer depends primarily upon this method, but rather, of late, on the less familiar form known as modus tolens, which goes like this:

> If Socrates is a god, then Socrates is immortal.
> Socrates is not immortal.
> Therefore, Socrates is not a god.

Karl Popper's philosophy of science uses modus tolens as the central method of disconfirming, or falsifying, scientific hypotheses. Scientists start with a current scientific theory and use the usual methods of deductive reasoning to derive specific conclusions, of which some are "predictions".

Strictly deductive reasoning is "truth preserving", that is, it is such that, if one starts out with "true" premises, one can only deduce "true" conclusions. Starting with a "theory" and deducing "predictions" can then be stated in the form of a premise:

> If the theory is true, then the prediction is true.

Popper shows that we cannot prove that a theory is true, but we can certainly show that a prediction is false. If the scientist tests one of these predictions and finds out that it is not true, he uses good 'ole modus tolens to conclude that the theory cannot be true.

> If the theory is true, then the prediction is true.
> The prediction is not true.
> Therefore, the theory is not true.

Of course, there is more to Popper's philosophy of science than this. We must consider what he means by a "scientific" theory, and by empirical, and how this method can apply to "parts" of theories as well as to "whole" theories, etc.

The key feature of Popper's perspective. as exemplified by the modus tolens argument is "critical testing". In order for critical testing to give valid results, the theory to be tested must be free from any "looseness"; Popper lists four criteria, or levels of evaluating, for determining whether a proposed theory is sufficiently "tight" to be admitted as a "scientific" theory.

We may, if we like, distinguish four different lines along which the testing of a theory could be carried out. First, there is the logical comparison of the conclusions among themselves, by which the internal consistency of the logical system is tested. Secondly, there is investigation of the logical form of the theory, with the object of determining whether it has the character of an empirical or scientific theory, or whether it is, for example, tautological. Thirdly, there is comparison with other theories, with the aim of determining whether the theory under examination would constitute an actual scientific advance, should it survive all our various tests. Finally, the theory is tested by way of physical and empirical applications of the conclusions which can be derived from the theory

Before looking at these four criteria let us see what Popper means, in general, by a theory. For Popper, a theory consists of a set of statements. Some of these statements are dependent upon other ones, and some are independent of other ones.

The statements can be classified according to "levels"; one statement is at a lower level if it can be derived from statements at a higher level. We will call the lower level statements, which depend upon, or can be deduced from, one or more of the higher level statements, dependent statements.

Higher level statements have the character of a "hypothesis" in relation to dependent statements at a lower level. Statements which are at the lowest level are called "basic statements", and correspond to specific singular facts (including "initial conditions") or predictions. Statements which are at the highest levels (universal statements), are related to the other statements in much the same manner that axioms are related to theorems in geometry. Statements at higher levels are typically used to "explain" dependent, lower level statements.

We have two different kinds of statements, both of which are necessary ingredients of a complete causal explanation. They are (1) universal statements, i.e., hypotheses of the character of natural laws, and (2) singular statements, which apply to the specific event in question, which we shall call "initial conditions". It is from universal statements in conjunction with initial conditions that we deduce the singular statement, which is, in other words, a specific or singular prediction.

The initial conditions describe what is usually called the "cause" of the event in question. The prediction describes what is usually called the "effect".

To give a causal explanation of an event, means to deduce a statement which describes it, using as premises of the deduction, one or more universal laws, together with certain singular statements, which we shall term the initial conditions

The first criterion Popper requires for a set of statements to be admitted as a theory, is that it must be internally consistent from a formal, and logical point of view. In order to meet this criterion the following conditions must be satisfied: (1.) The set of "axiom" statements must be independent and not contradict one another. (2.) There must be no dependent statements which contradict other dependent statements. (3.) None of the axiom statements are allowed have any manner of "built-in" inherent contradiction (i.e., self-contradictory). When a theory satisfies these conditions, then all of the "basic statements" of the theory can be deduced, in the strictly logical sense, from the axiom statements. This criterion is necessary to permit "falsification" to be extended to higher level theories.

By insuring that these criterion is met, we guarantee that: If a higher level statement is true, then an immediately lower level, dependent, statement is true. (Let us call this "Premise 1")

We can show what this means by two steps of modus tolens. Suppose we have a simple theory with one axiom, one "middle level" statement which, for this example, we will call the "hypothesis", and only one basic statement, the "prediction". Since Premise 1 is true for this theory, we can show relationships among the statements of this theory:

P1. If the axiom is true then the hypothesis is true.
P2. If the hypothesis is true then the prediction is true.

Suppose the prediction turns out false. Then we would say:

P3. The prediction is not true.

Argument 1

If the hypothesis is true then the prediction is true. (P2)
The prediction is not true. (P3)
Therefore, the hypothesis is not true. (modus tolens)

Argument 2

If the axiom is true then the hypothesis is true. (P1)
The hypothesis is not true. (from argument 1)
Therefore, the axiom is not true. (modus tolens)

Because the first criterion was satisfied in this simple theory, a false prediction was given, which when "carried through" proves the axiom of the theory false. If this first criterion had not been satisfied, this technique of using modus tolens could not have been used, and we would not know how a false prediction affected the theory as a whole.

In more complex theories, a false prediction might show only that a combination of axioms is inconsistent in regard to their consequences, but not which of the axioms is the one which caused the trouble. Because of this, a false prediction may cause the whole theory to be falsified, or only a part of it.

A theoretical system may be said to be axiomatized if a set of statements, the axioms, has been formulated which satisfies the following four fundamental requirements. (1.) the system of axioms must be free from contradiction (whether self-contradiction or mutual contradiction). This is equivalent to the demand that not every arbitrarily chosen statement is deducible from it. (2.) The system must be independent, i.e., it must not contain any axiom deducible from the remaining axioms. (In other words, a statement is to be called an axiom only if it is not deducible within the rest of the system.) These two conditions concern the axiom system as such. As regards the relation of the axiom system to the bulk of the theory, the axioms should be (3.) sufficient for the deduction of all statements belonging to the theory which is to be axiomatized, and (4) necessary, for the same purpose; which means that they should contain no superfluous assumptions.

In a theory thus axiomatized it is possible to investigate the mutual dependence of various parts of the system. For example, we may investigate whether a certain part of the theory is derivable from some part of the axioms. Investigations of this kind have an important bearing on the problem of falsifiability. They make it clear why the falsification of a logically deduced statement may sometimes not affect the whole system but only some part of it, which may then be regarded as falsified.

A second criterion concerns the logical form of the theory. Popper admits only theories capable of being tested by experience. If the form of a theory is such that its basic statements simply don't correspond to experience, or are otherwise not testable, then that theory does not qualify as empirical scientific. It may be some other kind of theory, but it is definitely not to be considered scientific. For a theory to be considered scientific, it must be testable.

The task of formulating an acceptable definition of the idea of a reproducibly observable "empirical science" is not without its difficulties. Some of the difficulties arise from the fact that there must be many theoretical systems with a logical structure very similar to the one which, at any particular time, is the accepted system of empirical science.

This situation is sometimes described by saying that there is a great number – presumably an infinite number – of "logically possible worlds". Yet the system called "empirical science" is intended to represent only one world: The "real world" or the "world of our direct experiences".

In order to make this understanding a little more precise, we may distinguish three requirements which our empirical theoretical system will have to satisfy. First, it must be synthetic, so that it may represent a non-contradictory, possible world. Secondly, it must satisfy the criterion of demarcation, i.e., it must not be metaphysical (fictional), but must represent a world of possible actual experience(s). Thirdly, The system under examination must be distinguished in some way from other such systems, as the one which actually best represents our world of direct experiences.

The third criterion concerns the comparison of one theory with another, and can be considered a decision process in selecting among theories which satisfy the other criteria. This is chiefly appropriate when one considers adding a new "axiom" to a currently accepted theory. Conditions which apply to this criterion include whether the proposed revision has fewer axioms (Stuart refers to this condition, which has been known through the ages as Occam's Razor, as "Popper's Chopper"), whether the proposed revision produces more and different basic statements (predictions), and, which is particularly significant in selecting from among competing theories, whether the new axiom produces basic statements (predictions) which contradict basic statements produced without that new axiom.

The fourth criterion again concerns testability, but of the empirical applications of the basic statements. If the theory is to be "about reality", its various statements must be directly correlated with the world of our direct observations. The "axiom" statements would need to correspond to "laws of nature", and the "basic statements" need to correspond to individual facts or predictions. Acceptance (tentatively) of a theory requires that it be highly "corroborated" by actual direct experience (empirical).

But how is the system that represents our world of experience to be distinguished? The answer is: By the fact that it has been submitted to various tests, and it has stood up to those tests. This means that it is distinguished by applying to it that deductive method of logic which it is my aim to analyze, and to describe.

"Direct Experience", in this view, appears as a distinctive method whereby one theoretical system may be distinguished from others; so that empirical science (physical evidence) seems to be characterized not only by its logical form but, in addition, by its distinctive method.

Several factors go to make up this criterion. No basic statement (fact or prediction) may be contradicted by experience. Some basic statements must have been tested with successful results many times. Not all basic statements must have been tested; they must simply be capable of being tested. Testability also means repeatability.

Popper says: "Indeed the scientifically significant physical effect may be defined as that which can be regularly reproduced by anyone who carries out the appropriate experiment in the way prescribed.

Additionally, Popper claims we must describe the rules of the method of science. What he means by that, is an accurate description of the context in which the given "science" occurs.

For example: Methodological rules are regarded as conventions. They might then be described as being the rules of the game of empirical science. Two simple examples of methodological rules may be given:

1. The game of science is, in principle, without end. . . . .
2. Once a hypothesis has been proposed and tested, and has proved its mettle, it may not be allowed to drop out without "good reason". A "good reason" may be, for instance, replacement of the given hypothesis by another, which presents superior testability, or by the falsification of one of the consequences of the hypothesis.

What happens if this methodology is used in conjunction with Popper's criteria, in particular, to the body of "scientific" knowledge, as a function of time? At any given time, there will be the "current (accepted) theory" for that time. As new hypotheses or "laws of nature" are proposed and tested, some will be, from time to time, refuted or falsified.

Along with the current theory, educators typically teach some of those theories which have been "shown to be false". The common view is that as we eliminate falsity, what we have left is closer to the truth. In other words, along with the current given theory, we have a list of previous theories or hypotheses, which have been discarded, which should never be entertained again, as they are inferior, or wrong.

It is as much of a mistake to think that these "refuted" hypotheses are "false" as it is to think that the "accepted" theory is "true".

Popper himself was aware of this: "...but those who uphold it dogmatically [a system] . . . are adopting the very reverse of that critical attitude which in my view is the proper one for the scientist. In point of fact, no conclusive disproof of a theory can ever be produced; . . . If you insist on strict proof (or strict disproof) in the empirical sciences, you will never benefit from experience, and never learn from it how wrong you are."

Wherein lies the flaw in our thinking? Different theories may equally account for the facts. If we were starting out with two theories to compare, our criteria would provide a means to select between them. What about when a sub-theory, or part of a theory has been "refuted", and we are seeking to find the best candidate among alternative sub-theories? In such a case, the selection is relative to the overarching part of the theory which was not in question. A single hypothesis could be disconfirmed within a sub-theory, but, a change to the main theory might prevent the falsification of the whole!

For example, a long time ago in a far away land, it was believed that Earth was the center of the universe. That was the accepted theory of the day. When Galileo's theory, the hypothesis that Earth was the center of the universe was falsified; the Sun then held that distinction. Newton's theory did not falsify this, but made it less tenable.

What has happened now that we have "relativistic" theories? One basic characteristic of general relativity is that the natural laws of the universe look the same to all observers, wherever they may be. There is no way to tell that any place is not the center of the universe!

Naturally, that Earth is the center of the universe is no longer false. Of course, neither is it verified. So, the hypothesis that Earth is the center of the universe was held to be true in early science, was "falsified" in later science, and is now no longer falsified. (Of course, this view is not held by many either.) As a practical matter, Earth is chosen as the center of the spatial coordinate system for many purposes.

In addition to this theoretical consideration, there is the matter of what Popper calls "inter-subjectivity". Inter-subjectivity operates at the level of our holding up some material object and agreeing to call it a "stone".

Such inter-subjectivity provides room for tacit and concealed assumptions about what the nature of the world is like. In many cases, the progress of science has required rethinking the most simple and thoroughly accepted so-called "basic facts".

Any significant revision of the expression of any of these "basic facts" will change the primary facts and must be accounted for by a valid theory. Such a change in basic facts could result in a previously disconfirmed hypothesis becoming, under the revised theory, no longer false. In such circumstances "falsity" seems to be no more "find-out able" than "truth".

However, time results in inevitable changes and expanded awarenesses and expanded comprehensions of our reality. Thus, we endeavor to depend only on what is reproducibly experiencable (empirical), in the moment, rather than to make what may be delicate or wrong theories.

REFERENCES

[1] K. R. Popper, *The Logic of Scientific Discovery*, Harper & Row, New York (1968); Routledge Classics, London and New York (2002).

[2] K. R. Popper *Of Clocks and Clouds*, St. Louis, Washington University (1966).

[3] A. Kanda and P. Apostoli, Logicism v.s Empericism, *International Seminar on the Structure in Cosmology and Theoretical physics*, Finish Physics Foundation Society and Finish Society for Natural Philosophy, Helsinki, 2010.

[4] A. Kanda, M. Prunescu, R. Wong and A. Biehl, The ghost of modality in theoretical physics, Chapter 7 in this book *Old Problems and New Horizons in World Physics*, Nova Science Publishers, New York (2019).

[5] A. Kanda, M. Prunescu, and R. Wong, Role of empiricism in physics, Chapter 8 in this book *Old Problems and New Horizons in World Physics*, Nova Science Publishers, New York (2019).

In: Old Problems and New Horizons in World Physics
Editors: Volodymyr Krasnoholovets et al.
ISBN: 978-1-53615-430-6

Chapter 10

EMPIRICAL SCIENCE: BACK TO REALITY!

Robert Neil Boyd*

Princeton Biotechnology Corporation, Barrington, NJ, US

ABSTRACT

The author argues that Science is based on the principle of Cause and Effect. We observe an effect. (Something happens, or some event is observed). Then we got a fundamental question: *What caused it?* Then, on finding out, we want to know, "What caused that?" going from cause to cause, looking for the actual origination of things. Cause and effect. This principle is used and applied by all Life and all forms of Consciousness, continuously. Knowing the cause of a thing or event is fundamental to existence. Nothing can live without applying this principle of cause and effect, at every opportunity. So learning about cause and effect is a natural condition, and is crucial to all Life. From that reasoning, the author criticizes some basic established theories, such as gravitation, ether, relativity theory etc.

INTRODUCTION

Science is based on the principle of Cause and Effect. We observe an effect. (Something happens, or some event is observed). *What caused it?* Then, on finding out, we want to know, "What caused that?" going from cause to cause, looking for the actual origination of things. This succession of asking the same question, at every point in the sequence, is a process which is apparently without end. (Though there may be some end-points reached eventually, in some regards).

* Author's Email: rnboydphd@comcast.net.

"What caused it?" is the *primary and most fundamental question in science.* What causes that observable event to happen? In fact, this question is primary to existence and to life experience. We are always asking questions in our lives, such as: Where did that come from? Why did that happen? What started it? Who started it? Why did they do that? What was the origination of that (event)?

These are all varieties of the same question: What caused it, actually? Because we know that until we find out what is the cause of some experience, the origination of it, we may be experiencing the same painful experience over and over again, until we learn to avoid the results of that cause, or cease performing that particular action. Or we may be missing the same pleasant experience, time after time, until we learn how to get it to happen again.

Cause and effect. This principle is used and applied by all Life and all forms of Consciousness, continuously. Knowing the cause of a thing or event is fundamental to existence. Nothing can live without applying this principle of cause and effect, at every opportunity. So learning about cause and effect is a natural condition, and is crucial to all Life. So, doing "science" (in terms of cause and effect) is a natural behaviour.

The proverbial tale of Newton and the apple that bonked him on the head (it never happened, really) leading to Newton's discovery of the Law of Gravitational Attraction, is an example of how exploring and examining our experiences in order to find what is the origin of the given observable event, results in reliable understandings about how Reality *actually works.*

PREMISES AND EMPIRICAL FINDINGS

Given a known cause, we can observe that a known effect is always resulting from that same cause. What makes something a "known"? "Known" means that it is a reproducibly observable fact. For example, "Every time I do this, that happens." Why is that?

For example, to state that the sky is blue, is a reproducibly observable fact. Moreover, this particular fact is the same for all observers who possess the same instrumentation, e.g., eyes. For an individual to state that the sky is some other color means either that they are imagining this, or that their instrumentation is faulty (e.g., color blindness). *Imaginings are not acceptable as scientific premises.* Flawed instruments lead to flawed measurements, which lead to flawed conclusions, relative to properly functioning instruments.

Now let's speak about facts. What is a fact? It is that which is *empirical.* It is that which is reproducibly experienced or observed, under exactly the same set of conditions, using exactly the same instruments in exactly the same way. For example, every time I look at the cloudless sky above me while I am on this planet Earth, I see that the sky is blue. That is observably true, unless the sun is in the process of setting, or unless the sun has gone down below the horizon, in which circumstances the observed colors are other than blue.

These *"unless"* conditions are also crucial. Why does the color of the sky change when the sun is near the horizon? And why does the color of the sky become progressively (predominantly) more black as the sun passes below the horizon?

So such *observations* lead to more questions. More questions lead to more explorations, through direct experiences and direct observations. More explorations lead to more discoveries and discoveries eventually add up to become realizations. And realizations lead to *reliable* cause and effect relationships (Laws), given numerous repetitions of the same sets of circumstances.

Let us keep in mind that the "known" is always based on the past. The "known" is what has already been experienced at some time in the past. The "known" then becomes a body of "knowledge" based on the past. However, *direct experience always exceeds the boundaries of "the known,"* since observations and experiences occur *only in the moment,* through the vehicles of the various senses and emotional sensitivities (which are best viewed as very reliable instruments).

Some "knowledge" is reliable as reproducible observations, such as the fact that the cloudless sky is blue when the sun is well above the horizon. Such observations may be considered "absolutes," being almost inevitably reliable.

Most knowledge is to some degree, provisional, or it may be *completely provisional.* This means that the given *concept* may be to some degree reliable, but it is not as reliable as a *fact.* So we put the concept in the "holding area" until it has received such sufficient and reliable confirmation so as to be accepted as a fact.

What this means is that, to all appearances, this or that might be going on, and this or that may be causing it. But we can't be absolutely certain yet that these particular observations, under these particular circumstances, will always result in that same effect, or that the same apparent event always has exactly the same sort of cause. So we put it in the "pending" status, as a provisional concept which is awaiting further observational evidence and further correlations and further corroborations.

An unfortunate habit that human beings develop is the tendency to categorize everything, and put everything into little boxes which comprise "the known" for them. What this leads to is another Whap! "Now where the heck did that telephone pole come from? It's not supposed to be here!?" As such categorizations tend to remove people from observations of the actual facts, in the moment.

Some of what is contemporaneously considered "knowledge" is not at all knowledge, but rather it is a set of *interpretations* of the observations. Interpretations are vast in variety and number, and cannot be relied on as being at all factual. Interpretations of the given facts often lead to dismal failures and obviously wrong conclusions, when such conclusions, being based on *interpretations,* are considered after the events have transpired. ("There's another one of those blasted telephone poles again, doggone it! Now how did that happen? Oh, I see. I was thinking about such and so and not watching the road.") This faculty is often called "hind sight."

Interpretations are almost completely unreliable, but it is still tempting to ask, "When this happens, that happens. *What could this mean?"* There in that instant of interpretation, we have gone into lands of fantasy and imaginings, where anything is possible, and where there are no rules, nor are there any requirements for any factual correlations or substantiations, when ones attention is purely preoccupied in fantasy-land.

Clearly, interpretations lead to infinite numbers of imaginings, fantasies, and flights of fancy, having nothing at all to do with the reproducibly observable facts. So, interpretations most often may be viewed as fantastical and imaginary whimsies, not even being classifiable as "provisional knowledge." *Interpretations often lead to beliefs.*

What is a belief? *A belief is an unsubstantiated imagining which is often treated as though it were a fact.* Needless to say, this sort of behavior leads to all sorts of problems, when confronted with the Real World of Actual Facts. "(Ouch! Hey! There's not supposed to be a telephone pole here! Where the heck did *that* come from!?!") Beliefs and opinions do nothing to alter the reproducibly observable facts, as our friend here found out.

Now, the word "belief" is synonymous with the word *"theory."* In other words, a theory is a sort of belief, or an opinion, or an interpretation of the facts. *A theory is never a fact,* and is never to be treated as such! Yet there are many who want to treat theories as though they were facts, when "theories" do not even represent a kind of "provisional knowledge." Theories are *not reliable,* and are certainly not as reliable as reproducibly observable facts!

The "scientific" blunders which have resulted from theoretical (belief) interpretations of the actual facts are countless and pitiful to behold! Such theoretically-based behaviors have led to many pains and difficulties in our world, just like our friend above who ran into the "unexpected" telephone pole. Well, if our friend had been actually watching the road instead of being preoccupied in the "Land of Oz" of his theoretical imaginings, he would have *seen* that telephone pole, and effortlessly avoided it.

Awareness (especially of the senses and sensitivities) and attention, are the keys to Actual Factual Reality, which is always and inevitably composed of reproducibly observable and reliable *facts.* (Think your senses are not reliable? Try driving a car for an hour without them, and see how far you get!)

A *Law* is a reliable physical relationship or a reliable interaction. (A theory is a belief). What causes such behaviors and such facts is often the source of conjecture, imaginings, and countless useless fantasy theories. The proper approach is to find out, *what causes that Law to be so?* This means that the processes of exploration and experiment and discovery must of necessity be endless. And it is good that it were so, else we would become bored. So we have suppositions, conjectures, imaginings, fantasies, beliefs, theories, and prejudiced fanaticisms. None of these conditions can be treated as reliable, nor as Laws. And regardless, exploration is an endless and wonderful process.

Scientific history is replete with instances where "anomalies" arise, which appear as exceptions to previously held convictions. History has shown that it is inevitably the case, that when "anomalies" appear, it is the "science" itself that is the *actual anomaly,* not the observed event. The unexpected appearances of "anomalies" are symptomatic of incomplete, or just plain wrong (!), scientific conclusions.

Actual Reality does not follow dogmas or beliefs or theories. The Universe does what it does, and is how it is, and it is up to us to make sense of it in some sort of reliable way. And beliefs and theories just don't cut it, by Actual Reality standards.

Here is one example of what has happened in quantum physics, due to various interpretations of the observable facts: It is commonly held that Heisenberg's Uncertainty Relation (a theoretical interpretation of the observable facts) is a factual relation, tantamount to a physical law. In actual fact, replicable experiments have demonstrated reliable evidence that Heisenberg's Uncertainty Relation does not hold under all circumstances. This means that this purely theoretical relationship has, at best, a very limited domain of usefulness and applicability.

Hans Dehmelt was awarded the Nobel Prize for demonstrating that the "Uncertainty Relation" does not hold for a single electron held exactly in one spot by Penning and Paul traps. Dehmelt was able to observe and measure both the momentum and location of the thus-fixed electron for spans of time as long as three consecutive months, with never a doubt about the validity of either of these simultaneous measurements. This means that the Heisenberg Uncertainty Relation is *not valid* in the context of an electron held fixed in a Paul trap. Thus, Dehmelt was awarded the Nobel Prize.

Later on, this writer discovered another exception to Heisenberg's Uncertainty (theory). Photons are not subjects of Heisenberg Uncertainty. For example, if one is observing coherent monochromatic light, at any point in time along the line A to B, one can easily can predict each and every one of these factors: *Wavelength, Frequency, Phase, Momentum, and Position* of the photons, at all times, with *complete certainty.*

For example, the only limitation to accuracy of the photon's position, is the time-of-emission accuracy, which is related to the accuracy of the timer. NASA has developed a timer system accurate to 10 femtoseconds, with projections of improvements into the .001 femtosecond regime. Emission time is then not an issue over a premeasured course and thus location of the photon is known to within the accuracy limits of the timer. (This is because we commonly consider that the speed of light is a "constant," even though this is not the fact of the matter. Repetitions of actual measurements of the speed of light, performed over several centuries, have demonstrated that the speed of light is not at all a constant, but is actually only an "average value." But this "average value" may be considered as reliable, for most practical purposes, under the most common of circumstances).

We can simultaneously know the momentum of the photon with complete certainty, because the momentum of a photon is directly related to its frequency. (If you know the frequency of an electromagnetic propagation, you know its momentum). Obviously frequency is not blatantly subject to the Uncertainty Relation, else radios and televisions and such, wouldn't work. The several other parameters mentioned above follow along similar lines.

Thus, contrary to Heisenberg, one can know both the momentum and the position (and many other physical properties and measurable conditions) of the monochromatic coherent photon simultaneously, and with absolute accuracy (for all practical purposes).

Thus, Heisenberg's Uncertainty Relation has a limited and very circumstantial domain of applicability and usefulness. One might well say that the Uncertainty Relation is rather extraordinarily uncertain!

This rather unfortunate sequence from the world of quantum physics is due to the circumstance of a theoretical interpretation (belief) taking the front seat over empirical and instrumented observations of reproducibly observable facts. Poor show! So what value is a theoretical interpretation (a belief, a conjecture) in the first place, except as trying to consider all possible alternatives, looking for that which best fits the facts?

But that's nothing compared to the vast array of fallacies and theoretical fantasies which accompany todays "officially accepted" versions of so-called "astrophysics"!

Long ago it was believed that the Earth was flat, and that if one sailed too far out to sea, one would "sail off the edge of the world." There was nothing other than the visible horizon for those folks.

Later, it was discovered that one would not sail "off the edge" of the Earth to a certain doom. This realization resulted in the belief that the Earth was the center of the universe. That was the accepted theory (belief) of the day.

With the acceptance of Galileo's telescopic observations, the prevailing belief that Earth was the center of the universe was falsified; the Sun now held that distinction. Subsequently, Newton's expressions did not refute this Sun-centered view, but simply made it much less tenable.

Then along came verified evidence that the Sun was not at all the center of the Universe, that there were vast numbers of other suns out there in the sky, called "stars," along with perhaps many other solar systems. But our solar system was now considered the center of the Universe.

Later, Einstein's flawed version of relativity theory became all the rage. This had folks in all the lands thinking that, as interpretations of Einstein's relativity theory had it, that the Universe must be a finite and bounded and limited sort of thing (and of course, centered on us). So it must have a center. And the center might very well be us.

Ecclesiastical interpretations of Einstein's version of relativity theory also had it that the entire of the Universe must have had some sort of "beginning," and then, that it must be doomed to have some sort of "an end." (So far there is no physical evidence to unambiguously support either of these assertions).

Later came evidence that our solar system is not very likely to be the center of the Universe, so now our Galaxy became the Center, as expressed by some proponents of the "expanding Universe" hypothesis. This kind of thinking is actually not so far removed from the views expressed by the (still existing!) proponents of the "Flat Earth Society."

The proverbial "fly in the ointment" is the fact that, given an infinite volume, any point in that volume may be considered as the center of the infinite volume, and/or that every point in the volume is *also* the center of the volume! (By definition, an infinite volume has an infinite number of centers, or any point in the infinite volume is the center of the volume).

There is physical evidence that the Universe is infinite in volume. This lies in the experimental fact that wherever and whenever instrumented measurements have been made, so as to ascertain the local value of the "curvature of space-time," all such measurements have met with the determination that space is asymptotically flat, meaning that there is no detectable "curvature of space-time." This fact infers one of two things: Either the Universe is infinite in volume (since an infinite volume will exhibit zero space-time curvature), or Einstein's version of relativity theory is completely without merit. Or, both these statements are true. (That both statements are true agrees with this writer's determinations).

So, if the Universe is infinite in volume, why do we need such unsupportable and untestable conjectures as some sort of "big bang" theory? In an infinite volume Universe, *Creation is a Continuous process,* rather than a once-in-forever event.

Some have suggested, on a relativistic basis, that perhaps there are many, more localized, "little bangs." Well, what do we need them for? To be sure, the proposal of "many little bangs" has more going for it than the concept of a single "big bang," because then there is some vanishingly small hope that we might eventually be able to see physical evidence which might lend some support for the notion of such "little bangs," by some kind of new and very advanced astronomical means. But we are still left with thousands of conundrums as to what happened *before* these various proposed kinds of "bangs" and "inflations," and what caused them and how and why they may have been caused.

Let us keep it in mind, that in all these speculative "inflationary" and "big bang" scenarios, according to the most common paradigm, all of everything was "crushed together" into a single "dot," which somehow managed to contain in it all the matter and energy in the observable universe.

This circumstance is forbidden by Classical Mechanics. It is directly forbidden by the Maxwell equations. It is also forbidden by Einstein's version of Relativity theory, and is completely forbidden by Quantum Physics. Still, the "big bang" proponents want to say that, "All the laws of physics, as we know them today, were *suspended* inside of the Initial Singularity."

Then they go on to say that by some undisclosed, indiscoverable, and unobservable means "a miracle happened," whereupon the "Singularity" "exploded" or "inflated," whereupon specific and selected various kinds of matter and the various forces and physical laws began manifesting themselves, in some sort of sequential order.

Of course all of this is sheer fantasy and has no empirical evidence supporting it. The proponents of this fantasy have been trying everything they can imagine to find some physical evidence, which might lend physical support to this fantasy, all to no avail, so far. (The attempt at the LHC facility to find the hypothetical "Higgs Boson," is just another of the many expensive attempts to try to force some mathematical fantasy fit with Actual Reality).

What is empirical science as opposed to theoretical science? Empirical science is based on reproducibly observable facts, as opposed to theoretical "science" which is based on sets of *believable beliefs*.

With "Occam's Razor," isn't it so very much easier to consider that all of the Universe has always been, more or less, the way we found it when we got here? Is it so difficult to conceive that matter and energy and the various forces are constantly being created and destroyed?

The entire concept of a "big bang" which is preceded by an impossible condition where all the laws of physics are "suspended," depends on the unwarranted assumption that the Universe is "expanding." The "expanding" Universe concept depends exclusively on the so-called "Doppler effect," which has light shifting towards the red end of the spectrum because the source of that light might be "moving away from us," as it passes across vast distances of space, thus shifting downwards in frequency, as demonstrated by Doppler with respect to sound waves.

This physically verifiable observation is then applied to light, without proper physical verification, and then interpreted(!) as meaning that the farther light has travelled, the more it must be shifted towards the red end of the spectrum. This is interpreted to mean that all the sources of light must be "moving away from us." This unjustified assumption is further interpreted to mean that this "red-shifting" must imply that the Universe must be "expanding."

In actual fact, there are at least a dozen additional ways that light can be red-shifted, in addition to the Doppler effect! This means that the so-called "Hubble's constant" is not a valid measure of interstellar distances, because many additional physical events and fields and forces may influence a given photon, in addition to the Doppler Effect. There are *many ways* that light can be shifted towards the red end of the spectrum, in addition to the "Doppler Effect," which is the origination of the "Hubble's 'constant.'" *In addition* to the Doppler Effect we have:

Gravitational red-shifting (See: Misner, Thorne, and Wheeler's "Gravitation")

Quantum fractional Hall impedance (Prof. R. M. Kiehn says, for example, "...the Action 1-form, A, has the physical dimensions of the flux quantum, h/e. The 2-form, G, has the physical dimensions of charge, e. The 3-form, A^G, has the physical dimensions of angular momentum, h, and the 3-form A^F, has the physical dimensions of spin multiplied by the Hall impedance, (h/e2) = h(h/e2) = h(Zhall)." He goes on to show that fractional

Hall impedance is a built-in function of the vacuum, which *shall cause* red shifting of photons).

The *spin field* can cause both red and blue shifting.

The *torsion field* causes red-shifting.

The *subquantum aether* can cause red-shifting by several mechanisms, particularly when there is *a density gradient in the aether gas/fluidic media,* which will directly result in a *change in the index of refraction* of the media, as seen by transiting E/M propagations. A related mechanism would be similar to Paul LaViolette's description, having to do with the subquantum superfluidic plenum creating a very small drag on photons over cosmic distances, causing them to loose energy.

6 & 7. Whittaker and Bateman both showed red-shifting mechanisms, caused by *strong electric and magnetic fields,* early in the last century. This is directly related to the red-shifting caused by gravitation, and is also related to Delta Grad E events.

Topological red-shifting, which can arise when a discontinuity is created in the vacuum by various topological means. Photons entering the discontinuity can lose energy in transit.

Bose-Einstein Condensates can red-shift the frequency of light.

Quantum Scattering (See http://www.rialian.com/rnboyd/quantum-redshift.htm).

Delta Grad E events: Changes in the gradient of an electric field result in instantaneous and directly measurable changes in the permittivity of the physical vacuum, and thus, instantaneous variations in the speed of light in the physical vacuum are measured. In addition, simultaneous, and instantaneous variations, are observed in the force due to gravitation, in the measured inertia of masses in the volume of the effect, and variations are observed in the pace of time. In addition, divergences are produced in the quantum potential. In addition, the electromagnetic radiation patterns that arise from electromagnetic transmission antennas are observed to vary dramatically under these conditions. (This statement has been tested as valid for normal E/M, complex E/M, and symplectic E/M transmission facilities). *Delta Grad E events are ubiquitous in interstellar and stellar plasmas, and are easily produced on Earth.*

Knowing some of these facts of optical physics, Hubble himself was completely opposed(!) to the notion that the Doppler Effect was the only thing that could shift the frequency of light. Then, despite poor Hubble's decades of protestations, this so-called "constant" was named after him, posthumously.

The "expanding universe" is a fraud, a fiction, and a lie. There is no physical evidence whatsoever to support the idea, given the fact that "Hubble's constant" is correctly seen as vastly flawed, and physically unreliable.

So if the Infinite volume Universe was never "expanding," why do we need any sort of physically impossible "big bang" or "inflation"? Creation is a local and ongoing set of events. In fact, everything we are capable of experiencing is direct evidence that all things are created as a result of local process, just as you were the result of quite localized Creation events.

It is also commonly and wrongly assumed that interstellar space is isotropic (having exactly the same properties everywhere). In actual measured fact, interstellar space is exceedingly anisotropic! (Meaning that the properties of space vary dramatically from place to place, and from time to time). This anisotropy gives us all the more reason to understand that many, or indeed all, of the possible ways that photons can be shifted towards the red end of the spectrum (or the blue end of the spectrum), might very well be directly influencing photons which are passing through vast distances and encountering vast numbers and kinds of anisotropies.

The fact that space is not "expanding," and the observable fact that Creation is a Continuous process, throughout the Infinite volume of the Universe, validate the "Steady State" cosmologies of Hoyle, Narlikar, Bondi, Gold, and many others, with the required caveat that Hubble's "constant" is a lie. And Hubble himself is thus vindicated in his strident opposition to this "constant."

In the altogether, Hubble is vindicated in his opposition to the co-called "constant" which was later attributed to him, despite his opposition to the notion. And the misnomer "Hubble's constant," is perfectly wrong. This "constant" is based entirely on a single and *intentionally narrow-minded* way of looking at cosmological red-shift behaviors. Since there are diverse additional and more reliable ways of producing red-shifts in the laboratory, the entire notion of an "expanding Universe" is perfectly ludicrous! The Universe is *Infinite* in volume, and is never "expanding."

So, did all of the observable Universe, and everything in it, happen by some blind random chance (?) due to some obscure, inert, and lifeless and mechanical mathematical considerations, such as some sort of imaginary "bang" of an impossible set of conditions, all of which conditions are perfectly denied by all of the empirically physically well-established and well-known Laws of Physics?

Such inert and mechanical concepts go back to Victorian times, when a "very predictable," purely mechanical, sort of "clockwork" Universe was very fashionable. In those days, the tiny and intricate pinions and gears of the fairly reliable wrist watch were the very pinnacle of advanced technology. Surely we are well beyond that state of affairs, in these days.

We've long known that this sort of "clockwork mechanical universe" paradigm is riddled with vast numbers of failings, being unable to properly describe the majority of the observable facts and being unable to make reliable predictions regarding the observable facts, particularly since the advent of quantum physics.

Well, if all this didn't happen in some sort of purely mechanical "clockwork" sort of way, then how did it manage to happen? Maybe it "self-organized" itself? How did it manage to do that then? Perhaps all the pinions and gears just kept randomly falling in an ever growing heap on the ground, where the ground itself was constantly moving, so as to keep the ever falling pieces and parts ever mixing together? Until such time as the pinions and gears and springs all "magically" happened to fit themselves together into a perfectly functioning, and very handsome, wristwatch? Very doubtful, wouldn't you think?

And there is also the issue of who designed these purported clockworks, and who managed to construct them, and by what means?

To make all these amazing things we see in the Universe all around us, several sets of guiding parameters must exist, which parameters must be guided in turn, by an inherent and Aware Intelligence which must exist at the very core of the All-That-Is, and simultaneously, pervade it everywhere. Intelligent Design has to be seen acting directly in the production of all Observables, where Creation- Information and Intention (the Quantum Potential) produces all of everything else.

Once a something has come into being, it seems that the given item, over the course of infinite time, is destined to decay and then eventually cease to be. But this natural process is not constrained to adhere to theoretical considerations which would calculate that the proton, for example, has a theoretical life- span of some 10e 33 years, based on expressions such as those found at http://hyperphysics.phy-astr.gsu.edu/hbase/particles/proton.html.

Not so! This is yet another example of how apparently air-tight theoretical results are so often completely falsified by empirical observations and the results of physical experiments, yet theoreticians are often found many years after the physical evidence has completely dismantled their beliefs, clinging desperately to their physically failed false imaginings as though they were life rafts, in a vast and empty ocean of dreams.

It was experimentally and empirically established by Gustave Le Bon of Belgium, as early as 1906, that all forms of atomic and subatomic matter cease to exist in rather short time spans, measured in spans of time between fractions of a second to mere centuries, through a process called photonuclear reactions. In a classical photonuclear reaction, when a gamma ray of exactly the correct resonant frequency of a given kind of atom, encounters that particular kind of atom, that atom vanishes, and completely and utterly ceases to be.

When the frequency of the gamma radiation is slightly off the "station frequency" of the given atom, the atom will dissociate in a shower of subatomic particles and various other kinds of radiations, and will also produce measurable events of the various kinds of forces.

Gamma radiations are ubiquitous in the Universe, especially resulting from stars and during various interstellar plasma behaviors. So atoms are being vanished by photonuclear reactions, constantly and everywhere.

Fortunately, this continual destruction activity is fairly well balanced by constant creations of new matter, everywhere. These processes are rather more complex than destructive ones, so we will give only one example of such processes here. During plasma physics studies which were designed to attempt to keep thermonuclear plasma species around long enough to produce fusion reactions here and there in the containment system, usually composed of magnetic field structures, it was realized early on that if any impurities got into the plasma, other than hydrogen or helium atoms, that the plasma would rapidly lose energy and drop below the fusion energy threshold.

Due to this circumstance it became common practice to pump the containment volume down to produce the best vacuum they could produce, so that no impurities would be lurking there, waiting to drop the energy of the plasma below the fusion threshold. As vacuum technologies improved, the ability to make better vacuums improved. When a certain threshold of vacuum was reached, the researchers discovered, much to their dismay, that hydrogen atoms began "showing up" in the vacuum chamber.

From that threshold and beyond, no matter what they tried, they could never get rid of all the hydrogen in the vacuum chamber. Eventually, through measurements, they realized that hydrogen atoms were "popping into existence," from out of nothing, as it were, obviously violating the theoretical tenets of the conservation of energy and the conservation of matter. Rather than announcing these results to the world, it was decided that they should prevent these discoveries from ever reaching the public domains. Nevertheless, it appears that the old saying, "Nature abhors a vacuum." is really true.

Matter and energy throughout the Universe are constantly being Created and constantly being Destroyed. Interestingly, this re-discovery of the Continuous Creation/Destruction process aligns perfectly with the Hindu pantheon of Brahma, Shiva, and Vishnu. This physically verifiable assertion falsifies entirely, the ecclesiastical notions of a "beginning" and an "ending" of the Universe. Of course, the various *individual items* and forces in the Infinite Volume Universe, *do* have "beginnings" and "endings." But this concept does not apply to the Infinite Whole, which was here before we got here, is now, and forever shall be.

The first key realization is in the fact that there do exist many varieties of faster than light events. The Mobius transformation solutions of the Maxwell equations lead to an understanding of various entities which can travel with any velocity, from zero velocity, to an infinite velocity. This understanding leads to the perception of an infinite fineness of the vacuum's structure, thus pointing to SubQuantum entities with the same propagation properties as those indicated by the nonlinear projective Mobius transformation solutions of the Maxwell equations, with individual propagations of these entities ranging from zero velocity to an infinite velocity. (As opposed to the more commonly understood Lorentz transformation solutions of the Maxwell equations, where propagation velocities are limited to the average speed of light, in the given media).

The second key understanding is the realization that matter is infinitely divisible, so that there is no end to smallness, where the Planck scale of 10e-33 cm., is just one level of a hierarchy of increasing fineness, which apparently continues into the infinitely small. Quantum fluctuations are direct physical evidence of the existence of the subquantum, as related to Brownian motion. In the subquantum model, increasingly fine subquantum entities and structures are resulting from composites of infinitesimal particles, which are the origination of the Entire of the physical Universe, and all of its specific components.

Such compositions are orchestrated by an Innate Sentience which is contained in the general Quantum Information Field, which Sentience orders and organizes all of everything else. This realization brings about a much deeper understanding of what are the actual fundamental building blocks of our Reality.

Information systems, operating in an infinite array of implicate orders, as described by Bohm, become pre-physical "stencils." Then as these "outlines" are filled in, there result all the observable physical manifestations. Such manifestations are resulting from the interactive and Intelligent presets and Sentient patterns which are already inherent in superluminal information-based connections and behaviors.

These patterns and presets are originating in the SubQuantum domains. Keeping in mind that all quantifiable energy is constantly coupled into the SubQuantum, by infinite velocity, infinitesimal particle-carried Information fields, we are becoming increasingly more familiar with the foundations of Information-driven phenomena, in the physical world.

Organizing functions originating in, and evolving through, the SubQuantum realms, are the inner workings of the, actually endless, Creation Process. All apparent Quantum "randomity," as related to observable forces and matter, is then actually due to an externally induced determinism, one which is reflecting the great overarching Harmony of All, in an infinite array of manifestation layers, proceeding from the small, into our normally familiar physical world.

In this understanding, Life, the brain, and all of everything else, are directly designed and constructed by adaptive morphogenetic SQ information-field influences, which are in turn, under the direct control of Harmonious Cosmic Intelligence, rather than through some Darwinistic "evolutionary sequence," which concept has been physically proven false by the observations and experiments of the Nobel laureate, Crick, and by many other information-related experiments.

The Cosmic Microwave Background (CMB) radiation is direct evidence that Creation is an ongoing process, based on subquantum determinations arising out of studies of the Kolmogorov vortices at the scale of 10e-60 cm. The advent of the subquantum microscope has already allowed the imaging of entities as small as 10e-96 cm., according to recent results out of Serbia. Such physical observations are validating the subquantum as the actual and foundational basis of all experiencible Reality.

What is empirical science as opposed to theoretical science? Empirical science is based on reproducibly observable facts, as opposed to theoretical "science" which is based on sets of *believable beliefs*.

Let's see how Einstein's version of relativity theory stacks up, relative to the empirical approach to the sciences, as illustrated by Popper's Criteria for Reliability in the Sciences, as paraphrased below:

Popper's 1st Criteria:

> "Is the scientific expression internally logically consistent? Are there any expressions therein which are obviously and instantly contradicted by previously well-proven and reproducibly observable substantiated facts?"

Einstein's version of relativity is not internally consistent, in several regards. Many of those inconsistencies are covered by Stephen Crothers in his previous writings regarding the various mathematical and logical inconsistencies and flaws contained in relativity theory, *here:* http://www.thunderbolts.info/thunderblogs/archives/guests08/090807_sjc.htm http://www.sjcrothers.plasmaresources.com/.

As to the 2nd part of the 1st Criteria, *never before in all of history* has any human being *ever directly observed* any "curved space," nor any "curved time," nor any of the even more ludicrous "curved space-time." So, this version of relativity is directly contradicted by thousands of years of empirical human experiences.

Popper's 2nd Criteria:

> Are these expressions capable of being **tested** by **direct experience, empirically**?

No, they are not. In fact Einstein himself once said, "Relativity theory can never be proved by any physical experiment." Why would he say that? Because Einstein knew full well that his version of relativity theory lives only in one's imagination, kin to the Disney production called "Fantasia," equally physically real, and equally physically unprovable. Einstein's version of Relativity theory is *perfectly non-physical!*

So it fails Popper's second criterion. Popper's 3rd Criteria (broken out by part)

> "Does the proposed revision have fewer axioms or more?"

Einstein's version of relativity includes *many more* axioms than Galileo's original version of relativity. In addition, none of these extra axioms have ever been proved to be physically real, nor absolutely required physically, nor physically useful, under any set(s) of circumstances which have been tested to date.

> "Does the proposed revision produce larger numbers of verifiable predictions?"

Einstein's version of relativity theory has produced no physically verifiable unambiguous predictions, about anything! In addition, 100 years of this version of relativity have resulted in zero innovations and zero new conveniences. (So what good is it?)

> "Do the given expressions produce alternative conclusions from a differing (relative to other viable expressions) set of basic statements (predictions) which appear to be more accurate to the observable facts?"

Alternative conclusions, yes. More accurate to the actual facts? *Not in any regard.* Actually, all the premises of Einstein's version of relativity require a complete denial of all the observable facts. Are there any other viable explanations? Indeed there are hundreds of years of other viable explanations, in addition to Galileo, such as model of the *actual cause* of gravitation published by the Marquis de Laplace, during the 1850s. When asked the question, "What is the cause of gravitation?" Sir Isaac Newton famously responded, "I frame no hypothesis." Laplace's model of gravitation was the first model to surpass Newton, in this regard, showing the actual cause of gravitation, a topic regarding which Einstein's version of relativity is equally as mute as was Newton.

> "Does the new axiom produce any predictions which **contradict** any existing predictions which have been made (and preferably **verified**) without taking the new axiom into consideration?"

Einstein's version of relativity theory directly *contradicts* Newton's long established Law of Gravitational Attraction, by empirical measurement. Any relativist will happily agree that Einstein's version of relativity theory does not agree with the direct and empirical observations of Newton, saying. "Rather, Einstein's version of relativity theory 'corrects' Newton." I don't buy it! Where is the physical evidence of this "correction"? Can't find any? Well then off with you and your "pig in a poke!"

Einstein's version of relativity theory fails to meet Popper's 3rd Criteria. So, on to Popper's 4th Criteria:

> "Is this scientific expression "**falsifiable**?" Can be confirmed or denied by experiments and observations?

Absolutely not. Again, Einstein himself said, "Relativity theory can never be proved by any physical experiment." That means it is impossible to confirm or deny, simply because it is imaginary, and lives only between ones ears, as all other imaginings do. That means it is not falsifiable, it is not testable, and since it has no relation *whatsoever* to the *actual physical universe,* no possible combination of experiments and observations can ever act so as to confirm any of these expressions.

The theory fails Popper's 4th Criteria.

Einstein's theory fails all of Popper's Criteria for Reliability in the Sciences.

Some well-respected mathematical physicists, such as Roger Penrose, are aware of these difficulties with respect to Popper's Criteria, and rather than admitting to these egregious flaws, would rather have Popper's Criteria white-washed away, calling them "overly stringent" and "harsh." [See his book, "The Road to Reality," pages 1020 and 1021, for example.] To this writer, it is *vastly more harsh* be forced to live in a completely unreliable world, made almost entirely of fantasies and illusions and delusions and imaginings, producing endless difficulties and uncertainties for the normal and sane person. This writer insists that Popper's Criteria should be applied with *unrelenting stringency,* resulting in some reliable understandings regarding Actual Reality, resulting in a world which contains some degrees of certainty and reliability.

Popper once said, if a theory fails to meet all of his criteria, then that line of investigation should most properly be abandoned for all time, as being perfectly invalid, and completely useless. I agree with Popper on this, with respect to Einstein's version of relativity. Forget about it! It must be added that there are many more theories that are presently "widely accepted" which also perfectly fail to meet Popper's criteria, such as "string theory."

Now, as to all the other fantasies which are based on this fantasy called relativity theory:

> "Black holes. Hey, wasn't that the show that came on TV right after "Fantasia" last year? Well, I thought they were both a bit entertaining. But Fantasia was much more fun! Besides, I didn't like the part where all those black holes went around eating all of everything else."

Gone away are all the other imaginary figments which have arisen from the mathematical fantasy called relativity theory. Gone then is "curved space-time." Gone then are the "black holes," the "big bangs," the "expanding (even accelerating!) universes," and the equally imaginary "neutron stars."

Gone as well, are such concepts as "Lorentz -Fitzgerald contraction," which has a massive object growing "thinner" as the "speed limit" of the speed of light approaches. Gone as well are the "twin paradoxes."

Gone are all considerations regarding light as the "highest possible velocity." Gone is the idea of "traveling backwards in time." Gone is the idea that time "slows down" as the speed of light approaches. Gone are "wormholes." *Gone is the idea that the speed of light is in any way physically related to Time.*

(Einstein relied on Minkowski's mathematical abstraction called "space-time," where a 4 dimensional, complex, pseudo-manifold ("pseudo" means fake or artificial or "looks like but isn't") is mathematically created, with the 4th dimension being time, which enters into the system as "*i*Ct," which term is called "imaginary time." This construction falsely links the pace of time, t, *exclusively* to the speed of light, C [This linking is a completely unwarranted assumption.], then enters time as a dimension of rank equal to the 3 normal Euclidean space dimensions, by way of the imaginary number *i*. This is the actual origination of the false assumption that nothing can exceed the speed of light, based on the unjustifiable, untested, and unquestioned assumption of "*i*Ct," which is linking of the pace of time, with the speed of light, along with a further unwarranted and unproven assumption that time is actually caused by, or linked to the speed of light, in any regard whatsoever).

Also gone are all the ridiculous and expensive experiments such as LIGO, which are still out looking for the legendary "dancing hippopotamus in the pink tutu," also known as "gravity waves."

In summation, there is absolutely nothing right about Einstein's version of relativity. Whether we are considering GR, or SR, every supposition that is supposed to be deriving from any completely flawed scientific expression, is equally flawed, and is more likely to be an even worse flaw, which is stacked up on top of all the previous flaws. What we're left with is a completely useless stack of garbage.

Appropriate to these circumstances, an old saying from Computer Science comes to mind, "Garbage into the system, leads to garbage out of the system!," often abbreviated "GIGO," for "Garbage In, Garbage Out."

Back to the empirical method! Back to Reality!

In: Old Problems and New Horizons in World Physics
Editors: Volodymyr Krasnoholovets et al.
ISBN: 978-1-53615-430-6

Chapter 11

CANTOR'S CONTINUUM IN A VACUUM

Arnold Gorgels*
Mathematical Physics, Institute in Potsdam, Germany

ABSTRACT

A symmetrical French garden and a seemingly disorganized English garden lay adjacent. A vase is located in the English garden. Its location seems to be randomly chosen and fully incomprehensible for visitors. Only if looking from the French garden, it can be seen that the vase's location fits perfectly with the allees of the French garden. Take the speed of light symbolically as the vase and its value in the universe as random according contemporary knowledge. Is the speed of light indeed random? Or is it possible to look at it from another angle? That is the idea of this chapter for going after a new logic interpretation of nature. It is a view from another angle; from the angle proposed by Georg Cantor's mathematics and applied to the vacuum.

This study describes the consequences of assuming Set Theory, based upon sets of natural and real numbers, as the most appropriate mathematical tool to describe nature. It is a qualified suggestion meant to present an alternative logical possibility in the universal search for a unified theory. The authors believe that the axioms of Zermelo-Fraenkel (1907-1930), forming the standard form of axiomatic Set Theory (including Cantor's theory of infinities (1877)), allow for the most accurate description of the universe; natural laws and nature's constants should be derivable from it. Set Theory allows real and virtual elements of nature to be arranged either individually or in groups. It assumes the existence of any elements to appear within a mathematical environment that is called "the set of those elements," in certain cases sets with (actual) infinite number of elements in which case Cantor's mathematics of infinities applies. The consequences of this (infinite) approach are found both in the furthest corners of the universe as well as in the smallest nuclear particles. The first part of this article deals mainly with the universe and the second part with light waves, positrons and electrons.

---

* Member DPG; Author's Email: arnogorgels@gmail.com.

Keywords: void, ether, vacuum, space, set theory, aktual infinity, continuum, quantum, quaternion, electron, positron, proton, quark, relativity, particles, fields, gravitation, EM-waves, philosophy, religion

INTRODUCTION

On the 7th of July 2005 the undersigned published together with Mrs. Shevkinaz Bulut (Cologne, Germany) a document that we called SQT "Superquantum Theory, candidate for "Theory of Everything." At this time, in the year 2005, the "Einstein Year," Albert Einstein's search for a "Weltformel," for a mathematical description that should combine all natural forces, was discussed as part of his heritage. Physicists called Einstein's quest the search for the "Theory of Everything." Physicists are still searching it in 2018, relying on mathematical subtleties like string and M-theory. We, however, looked first of all for an overall mathematical theory and found it in Cantor's set theory that enables connection of all elements, facts, events and formulas. We called it: Cantor's Universe, because Cantor's Continuum was taken as starting point of our work. Cantor's theory is generally considered to constitute the backbone of mathematics out of which all mathematical derivations and relations are inferable. The existence of the continuum was proven (to be rephrased) by the author (see references).

Since the sought-after theory should be a "theory of connecting it all" its characteristics should be holistic and should contain an element of infinity. Cantor's theory containing an infinite number of elements and his genial continuum formulas appeared to us to be the most suitable mathematical theory available. We decided in August 2004 to base our work upon this theory. However, a stumbling block appeared: all things with a material character, whether molecules, atoms, quarks or muons appeared to be unsuitable to cover the property of numerical infinity, no matter how big the number of elements. Whilst looking for a substance or medium where infinity and its mathematical formulas could be applied, we selected vacuum as the only possible candidate. Thus, we concluded that the vacuum was to be quantified. We called the quanta: superquanta (to be in line with superstrings), and also vacuumquanta. Furthermore, expressions like observer and infinity play essential roles in the present work; the position and distance of a virtual observer and a virtual object is decisive for his interpretation and its consequences, as will be seen.

METHODS

At this point, it is necessary make an side observation: had we known, that physicists of the 20th century had abandoned the idea of an ether (a structured vacuum) and had done their utmost to proof this, even with an experiment, the Michelson-Morley experiment, we may have not continued our investigations. *Our* philosophical starting point was different: we considered that stability of the universe, if lacking overall comprehensive logic that culminates in a profound mathematical description, couldn't be supported; its existence in the next fraction of a second then not secured. Mathematics as fundament of stability of all facts and events is a *conditio sine qua non*.

Contemporary philosophers supported however the physicists and encouraged them to limit their work to visible and measurable facts. Those philosophers earned fame too, who discarded mathematics as a non-fundamental semantics tool, i.e., a language for mutual communication only: no Nobel-prize for mathematics. We, however, didn't know or rejected these objections. Was not the imaginary number j and Leonhard Euler's theorem in the 18th century the foundation of all electro-magnetic calculations in the 19th century? Unimpressed, we continued, because Cantor's Set Theory projected on the vacuum seemed to work too well as universal fundament to be non-applicable. We continued our research, even when it produced ever more results that contradicted mainstream knowledge. We looked for fundamental order and noted that order is a subset of disorder (chaos).

The idea of a comprehensive mathematical order urges that the theory should apply without limitations, i.e., in accordance with Cantor's continuum hypothesis. Consequently, we accepted the existence of power sets in vacuum. In the course of further considerations, we understood, that the superquanta are quaternia, the mathematical-physical (x, y, z, i, j)-construction of three-dimensional space combined with two imaginary axes, from which Maxwell's equations were derived. Quaternia or Hamilton-numbers, called after its inventor William Rowan Hamilton, were already known since 1843. Well, since Cantor's continuum should apply, further spaces can be constructed from the superquanta of the "visible" space, that we call Hubble-space for the sake of identification. These follow out of the axiom of power sets. These subspaces are firmly related through power-set-relation to the Hubble-space. This view of subspaces gave raise to a suggestion concerning the true source of the phenomenon of background radiation: if the Hubble-vacuum is the source of the visible EM-waves, then its first power set, i.e., the next space, may well be the source of the cosmic microwave EM-wave background.

Stability of the vacuum and its elements is imminent. Stability is the subject of the next investigations. The vacuum and its elements can be measured by its length, height and width. It possesses a fundamental stability, since it's the carrier of all facts and events. This fundamental stability allows formulation of mathematical consequences. The modelling leads to the formulation of the following three criteria: Each vacuum quantum must be subject to:

Criterium 1) Stable diameter: d
Criterium 2) Stable surface: d^2
Criterium 3) Stable volume: d^3

These three stability criteria of the vacuum quantum should allow the formulation of fundamental formulas and further derivation of natural laws and constants. How? A fixed length d can be described mathematically by a standing wave, i.e., the sum of two opposite running waves. This is the basis of further calculations, amounting into Maxwell's equations of electromechanical waves and Einstein's Special Relativity. In fact, SRT describes One Stone of many.

The theory of the stability of superquanta however shows that the wavelength and frequency of lightwaves depends on the size of the superquantum.

Possible conclusion: Cantors Universe allows a steady state universe and proposes galaxies that possess vacuum quantum of several diameters d1, d2, d3 etc.

See the illustration of a steady-state universe:

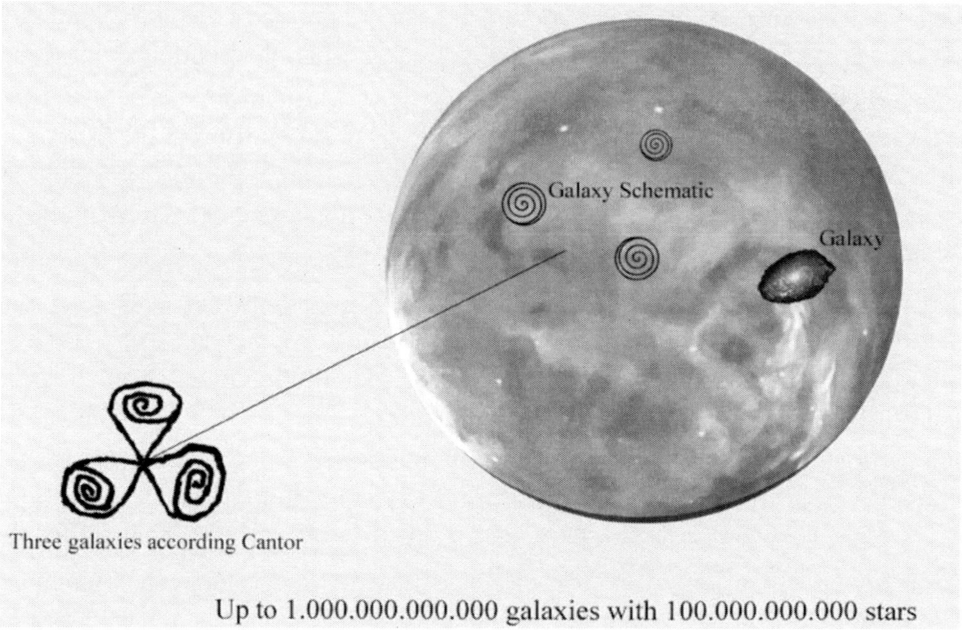

The result is a universe of disorder (chaos, unordered) with islands of order (galaxies). Each single island starts off as single quaternion, then it expands (arithmetically-virtual-real as per Dedekind-Peano-Axioms) as volume of superquanta, identified as a set of infinite number of single quaternions. Their powersets constitute (inside) further inner spaces of (smaller) quaternions, whilst the elaboration of the Maxwell-equations out of quaternions lead to the occurrence of photons/light as per well-known derivations. New in this respect is the Cantor-based application of inner spaces that can clarify how waves propagate in vacuum. Since quaternions are grouped in space as mathematical numerical units, their inner construction follows the elasticity known from number theory; the stabile elements are quaternions by themselves. This way the ordered galaxy has a firm logical inner structure which is a condition sine qua non for its unshakable stability. Each particle has an inner body and an outer environment.

If indeed the universe has the shown structure of disorder with individually ordered corpora (galaxies), of which a hundred to a thousand billion appear to exist, ideas like dark energy, the introduction/deletion/re-introduction of the cosmological constant, the expansion of the universe, the idea of an expanding ether, the idea of galaxies flying faster than the speed of light become obsolete. At the other hand, the phenomenon called dark matter, though, is a different issue; set theory allows a simple qualitative explanation by introducing a local topology with a vertical component; i.e., in addition to the well-known inter-mass effect a set-theoretical space-time gravitational effect is to be reckoned with, especially where the rotational speed of heavenly bodies is under discussion. The reason is that the inner spaces may play a (small) role, whereas certain particles may be related to the inner spaces and less to the Hubble-space.

It should be possible to investigate the inner order at the level of protons in greater depth, photons and em-waves, then electrons and positrons, of protons/quarks/neutrons, finally hydrogen leading to the hydrogen gas clouds in space. It is suggested that protons are built up from positrons surrounded by second space elements that are called eons and peons in our research.

Set theory may allow a special construction of the neutron/quark/proton as shown here. It resembles the atomic model of Niels Bohr.

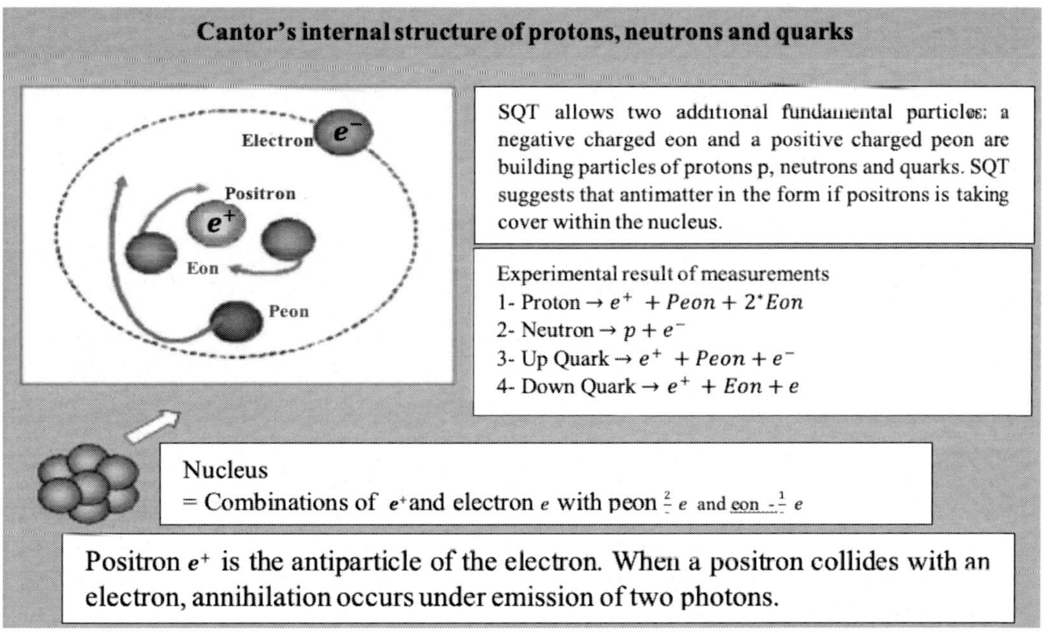

The Atom

In the following paragraphs, the three-dimensional massless region which traditionally is called ether or vacuum, void or space, an attempt is undertaken to introduce a pure mathematical-physical approach. This is based upon the set theory of infinite sets developed by Georg Cantor in the nineteenth century. Thus, an interesting new ether-model is developed. For this document, the expressions ether, vacuum and space are used to describe different properties of the phenomenon void.

Introduction

For the purpose of the following explanations, we call the three-dimensional massless region ether. For reasons of stability of the universe, set theory of infinities is taken to be the adequate description of the vacuum. Elements of natural numerical sets are representative of vacuum quanta. Power sets play an important role in as far as these allow a stabile construction of numerous interdependent solid layers of vacuumquanta separated by dynamic borders. Of these numerous layers, only a few layers are of direct access to experiments. The existence of sets and power sets invite acceptance of a hierarchical overall setup. This will demonstrate that the vacuum which, in line with experiments, introduces forces between particles, will also be their origin and their existential environment. Whilst peering only the first two layers, vacuumquanta have an inner corpuscle and an outer environment. Whilst experimental physics seeks the particle's *outer* environmental conditions, this paper attempts to investigate and document their *inner* structure. Investigation of the outer and inner space layers may raise possibilities to adequately allocate the origins of strings, quanta, fields, photons, entanglement, tunnel-effects, dark matter, dark energy, natural constants, time, dimensions and alike or reject them as found appropriate. The quantum-mechanical provision of infinite chaotic on-off possibilities and the creative (selective) observation thereof are basic tools of ontological awareness. Provision and selection must coincide if stable existence is to be involved. This resembles the two waves general principle that leads to standing waves of facts and events, i.e., reality.

Methods

In the vacuum, for a start, at virtual level, only location and distance plays a role; that is to say, location and distance are initially the only observation at virtual (vacuum-quantum) level; sharp (close-by) and vague (at distance) views can be identified. All facts and events at vacuum level are a consequence of awareness of such distance, accompanied by a mathematical-physical contract (formula(`s)) giving visible and durable expression to them. The necessary tool of dynamism will be identified herein.

Questions like "what is the reason that the absolute values of the charges of quarks are precisely 2/3*e or 1/3*e " or, "what is the true origin of natural constants like for example the velocity of light in vacuum or the gravitational constant *G*, Planck's constant *h*, the electric constant ε_0, and the elementary charge *e*" are unresolved in spite of good experimental knowledge. Experimental science has lately accumulated some piles "unresolved". Unresolved questions are traditionally given names which allow detailed resolution at a later point in time. In the last ten years dark matter and dark energy have been added. Historically, this name-giving practice can be successful as happened with expressions like atom, entropy as measure of disorder or Leibniz' monads as metaphysical basic substances. Monads are factually virtual particles.

The last 100 years are marked by continued bookkeeping of experimental results along with invention of suitable mathematical descriptions. At the time, mathematics was considered merely a human invention of a logic language, without profound significance. Lately, an new awareness is coming up: logic is available at the cradle of particles. This perspective renders mathematics a more basic position in nature.

The string theory is nowadays a wanted candidate for explaining everything. Efforts are being done to reconcile string theory with field theory thru the M-Theory. In this paper however, the vacuum as construction of space layers with a proposed mathematical background is given stabile structural background as drumhead, using, inter alea, following logic.

Whilst a string is defined as a hypothetical vibrating one dimensional sub-atomic structure (...), the obvious question arises: "What exactly is it within a string that vibrates?" Now, it will be shown that this question can be answered by giving the vacuum a mathematical-physical structure of vacuumquanta; these vacuumquanta have an inner structure of infinitely numbered small space quanta as well as an outer structure of vacuum quanta of same size. The concept of stable, solidly structured vacuum quanta allows a new perspective point of nature. The expression "ether" is reserved for the whole of space related elements and structures. If vacuum quanta are understood to be spheres, it is seen that between them gaps exist that need to be filled in (See Illustration A).

Filling in the gap that occupies almost half of the available realm can only be done by ever finer space structures. This gap-filling process is to be continued, theoretically, until infinity, but only two three layers are factually of immediate importance (measurable consequences) to physics. A significant dynamic process of mutual collaboration of a provider function (generality) and an observer function (individuality) is identified. This principle has mathematical, physical but also obvious philosophical significance. Elaboration of the latter is not subject of this paper.

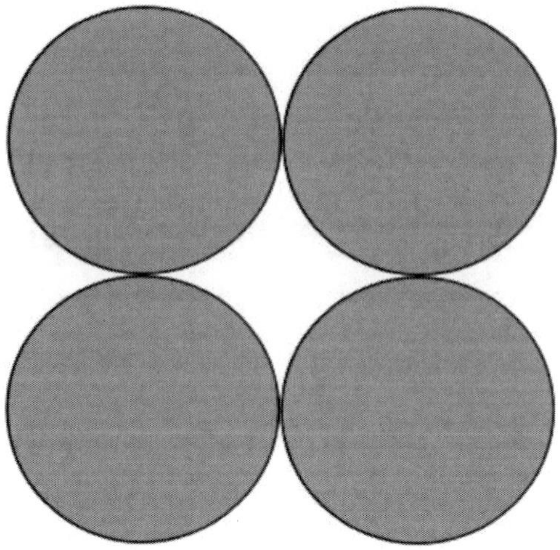

Illustration A.

Brief History of the Vacuum

The idea of "vacuum" has been subject to different views through history: from being a source of everything, despite its nature of being nothing, to being a simple geometrical platform, a playing room for mathematicians, many varieties have since been considered. The Buddhist term *Śūnyatā* means emptiness. This was considered as a platform of interdependent origin of all things that also provides for causality. Approximately 150 years later, the Greek philosopher Plato focussed on the geometrical features of "void" (Greek: kenos). Paparazzo describes the appearance of a twin-vacuum idea given by the atomists Leucippus and Democritus who define body and void. This view has hold into modern times: separation of void and material. Void is then defined as "absence of material". The void, as per the modern view, is the playing ground for particles to interact. Aristotle, in line with Paparazzos summary, seems to carry first thoughts of quanta when he writes about "cyclical replacement" referring to vacuum properties allowing movement of things. Whilst subjecting the vacuum to metaphysical questions, Aristotle concluded that vacuum lacks existence in ontological sense. He does this whilst giving some goodwill to surface science as opposed to space science since objects become identified in space through their surfaces, i.e., not by their space-occupation. In the middle ages Islamic scientist Abū Rayhān al-Bīrūnī concludes according to Ahmad Dallal that "there is no observable evidence that rules out the possibility of vacuum." First experiments with vacuum are reported to have been conducted during the Islamic Golden Age.

In 1644, French philosopher René Descartes proposed a geometrically based theory of atomism, avoiding the problematic nothing–everything oppositeness of void and atom. Where he set vacuum and matter to be equal, Descartes was critisized at his time and still is today. Absence of matter is substantial to the actual understanding of the abstraction "vacuum." Yet, the idea of a background of matter within a mathematically structured ether is subject of this study. Descartes' idea has parallels to the here proposed visible universe that finds its origin in Cantor's Continuum, positioned as space in space and multiple space-layers and basis of an "embroidery of light-waves" that constitutes matter and material.

Electromagnetism, quantum electro dynamics and quantum chromodynamics have given rise to more precise definitions of the vacuum. First of all, for experiments of electromagnetism (19th century) a so-called "classical vacuum" was supposed to allow radiation to travel at the fixed speed of light. Secondly the classical vacuum brings forth the values of constants of electrical permittivity and magnetic permeability. This paper is written to lay the foundation of understanding the reason for these constants. For quantum electro dynamics the vacuum is a state of lowest possible energy which can be transformed to the classical vacuum when it's cornerstone, the Planck's constant, is taken to be zero. Liberating the realm of quarks lead to the quantum chromodynamics vacuum. Naturally, several physical descriptions allow, when it its properties concerning matter/energy are brought to zero to corresponding vacuum descriptions.

An interesting pure mathematical approach to describing a "Mother-space" as a source and generator of matter and physical laws was given by Bounias in 2000. The illustration A given above can serve as its introduction if the gaps shown are neglected as well as the necessary awareness that these need to be filled in so as to form a complete ether without gaps. Such an ether will be seen to possess a hierarchic structure. Bounias' hypothesis was elaborated by Volodymyr Krasnoholovets [2]. Whilst results like differential geometry were recorded as validated by his approach, the development of an abstract empty set into the field of numbers was an achievement of Giuseppe Peano (19th century) who goes unnamed in said paper. Peano's logical unfolding of the empty set leads to the concept of the set of natural numbers out of an empty basket. At this point the paper at hand differs from Krasnoholovets elaboration. As from here, the full stability of set theory and its axioms should be valid which Krasnoholovets misses; with missing stability the abstractions can not lead to stabile constructions like elementary particles. Besides the obligatory unlimited adoption of Fermelo-Fraenkel axioms into all considerations of fundamentally combining mathematics and physics, special attention is to be given to investigating the ether as a combination of interdependent space layers. As can be seen from illustration A, consideration of the set of natural numbers as per Krasnoholovets must be complemented with gap filling spaces of higher order. The apparent mathematical interdependence between the main set and its gap-filling subsets must be a structure of power sets since there is no other stabile theoretical vehicle available; this allows full scale structured infinity in a chaotic overall environment: the root of the tree of physical reality is embedded in power sets within ether. Finally, we wish to mention, that Krasnoholovets' elaboration doesn't address how to treat the limitless limitation called "infinity" in the mathematical-physical context under his investigation. These are reasons why a further approach is imminent. Sets, axioms, infinities, power sets, stability, physical-mathematical formulas all these properties can be found back in a the Set Theory of George Cantor when applied as descriptive of the ether. the application of Cantor's Continuum as structure of the ether is subject of this document.

PROPOSED MATHEMATICAL-PHYSICAL ETHER

The theory of sets of natural and real numbers postulates a continuum of interdependent infinities that have cardinal numbers that are called A0 A1 A2 A3 etc. A0 and A1 are related through the theory of numbers where A0 is the cardinal number of the set of natural numbers and A1 the cardinal number of the set of real numbers which includes over and above natural numbers irrational numbers [1-2]. The (basic) sets of natural and real numbers with cardinality A0/A1 form a body or corpuscle. Further corpuscles arise by using the axiom for creating power sets. The cardinal numbers between the several power sets can be shown to have the following relation [1]:

$$A^{n+1} = 2^{An} \tag{1}$$

The mathematical principles of numerical set theory are based upon Zermelo-Fränkel-axioms and upon the vast number theory as from Pythagoras via Plato, Euler, Gauss and many others. In the nineteenth century so-called aktual infinities were added by Georg Cantor which made the infinite set of natural/real numbers to a field with limitations that we define on one side by elements $1/A_0$ and on the other side by actual infinity A_0. The corresponding corpuscle is bordered by its first power set with cardinality 2^{An}. This gives a next corpuscle. Georg Cantor introduced the continuum as the range of power sets of the infinite set of natural numbers. This paper relates the corresponding infinite sets (mathematical corpora) to as many spaces (virtual-physical corpora): Cantor's mathematical continuum describes virtual-physical layers of spaces.

These corpora are perfect interdependent mathematical entities. They contain any descriptive mathematical possibility, i.e., can be seen as a drum filled with executable possibilities.

So, this study postulates the corpuscle of the field of numbers to be congruent with the vacuum of a galaxy as a basis for all that exists. In doing so, the basis for a experimentally proven, philosophically acceptable, logically functioning nature is given, in as far the mathematical origin of quanta, particles, atoms etc. can be tracked back within its vacuum, described and more phenomena be found respectively explained. To achieve this is the aim of this study. Infinities, expressed with the cardinal numbers $A_0 ... A_n$, are the highest numbers available within the layer of spaces. All natural constants (but the velocity of electromagnetic waves) are related to very small figures. In contrast, the velocity of electromagnetic waves (in vacuum) is related to a very large number. Therefore, first of all,we postulate that the speed of electrical-mechanical waves which is the highest relation between meter and second in the relationship "speed" is related to A0/A1 being the highest number available within the vacuum. More generally spoken, we postulate that the speed of electrical-mechanical waves in a particular space(n) is related to the corresponding cardinality of its space(n) A_n. So we can write:

$c :: A_0$ and $c1 :: 2^{A_0}$

Experiments show two distinct fields of natural EM-waves: the realm around visible light from radio waves to gamma rays. Secondly, the cosmic microwave background. It is known that the latter is disputably related to events around the postulated big bang.

SPACES

Cantor's theory of the continuum if applied to the vacuum results in a structured, interdependent, in principle infinite number of spaces. The first space, the vacuum, is equal to Einstein's block universe. This paper is about the consequences of the vacuum as well as the physical influences of the layer of spaces as per Cantor's continuum. It is to be noted that the second space should not be considered to be a consequence of the vacuum; but, the structure of space 2 *as related to* the vacuum is rather to be seen as cleverly adapted; that's to say: within the layer of spaces Einstein's block universe (the vacuum) is firmly rooted, where the term "firmly" means: mathematically and logically sound for which the set axioms are valid. This paper supports that Cantor's Continuum is the only backbone structure that provides a completely sound logic foundation.

OBSERVER. INNER AND OUTER AWARENESS

Studying nature as a consequence of such an abstract basic empty structure like a quantized ether/vacuum requires the ability to awareness of the availability of multiple/infinite possibilities along with a selective mechanism of those possibilities that are factually to be realized; in fact, this is daily human experience. Said attributes can be brought back to a provider-function and an observer-function. In a abstract/metaphysical field of well-ordered vacuum quanta the observer function is virtual whereas, in a physical field, the observer function is real. In both cases the observer function is propelled by these two questions "What am I? Where am I?" The Provider function is given by "I am." The *virtual* observer has one single physical awareness: distance, the ability of specifically discerning between near and far or, in wider sense, clear and vague. The observer in the natural field (the world's physical environment) has senses (sometimes supported by tools) for identification of "who/what and where?" Facts (elementary particles and "is-realizations") and events (complex sequences involving facts) are the result of a contract between provider and observer involving the parameter "time." With this physical and metaphysical understanding the ether will be discussed more in detail (See the following Figures 1 through 3).

Figure 1 shows the massless vacuum as lowest-energy state but allowing quantum effects. Figure 2 shows a quantized vacuum filled with vacuum-quanta as actual transporters of quantum effects Figure 3 shows two vacuum quanta with an inner subdivision into an infinite number of space quanta (yellow) rendering them stability (for the observer).

Figure 1. Vacuum

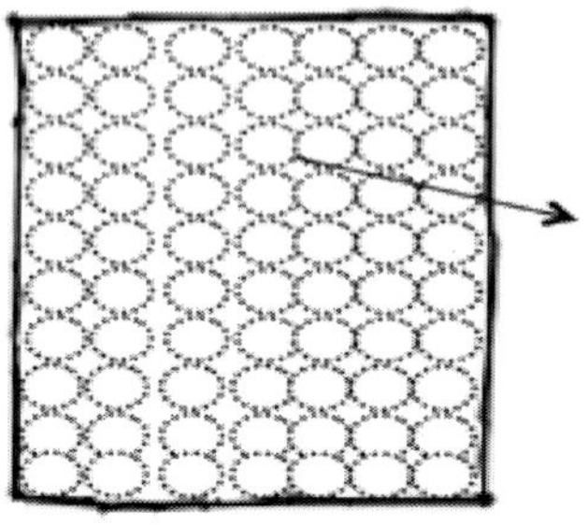

Figure 2. Vacuum quanta.

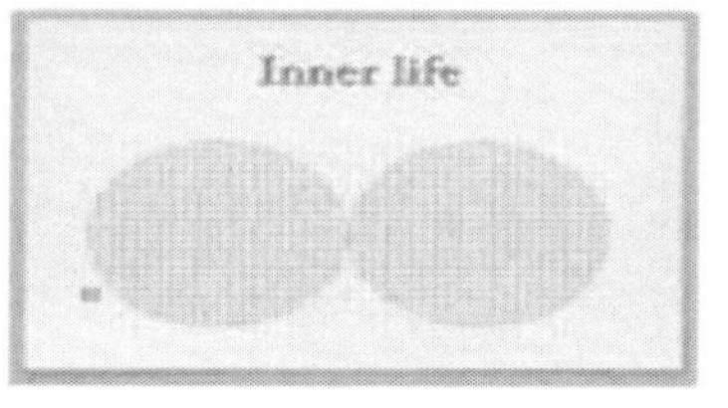

Figure 3. Vacuum quanta inner structure.

Vacuum quanta shown in Figure 3 have an inner reference and an outer reference. In order to investigate these both relationships we will take the virtual position O of one of the inner space quanta being placed on the surface of the vacuum quantum as shown in Figure 4.

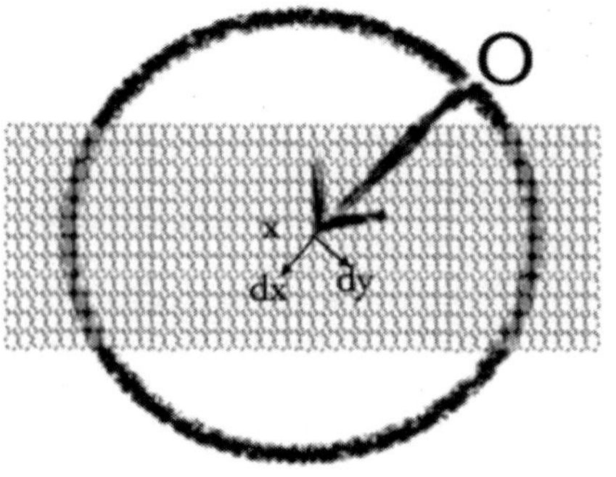

Figure 4. Vacuum Quantum = SQ1; O = random observer located on the surface; of the vacuum quantum. "What am I?; Where am I?"; x = inner actual infinity.

The vacuum quantum is a mathematical structure with vibrational inner elements which are spacequanta. For reasons of distinction, the elements of quantizing the vacuum will be called vacuumquanta and the elements of inner quantizing of the vacuum quantum will be called space quanta. The space quanta form, for reasons of countability rendering required stability, a field of numbers mounting to aktual infinity at x (seen from the standpoint of O) or vice versa.

Observer O located at the surface of SQ1 recognizes elements of own size within SQ1 until being confronted with such large inner distances x for which is he unable to make the distinction between x respectively x + 1, x + 2 etc. In order to make sense of the vague object formed by x, x + 1. x + 2, x + 3 etc., O groups those in a corpuscle in which he looks at each element x, x + 1, x + 2, x + 3 etc. one by one in a strict sequence. O will distinguish two different patterns: one along the direction dx he is looking at and one in the direction yd perpendicular on the direction he is looking at but at the distance of aktual infinity x. The corpuscle will be defined when x+n is at a distance where O is unable to distinguish the element x + n. As from there, O continues ordering elements in the given sequence but now with decreasing distance x + n, x + n-1, x + n-2 etc. until x is reached again: one period of the corpuscle is completed. In dy direction O will not have this difficulty, the distance x does't change; O will observe a circular movement whilst following the same sequential ordering of elements. To make a long story short: O observes waves in longitudinal and transversal direction. The wavelength in longitudinal direction dx must be equal to the wavelength in transversal direction yd which equals 2π*x. x equals A0 as has been shown above. When all this should be the origin of an electromechanical wave which it resembles.

Remark: Note that as long as nothing else happens, O will experience a stable environment in which all the waves he can imagine at distance x are compensated by left and right running waves such as to form standing waves. From the standpoint of an observer the inner structure exist of a solid structure plus, at infinity, of standing waves. The SRT formulas (Lorentz equations) describe this stability, not more and not less.

The picture we described above looks as follows:

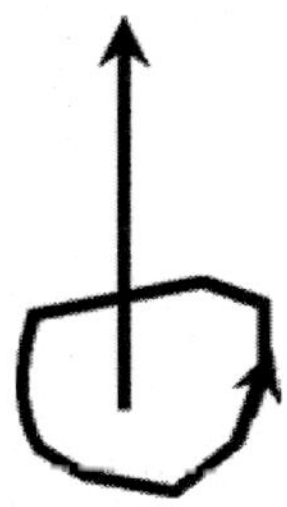

It is to be noted that the longitudinal sinus is perpendicular on the real space formed by near space quanta. A midpoint can be allocated. This sinus can easily be associated with the electrical field and with the electron.

Similarly, the transversal sinus is seen perpendicular on the electrical field but circling in a infinite path on which no starting point or sourcing can be identified. The field seems parallel to the volume whilst a double perpendicularity can be identified turning it back to being parallel. I doubt that this is the case; it seems more like a non-touch situation "at an inner distance x." O recognizes a magnetic plane "within." Medical applications have been reported. Themes like memory, brain and subconscious or communications on distance can be scientifically approached.

Remark: the shown magnetic property is a property of space-quanta, i.e., to space! Whatever the implications of this understanding of a magnetic medium may be, it will not be directly related to a specific material: to the contrary, a connection at - what can be called - a magnetic oriented space layer is underlaying and substratum to any material.

Figure 6 has far-reaching implications for the understanding of physics and metaphysics. In the centre of awareness is the observer O positioned at a decisive point of discrimination between "inside" and "outside." Both "outside" and "inside" reflect infinities. Figure 6 shows the inside perspective. The virtual observer O (What am I? Where am I?) acknowledges at $x = A_0$ (i.e., aktual infinity) vagueness that he can overcome by grouping the space quanta elements identified nearby. He identified the nearby space quanta between 0 and x by counting them which impressed as a field of natural numbers. At $x = A_0$ (aktual infinity) he needs a new instrument: grouping of spacequanta for the sake of achieving an identifiable corpus of sufficient size. Along his vista, by introducing an ordering one by one observation, he observes a longitudinal pattern. In the same period, and perpendicular on his vista, he observes a transversal wave-pattern. Effectively, he observes a electrical and a magnetic property. He acknowledges a potential (but extinguished) corpus existing of electrical and transversal waves. The corpus has, along the x-axis, electrical capacity related to distance to $x = A_0$. The measure of electric capacity is generally known as the vacuum permittivity ε_0 expressed in Farad/meter. Similarly, the corpus has, perpendicular on the x-axis but at $x = A_0$, magnetic capacity related to distance to $x = A_0$, amounting to what, in modern terms, is called magnetic constant, the former magnetic permeability μ_0.

This picture changes promptly when the random observer O, by impulse, jumps into position $x = A_0$. Corpus and observer change places; a movement in m/s is given and its speed is directly related to the observations ε0 and μ0. For reasons of symmetry, the distance A_0 is close to or equals 1/2d whilst d is the diameter of the vacuumquantum under observation. The realized agitation causing the above described em-corpus involved the whole vacuumquantum; not more and not less. The ejected em-corpus leaves a gap in the vacuum quantum in which only this particular corpus fits: em-corpus and vacuumquantum remain connected. It agitates vacuumquanta as it passes thru the vacuum. Note that observer and corpus have become identical. The direction of movement is along the x-axis: i.e., perpendicular on the vacuum inner volume represented by vacuumquanta. In 1678, Huygens proposed that every point which a luminous disturbance reaches becomes a source of a spherical wave. Whatever this implies, Huygens' view compares very well with the Observer-Provider view of Cantor's Continuum. The distance between location x and Observer O is 0. The medium for EM-fields is space quanta; it will be seen that the main medium for gravitation is vacuumquanta. A small portion is attributed by spacequanta.

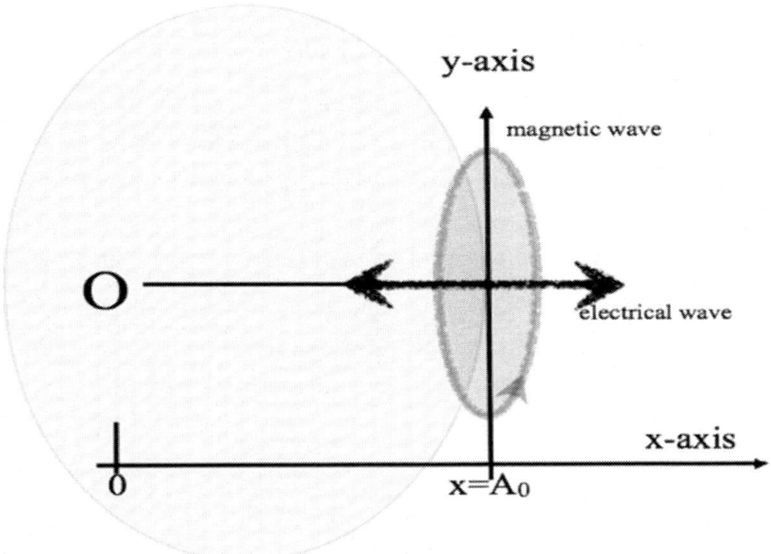

Figure 6. Inside perspective

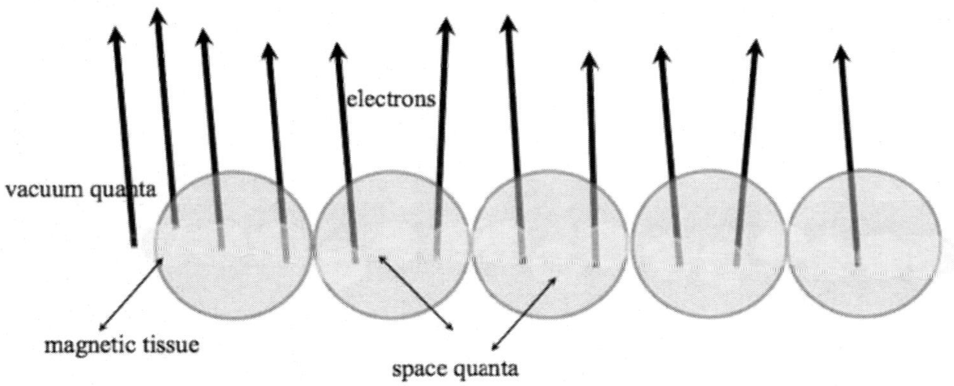

Figure 7. Vacuum and space schematic.

The electrical capacity stands perpendicular on the vacuum volume. The magnetic function is parallel to the vacuum volume but as an inner tissue. As long as em-waves exist as such, the magnetic tissue within the space quanta doesn't become apparent. However, when, for colliding reason, two em-waves interfere to become a positron and an electron, their magnetic components become loose elements of the magnetic tissue constituting the magnetic character of the space quanta at this level. The tail of the electrical vector is rooted in the magnetic tissue. Its direction is perpendicular on the magnetic tissue, i.e., parallel to vacuum. "One sits on his electrons" is a saying; it seems to reflect truth.

Electrical currents, vacuum level reality, electrons and magnetic tissue remind of conscious and subconscious brain levels. Set theory can assist in the search for physical components that allow interaction between consciousness, subconsciousness at a sub-level of space that is common substratum of all material and vacuum/space alike within a realm of collective subconsciousness.

Moving one step on, ideas can easily be developed describing an inner ether body of structured space layers with an outer material hull, suitable for life on earth.

These deductions and conclusions are possible upon positioning the observer O at the appropriate position between an ontological abstract of volume quanta with inner and outer infinity. For reasons of stability, The observer O requires the well-ordering of quanta applying the corresponding axioms. In the absence of O, order will not be established; all is brought back to multiple possibilities, i.e., chaos, which can be compared to the universal property of entropy: action and for that reason energy is required for the sake of maintaining order.

Furthermore, in case O observes empty sets of same definition in inner or outer vista, then, Cantor's Continuum requires a sound mathematical relationship. For example: vacuum quanta and space quanta are constructed as set and power set. Or: gravitation, electrical field and magnetic field as examples of the empty set Ø(field) are constructed as to satisfy the relation power set too. This interdependence between natural fields allows the necessary stabilizing logical structure within the set with elements "fields."

Apart from these considerations and opening doors to physically understanding other disciplines at this inner level, the proof of its connection to experimental physics is to be given. For this reason we do one step back and analyze the possible relation between the ether mathematics of Cantor's Continuum, and what is called the "classical vacuum," represented by Maxwell's equations.

Finally, wave theory shows that the above descriptions that base upon considerations/observations of Cantor's continuum lead to Maxwell Equations as follows:

Two opposite running functions f(x+ct) and g(x-ct) are given as sum u(t,x):

$$u(t,x) = f(x + ct) + g(x-ct)$$

These are solutions of this differential equation:

$$1/c2 * \partial 2u/\partial t2 - \partial 2u/\partial x2 = 0$$

One of Maxwell's four equations is already given by the above deduction of a pole-less magnetic field:

$$\vec{\nabla} \cdot \underline{\vec{B}} = 0$$

The other equations

$$\vec{\nabla} \cdot \underline{\vec{D}} = \rho$$
$$\vec{\nabla} \times \underline{\vec{E}} = -j\,\omega\,\underline{\vec{B}}$$
$$\vec{\nabla} \times \underline{\vec{H}} = \underline{\vec{J}} + j\omega\underline{\vec{D}} = (\sigma + j\,\omega\,\varepsilon)\ \underline{\vec{E}}$$

will follow upon breaking up the standing waves into two electromechanical waves or photons. Physically spoken this can happen when the observer O for some reason suddenly appears at position x, as shown before. An impulse should trigger this which is subject to future mathematical investigation.

In another collision two photons collide to become electron/positron pair. This is shown in the following picture that shows that the antimatter-particle positron is separated from its partner electron by the magnetic tissue. It is propelled into a further space layer where it remains. Schematically, this is shown in the next Figure 8, but the related interference formula is yet to be determined.

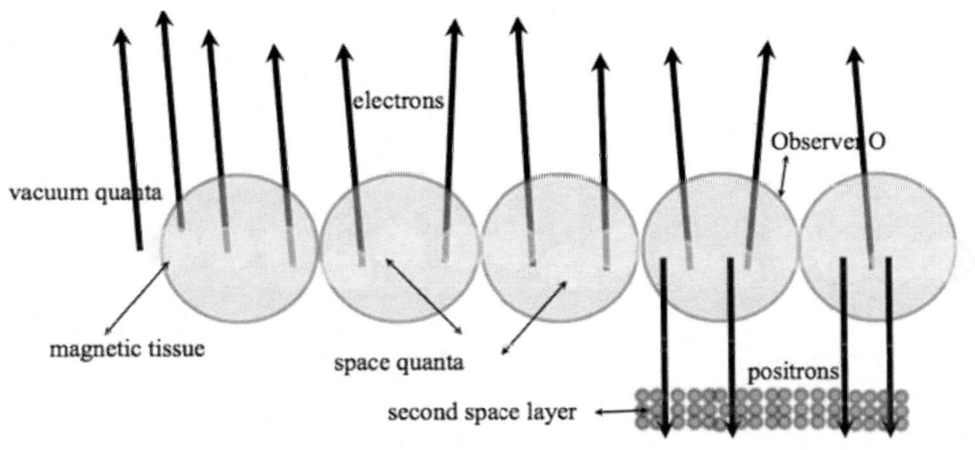

Figure 8. Vacuum and space schematic.

The second space layer is the consequence of quantizing of the space quanta into a further infinity that is a power set in the continuum. In this second space layer that is beyond the direct access of the observer O the positrons enter. The qualitative investigation of this second space layer starts with listing phenomena as follows: the second space layer is the home of

1) electro-mechanical cosmic microwave background.
2) particles with electrical charge +2/3*e and -1/3*e (peon/eon)
3) three families of quarks
4) shelter for positrons
5) the strong interaction
6) the tunnel effect, entanglement plus

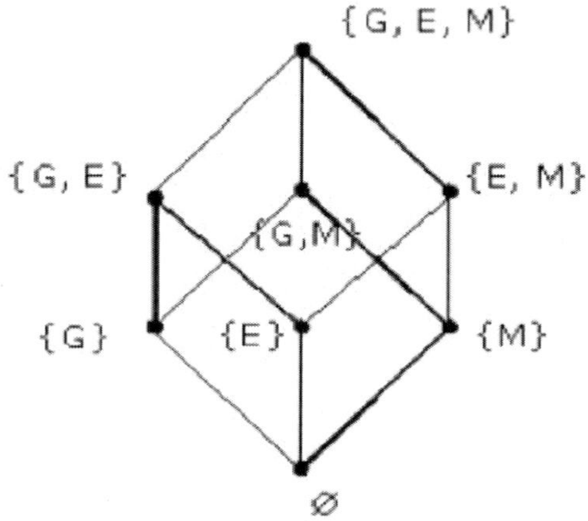

Figure 9. The Cube as image of the natural forces carrying the material world.

Hypothesis: the following logic is adapted which is borrowed from Niels Bohr atomic model: interference between cosmic microwaves causes the production of three hypothetical particles per each collision: two particles with electrical charge -1/3*e and one particle +2/3*e. We propose the names eon and peon for these particles. In analogy to Niels Bohr's atom model these hypothetical particles group in orbits around the positron to form the proton and the quarks as follows:

proton = positron + 2*eon + peon with eons and peon in second space orbits and the positron in the centre

The complicated resulting construction is observed at vacuum level as building stone of the nuclei. The cloud of loads thus available within the proton causes several experimental results for the same construction: up-quarks, down-quarks and protons. This could explain why quarks can't be brought into isolation. Quark-families can be understood to be excited protons that, as it is with excited atoms, are unstable, short-living protonic status.

When a electron is added in an orbit at vacuum level (the vacuum space numbering system applies), the observer at vacuum level observes a neutron. At protonic level the numbering system of the second space layer applies. The several numbering systems are stabile by power set interdependency. Conclusively, the consequences of the second space layer are hidden, along with antimatter, in the deepest core of the atom's nucleus. Eons, peons, positrons and electrons and their derivates attribute to mass and gravitation. Because of the mass attribution in several space layers, a simple overall causal gravity theory that is more than a mathematical formula describing experimental results (Newton) seems difficult to be formulated. The position of the source of em-waves having mass at rest in their backpack must play a role.

It is suggested that the processes qualitatively analyzed above do cause the observation of cosmic rays as well as the vast clouds of hydrogen clouds in the universe.

The three vacuum/space fields, grouped within the set of fields, with elements gravitational, electrical and magnetic, should become visible as the fields' power set in the second space layer, i.e., with 2^3 fields. Compare the theory quantum chromodynamics. Applying set threory, the classical theory of nine gluons should be corrected to eight. This corresponds to results of actual experiments. In the second space layer the impact of the natural forces can be shown in the above shown Figure 9, in which G is the gravitational force, E means electrical force and M means magnetic force. The power set of these three forces of the vacuum can be illustrated as a perfect cube. Its significance is the presentation in one picture of all known forces relevant to the construction of the material world that are effective in vacuum, space layer and the second space layer. It reminds of the instructions given to Moses at the Sinai Mountain as to the construction of the Tabernacle that was instructed to be a perfect cube. No cube is more perfect than a mathematical cube.

A religious commentator of the Tabernacle construction, Müsä ibn Maymün, who is reported to be strongly influenced by Islamic Sciences and Culture, comments Moses' Exodus with the remark that the Tabernacle was representative of the microcosmos and its elements symbolize the major elements of Creation. Compare: The Kaaba or Kaaba, Arabic al-Kaʿbah "The Cube."

In the next final section of this multi-disciplinary model study, the observer O focusses on his outer vista.

Earlier, the virtual Observer O focussed on the infinite (infinitesimal) inner structure of the corpus of the vacuumquantum of which he is give to be one random surface spacequantum. Now, we will have the Observer focus on his outer surroundings as follows: looking outward, O is surrounded by a field of vacuumquanta that he can count 1,2,3,...n.

This is per Cantor's definition the field of numbers. Note that, also in this case, distance is the observer's decisive factor; it is an issue of topology. The field of numbers is known to be congruent with the gravitational field. So, the outer awareness of the virtual observer O is that of a gravitational field. Gravitation is a topological issue with an strong outer impact and a very small inner impact the latter possibly being the origin of dark matter.

Conclusion

It is seen that the attempt of introducing Cantor's Continuum as fundament of the universe has severe complications. Its undertaking is a step into darkness but its results may shed new light on science.

First of all, its introduction leads to multi-disciplinary issues for which in each discipline answers and credos already have been formulated. After all, the nature of set theory is crossing borders since it connects all together. But then, wasn't this connecting the quest right from the start? What else can be expected when going after Einstein's quest looking for a theory of everything? So, not only mathematical and physical problems but also sensitive philosophical and even religious ideas were touched during the ongoing research of the last fifteen years.

Secondly, whilst arguing along the lines of Cantor's logic, very different answers to the list of open questions of contemporary physics were formulated. Whatever value these new answers have, they clearly contradicted some contemporary positions. Fortunately, these are convictions concerning issues at the very borders of our scientific knowledge, i.e., looking outward into far space or looking inward into smallest particles.

Whether the approach introduced in this chapter will be able to withstand apllication of Occam's Razor, future may tell. As per Razor's definition, the solutions offered should be simpler than the one existing. Sir Karl Popper postulated falsification. The sudden need among physicists for dark energy should have been an example of falsification. But it wasn't; an expression was invented but no new explanation.

Maybe this experiment could help: a possible experiment is the production of hydrogen in vacuum through interference of em-waves (laser beams) possessing suitable frequencies, such as waves of visible frequency should cause interference with em-waves of cosmic background radiation. Maybe, nature already is conducting this experiment by the creation of fast moving protons at the border of our atmosphere. But… that would be another falsification.

Postscript. Cantor's Continuum in Vacuum is a refreshing specific mathematical view on nature and physical facts and events, like a Theory of Everything. For that reason, further research needs to be done to prove its validity and value for future progress of theoretical and material knowledge. The following list shows some of the topics the author would propose to be investigated:

Compare and reconcile cantor's continuum with competing theories.

Experiment: make electrons and positrons thru interference of EM-Waves in vacuum chamber according to the mathematics of solitons.

Formulate and solve the mathematical trigger for the status change quaternion-electromagnetic wave

Describe the earth magnetic field against the background of interdependent electrical, mechanical and gravitational forces in accordance with the cube of natural forces

Explain the congruence of the formulas of gravitational and electrical forces

Research the inner structure of protons, quarks and neutrons in accordance with the new atomic nuclear model.

Investigate the origin of planets against the background of emerging fast protons outside of the atmosphere.

Describe the tunnel effect of photons and electrons in the continuum.

Reconcile the expressions x, dx, dx^2 with the first three spaces of the continuum.

Apply Cantor's Continuum to photon entanglement. Investigate possible experiments

Review the position of stars and planets when locally emerged thru interference of light waves. Compare the vase in the English/French garden.

Cantor's continuum and the Egyptian pyramids

Cantor's continuum and indian infinity philosophy

Cantor's continuum and western religion

REFERENCES

Aczel, Amir D. (2000), *The Mystery of the Aleph: Mathematics, the Kabbala, and the Search for Infinity,* New York: Four Walls Eight Windows Publishing. ISBN 0-7607-7778-0. Pages 226, 227, Does the continuum really exist?

Dauben, Joseph W. (June 1983), "Georg Cantor and the Origins of Transfinite Set Theory," *Scientific American*, 248 (6): 122–131, Bibcode: 1983SciAm.248f. 122D, doi:10.1038/scientificamerican0683-122.

Havva Shevkinaz Bulut und Arno Gorgels (July 2005), *The Superquantum Theory*, ISBN 3-00-016716-1, pages 60 – 63 solution of Cantor's continuum problem.

Michel Bounias, A theorem proving the irreversibility of the biological arrow of time, based on fixed points in the brain as a compact, Δ-complete topological space, *Amer. Inst. Physics' Con. Proc., Symposium* 8, 20–24 (2000).

Michel Bounias, The theory of something: a theorem supporting the conditions for existence of a physical universe, from the empty set to the biological self, Ed.: Dubois, D. M., *Int. J. Comput. Anticipatory Systems* 5-6, 1-14 (2000).

Volodymyr Krasnoholovets, The tessellattice of mother-space as a source and generator of matter and physical laws, in: *Einstein and Poincaré: The Physical Vacuum*, Ed.: V. V. Dvoeglazov (Apeiron, Montreal, 2006), pp. 143-153.

In: Old Problems and New Horizons in World Physics ISBN: 978-1-53615-430-6
Editors: Volodymyr Krasnoholovets et al.

Chapter 12

STATIONARY SPIRAL STRUCTURE AND COLLECTIVE MOTION OF THE STARS IN A SPIRAL GALAXY

Ying-Qiu Gu*
School of Mathematical Science,
Fudan University, Shanghai, China

ABSTRACT

Most fully developed galaxies have a vivid spiral structure, but the formation and evolution of the spiral structure are still an enigma in astrophysics. In this paper, we derive simplified dynamics for a spiral galaxy at first, and then solve approximate analytic solutions for a spiral galaxy. With these solutions, we can obtain manifest relations between mass density and speed, and arrive at some interesting conclusions: (I) The spiral pattern is a stationary or static structure of a density wave, and the barred galaxy globally rotates around an axis at a small angular speed. (II) All stars in the disc of a barred spiral galaxy move in almost circular orbits. (III) In the spiral arms, the speed of stars takes minimum and the stellar density takes maximum. (IV) The mass-energy density of the dark halo is compensatory for that of the disc, namely, it takes minimum in the spiral arms. This phenomenon implies that the stream lines of the dark halo are quite different from geodesics, which may be explained by nonlinear spinors.

Keywords: spiral structure, gravitomagnetic field, density wave, gravitational retarded potential

* Email: yqgu@fudan.edu.cn.

1. INTRODUCTION

Most fully developed galaxies have a vivid spiral structure, which has profound influences on the generation and evolution of stars. During the last 60 years, many efforts have been made to reveal the nature of the spiral structure. Despite the progress made in understanding the amplification mechanisms of the spiral arms and the role of the spiral arms in the dynamics of galaxies, the formation and evolution of the spiral structure are still an enigma in astrophysics [1, 2, 3]. We do not even know if the spirals are a long-lived phenomenon, or are short-lived and regenerated many times during the galactic evolution. C. C. Lin and F. H. Shu suggested that the spiral structure should be a quasi-stationary and stable density waves, which are excited by over-reflection at co-rotation due to some nonlinear effects [4, 5]. But some other opinions favor short-lived or recurrent spiral patterns which are developed in the galactic disks via swing-amplification in a rotating disk with shear [6], or by an external force [7]. The recent observational results seem to favor a long-lived pattern. In contrast the nearby galaxies with the distant ones observed through the Hubble Space Telescope and from the ground [8, 9], the results show that there is only mild evolution in the relationship between radial size and stellar mass for galactic disks within the redshift region $0 < z < 1$.

In the noncentral potential, the orbit of a star is usually an unclosed and complicated spatial curve, which sensitively depends on the initial speed and direction [1]. Therefore, the stellar system should be more reasonably treated as fluid rather than mass points. Although the dynamics for the two models is essentially the same Newtonian mechanics, the initial and boundary conditions are different. In the case of fluid, the stream lines are a consistent field, which are a natural result of the generation of stars from the nebulae.

W. Dehnen and J. Binney fitted the mass distribution within the Milky Way to the observational data, such as the rotation curve and Oort's constants [10], but they found the fitting mass distribution was ill determined. In [11, 12], the authors made numerical simulations for 2-dimensional stellar hydrodynamics of a flat galactic disc embedded in dark matter halo, and the model solveed the Boltzmann equations up to second order moment [1]. A dynamical approach to explain the formation of both spirals and rings in barred galaxies was proposed in [13, 14]. It is based on the orbital motion driven by the unstable equilibrium points of a given rotating bar potential, and then the spirals, rings, and pseudo-rings are related to the invariant manifolds associated to the periodic orbits around these equilibrium points. By adjusting dynamical parameters of the host galaxy, we get the spiral and ring like structures.

If arms of a spiral galaxy were constructed by fixed material, the arms would become more and more tightly wound, since the matter nearer to the center of the galaxy rotates faster than the matter at the edge of the galaxy. The arms would become indistinguishable from the rest of the galaxy after only a few orbits. In order to avoid this winding problem, Lin and Shu proposed the density wave theory [4, 5]. They suggested that the arms are not constructed by stationary materials, but instead made up of areas of greater density similar to a traffic jam on a highway. More specifically, for some dynamical reasons, the luminous materials such as stars intend to move slowly and spend more time inside the spiral arms and move rapidly outside the spiral arms. Then, an inhomogeneous and quasi-stationary distribution of mass density wave is developed, which forms the spiral arms. In the sense of morphology, this theory is quite successful in some aspects. However, the dynamical explanation for the formation of the density wave is still unclear, and almost all dynamical calculations overlooked the retarted potential of gravity for the large scale structure. The analyses of this paper attempts to give a clear dynamical explanation for the density wave.

According to the following observational facts, we find the problem can be simplified and described by an idealized fluid model, and then the nature of the spiral structure might be disclosed by dynamical approach.

1. By observation, except for the center part, all stars in a spiral galaxy are mainly distributed in a thin disc, so the spiral structure should be a visible display of collective motion of these stars, and the collective motion can be approximately described by the 2-dimensional hydrodynamics.
2. The collision among stars rarely occurs, so the fluid of star flow is zero-pressure and inviscid, and the stars move along geodesics.
3. All stars are driven by the average background gravity of total mass-energy distribution in the galaxy, and the background mass density and gravity can be constrained by some empirical data such as the flat rotation curve.
4. These are some basic assumptions of the following calculations. The results show that, the stationary solutions to such system indeed exist, and the distributions of mass density and speed do have stable spiral structure. So this idealized model and the solutions might shed some lights on the enigma of galactic structure.

2. Dynamical Equations for the Galaxy

At first, we establish the dynamical equations to describe galactic systems from Einstein's relativistic dynamics by weak-field and low-speed approximation. Some fundamental contents can be found in [15], but here we make a more systematic and detailed derivation for galactic dynamics.

In the context of general relativity, the whole dynamical equation system for galactic evolution should be the Einstein's field equation

$$G^{\mu\nu} \equiv R^{\mu\nu} - \frac{1}{2} g^{\mu\nu} R = -\kappa T^{\mu\nu}, (\kappa = \frac{8\rho G}{c^4}), \tag{2.1}$$

combined with the energy-momentum conservation law and the equation of state of the gravitating source. In most cases, the total classical approximation for energy-momentum tensor takes the following form [16, 17, 18]

$$T^{\mu\nu} = (\rho_{\text{tot}} + P)U^\mu U^\nu + (W - P)g^{\mu\nu}, \tag{2.2}$$

where W corresponds to nonlinear potentials, which acts as negative pressure and leads to the deviation from geodesic. According to the energy-momentum conservation law or Bianchi identity $T^{\mu\nu}_{;\nu} = 0$, we can derive the continuity equation $U_\mu T^{\mu\nu}_{;\nu} = 0$ and the equation of motion for the source as follows

$$U^\mu \partial_\mu(\rho_{\text{tot}} + W) = -(\rho_{\text{tot}} + P)U^{\mu\nu}_{;\nu}, \tag{2.3}$$

$$(\rho_{\text{tot}} + P)U^\nu U^\mu_{;\nu} = (g^{\mu\nu} - U^\mu U^\nu)\partial_\nu(P - W). \tag{2.4}$$

In case with $W \sim \rho_{\text{tot}} \gg P$ such as for the nonlinear spinor, we find the stream lines are quite different from the geodesics $U^\nu U^\mu_{;\nu} = 0$. So a fully relativistic simulation for the dynamics of galaxy should include such terms. However, in the following non-relativistic approximation, the effects of (P, W) are merged into one effective mass-energy density ρ, it becomes much simpler.

For convenience, we take $c = 1$ as unit of velocity. Noticing the facts that the collisions among stars rarely occur, and the trajectories of the ordinary matter such as electrons and baryons are almost geodesics, so for the stars, the following zero-pressure and inviscid energy-momentum tensor holds

$$T^{\mu\nu}_s = \rho_s U^\mu U^\nu, \tag{2.5}$$

in which ρ_s is the comoving mass density of the stars, and U^μ is the 4-vector speed of the stellar flow. Since the ordinary matter satisfies the mass-energy conservation law independent of the dark halo, we have $T^{\mu\nu}_{s;\nu} = 0$. Expressing it in the form of equations of continuity and motion, we get the dynamical equations for the stars

$$U^\mu \partial_\mu \rho_s + \rho_s U^\mu_{;\mu} = 0, U^\nu U^\nu_{;\nu} = 0. \tag{2.6}$$

The total energy-momentum tensor of the galaxy is still given by (2.2), and satisfies the dynamical equations (2.3) and (2.4). By (2.1) and (2.2), we get

$$R = \kappa(\rho_{\text{tot}} + 4W - 3P), \tag{2.7}$$

where $R = g_{\mu\nu}R^{\mu\nu}$ is the scalar curvature. Substituting (2.7) into (2.1), we get

$$R^{\mu\nu} = -\kappa(\rho_{\text{tot}} + P)U^\mu U^\nu + \frac{1}{2}\kappa(\rho_{\text{tot}} + 2W - P)g^{\mu\nu}, \tag{2.8}$$

where U^μ is the average 4-vector speed of all gravitating source.

In order to make weak-field approximation, we choose the harmonic coordinate system, which leads to usual Cartesian coordinate system when making linearization of metric. Then we have the de Donder coordinate condition

$$\Gamma^\mu \equiv g^{\alpha\beta}\Gamma^\mu_{\alpha\beta} = -\frac{1}{\sqrt{g}}\partial_\nu\left(\sqrt{g}g^{\mu\nu}\right) = 0, \tag{2.9}$$

where $g = |\det(g)|$. Denote the Minkowski metric by

$$\eta_{\mu\nu} = \eta^{\mu\nu} = \text{diag}\,(1,-1,-1,-1). \tag{2.10}$$

For weak-field approximation, we have the linearization for the metric

$$g_{\mu\nu} \equiv \eta_{\mu\nu} + h_{\mu\nu}, g^{\mu\nu} \triangleq \eta^{\mu\nu} - h^{\mu\nu}, \tag{2.11}$$

$$h^{\mu\nu} = \eta^{\mu\alpha}\eta^{\nu\beta}h_{\alpha\beta}\,, h = h^\mu_\nu = \eta^{\mu\nu}\eta_{\mu\nu}, \tag{2.12}$$

$$g \triangleq 1+h, \sqrt{g} \triangleq 1+\frac{1}{2}h. \tag{2.13}$$

In what follows, we directly use = to replace $\triangleq$ for convenience. By straightforward calculation, we get the linearization for other parameters

$$\Gamma^\mu_{\alpha\beta} = \frac{1}{2}\eta^{\mu\nu}\left(\partial_\alpha h_{\nu\beta} + \partial_\beta h_{\alpha\nu} - \partial_\nu h_{\alpha\beta}\right), \tag{2.14}$$

$$\Gamma^\mu = \partial_\nu\left(h^{\mu\nu} - \frac{1}{2}\eta^{\mu\nu}h\right), \tag{2.15}$$

$$R_{\mu\nu} = \frac{1}{2}\partial_\alpha\partial^\alpha h_{\mu\nu} - \frac{1}{2}(\eta_{\mu\alpha}\partial_\nu\Gamma^\alpha + \eta_{\nu\alpha}\partial_\mu\Gamma^\alpha), \tag{2.16}$$

$$R^{\mu\nu} = \frac{1}{2}\partial_\alpha\partial^\alpha h^{\mu\nu} - \frac{1}{2}(\eta^{\mu\alpha}\partial_\alpha\Gamma^\nu + \eta^{\nu\alpha}\partial_\alpha\Gamma^\mu), \tag{2.17}$$

$$R = \frac{1}{2}\partial_\alpha\partial^\alpha h - \partial_\alpha\Gamma^\alpha. \tag{2.18}$$

In which $\partial_\alpha \partial^\alpha = \partial_t^2 - \nabla^2$ is the d'Alembert operator.
In the harmonic coordinate system, we have

$$\Gamma^\mu = \partial_\nu \left(h^{\mu\nu} - \frac{1}{2}\eta^{\mu\nu} h \right), \tag{2.19}$$

$$R_{\mu\nu} = \frac{1}{2}\partial_\alpha \partial^\alpha h_{\mu\nu}, R^{\mu\nu} = \frac{1}{2}\partial_\alpha \partial^\alpha h^{\mu\nu}, \tag{2.20}$$

$$R = \frac{1}{2}\partial_\alpha \partial^\alpha h, G^{\mu\nu} = \frac{1}{2}\partial_\alpha \partial^\alpha (h^{\mu\nu} - \frac{1}{2}\eta^{\mu\nu} h). \tag{2.21}$$

By (2.19) and (2.21) we find if $\Gamma^\mu = 0, \partial_t \Gamma^\mu = 0$ at any given time $t = t_0$, it will always hold due to the Bianchi identity $G^{\mu\nu}_{;\nu} = 0$.

In order to compare with electromagnetism and to understand the physical meaning of the parameters, denote

$$\Phi = \frac{1}{2}h_{tt} = \frac{1}{2}h^{tt}, \vec{A} = (h^{tx}, h^{ty}, h^{tz}) = -(h_{tx}, h_{ty}, h_{tz}), \tag{2.22}$$

$$H = (h_{ab}) = (h^{ab}), (\{a,b\} \in \{1,2,3\}), \vec{B} = \nabla \times \vec{A}. \tag{2.23}$$

In the International System of Units, we have the order of magnitude for the components of metric

$$c^2 |h^{ab}| \sim c|A_k| \sim |\Phi| \ll 1, (a \neq b), \tag{2.24}$$

which means $|h_{ab}| \ll |A_k| \ll |\Phi| \ll 1$ if taking $c = 1$ as unit.

For the present purpose, we define the stellar speed $\vec{V}$ by

$$\vec{V} \equiv \frac{1}{U^0}(U^1, U^2, U^3), \tag{2.25}$$

which is approximately equivalent to the usual definition. For galaxies, we have the following order of magnitude

$$|\vec{V}| \sim 300 \text{ km/s} = 10^{-3}\, c, \vec{A} \sim \kappa \vec{V}, h_{ab} \sim \kappa\, |\vec{V}|^2, (a \neq b), \tag{2.26}$$

in which the coefficient κ is also a number of little value. Then according to

$$1 = \sqrt{g_{\mu\nu} U^\mu U^\nu} = (1 + 2\Phi - 2\vec{A} \cdot \vec{V} + g_{ab} V^a V^b)^{1/2} U^0, \tag{2.27}$$

by omitting $O(V^2)$ terms, the low-speed assumption gives

$$U^0 = 1 - \Phi + \vec{A} \cdot \vec{V}. \tag{2.28}$$

Substituting (2.25) and (2.28) into (2.6) and omitting the high order terms, we get the continuity equation and motion equation for stars

$$\left(\partial_t + \vec{V} \cdot \nabla\right) \rho_s = -\rho_s \left[\nabla \cdot \vec{V} + (\partial_t \Phi + \nabla \cdot \vec{A})\right], \tag{2.29}$$

$$\left(\partial_t + \vec{V} \cdot \nabla\right) \vec{V} = -\nabla\Phi + \left(-\partial_t \vec{A} + \vec{V} \partial_t \Phi\right) + \vec{V} \times \vec{B} + \vec{V} \cdot \partial_t H. \tag{2.30}$$

In (2.29), we used the de Donder condition $\Gamma^0 = 0$ in the form

$$\frac{1}{2} \partial_t \left(h_{xx} + h_{yy} + h_{zz}\right) = -(\partial_t \Phi + \nabla \cdot \vec{A}). \tag{2.31}$$

The equation of motion (2.30) has a similar structure to the electrodynamics. From it we learn that, Φ gives the Newtonian gravitational potential, and $\vec{A}$ leads to gravimagnetic field $\vec{B}$.

By (2.7) and (2.21), we have

$$\partial_\alpha \partial^\alpha h = 2\kappa(\rho_{\text{tot}} + 4W - 3P). \tag{2.32}$$

By (2.32), (2.8) and (2.20), we get the dynamical equations for $h^{\mu\nu}$

$$\partial_\alpha \partial^\alpha h^{\mu\nu} = -2\kappa(\rho_{\text{tot}} + P) U^\mu U^\nu + \kappa(\rho_{\text{tot}} + 2W - P)\eta^{\mu\nu}, \tag{2.33}$$

$$\partial_\alpha \partial^\alpha \chi^{\mu\nu} = -2\kappa \left[(\rho_{\text{tot}} + P) U^\mu U^\nu + (W - P)\eta^{\mu\nu}\right], \tag{2.34}$$

where $\chi^{\mu\nu} = h^{\mu\nu} - \frac{1}{2}\eta^{\mu\nu} h$. If the average speed of the dark halo is also small, omitting $O(\vec{U}^2)$ from (2.33) we get $h^{xx} = h^{yy} = h^{zz} \equiv 2\Psi, h^{ab} = 0, (a \neq b)$ and

$$\partial_\alpha \partial^\alpha \Phi = \frac{1}{2} \partial_\alpha \partial^\alpha h^{00} = -4\pi G \rho, \tag{2.35}$$

$$\partial_\alpha \partial^\alpha \Psi = \frac{1}{2} \partial_\alpha \partial^\alpha h^{kk} = -4\pi G \tilde{\rho}, \tag{2.36}$$

where ρ and $\tilde{\rho}$ are the effective mass densities with little difference. Their zeroth order approximation gives

$$\rho = \rho_{\text{tot}}[(U^0)^2 - 1] - 2W + P[2(U^0)^2 + 1] \triangleq \rho_{\text{tot}} - 2W + 3P, \tag{2.37}$$

$$\tilde{\rho} \triangleq \rho_{\text{tot}} + 2W - P. \tag{2.38}$$

In the following discussion, only the zeroth order approximation of (2.29), (2.30) and (2.35) are involved.

3. Hydrodynamics for Stars in a Spiral Galaxy

As discussed above, a fully dynamical approach to the galactic structure in the context of general relativity depends on the properties of the dark halo, which is unrealistic at present [16, 19]. Since the gravity in a galaxy is very weak except for the region near the center, the Newtonian gravitational theory is accurate enough to simulate the spiral structure. However considering the radius of a galaxy is about 10 ~ 100ly, the retarded potential should be included.

By (2.35), the zeroth order approximation gives the dynamical equation for Newtonian gravity

$$\partial_\alpha \partial^\alpha \Phi = -4\pi G \rho. \tag{3.1}$$

For the stars moving in the gravity, omitting the gravitomagnetic effect, the dynamical equation of the flow is given by (2.30)

$$(\partial_t + \vec{V} \cdot \nabla)\vec{V} = -\nabla\Phi. \tag{3.2}$$

Theoretically, the complete dynamical equation system should include the continuity equation and the dynamics of the dark halo. But these equations strongly depend on the equation of state of dark matter and dark energy, which is not clearly discovered. However, as shown below, the lack of this knowledge can be compensated by other empirical data such as the flat rotation speed curve, and then some important information of galaxy such as the total mass density and the spiral structure, can be inversely derived. That is to say, we do the inverse treatments as done in [10, 13, 14].

For convenience we rewrite (3.1) and (3.2) in the spherical coordinate system (t, r, θ, ϕ). By 3-dimensional tensor operation, we get the following equations

$$c^{-2}\partial_t^2\Phi - \left[\partial_r^2 + \frac{2}{r}\partial_r + \frac{1}{r^2}\left(\partial_\theta^2 + \cot\theta\partial_\theta + \frac{1}{\sin^2\theta}\partial_\phi^2\right)\right]\Phi + 4\pi G\rho = 0, \tag{3.3}$$

$$(\partial_t + V_r\partial_r + V_\theta\partial_\theta + V_\phi\partial_\phi)V_r - rV_\theta^2 - r\sin^2\theta V_\phi^2 + \partial_r\Phi = 0, \tag{3.4}$$

$$(\partial_t + V_r\partial_r + V_\theta\partial_\theta + V_\phi\partial_\phi)V_\theta + \frac{2}{r}V_rV_\theta - \sin\theta\cos\theta\, V_\phi^2 + \frac{\partial_\theta\Phi}{r^2} = 0, \tag{3.5}$$

$$(\partial_t + V_r\partial_r + V_\theta\partial_\theta + V_\phi\partial_\phi)V_\phi + \frac{2}{r}V_rV_\phi + 2\cot\theta V_\theta V_\phi + \frac{\partial_\phi\Phi}{(r\sin\theta)^2} = 0 \tag{3.6}$$

where the velocities are defined in the form of 3-dimensional contra-variant vectors

$$V_r = \frac{dr}{dt},\ V_\theta = \frac{d\theta}{dt},\ V_\phi = \frac{d\phi}{dt}. \tag{3.7}$$

(3.3)-(3.7) are just the Newton's second law for the motion of the stars.

For the stationary spiral structure globally processing around *z*-axis at constant angular speed Ω (namely the pattern speed), under the coordinate transformation $\varphi = \phi - \Omega t$, the solution will be static in coordinate system (t, r, θ, φ), namely independent of t. Then we get the following equations for such galaxy

$$c^{-2}\Omega\partial_\varphi^2\Phi - \left[\partial_r^2 + \frac{2}{r}\partial_r + \frac{1}{r^2}\left(\partial_\theta^2 + \cot\theta\partial_\theta + \frac{1}{\sin^2\theta}\,\partial_\varphi^2\right)\right]\Phi + 4\pi G\rho = 0, \tag{3.8}$$

$$(V_r\partial_r + V_\theta\partial_\theta + V_\varphi\partial_\varphi)V_r - r\,[V_\theta^2 + \sin^2\theta\left(V_\varphi + \Omega\right)^2] + \partial_r\Phi = 0, \tag{3.9}$$

$$(V_r\partial_r + V_\theta\partial_\theta + V_\varphi\partial_\varphi)V_\theta + \frac{2}{r}V_rV_\theta - \sin\theta\cos\theta\left(V_\varphi + \Omega\right)^2 + \frac{\partial_\theta\Phi}{r^2} = 0, \tag{3.10}$$

$$(V_r\partial_r + V_\theta\partial_\theta + V_\varphi\partial_\varphi)V_\varphi + \left(\frac{2}{r}V_r + 2\cot\theta V_\theta\right)\left(V_\varphi + \Omega\right) + \frac{\partial_\varphi\Phi}{(r\sin\theta)^2} = 0. \tag{3.11}$$

In (3.9)-(3.11), some terms have manifest physical meanings, $r\,[V_\theta^2 + \sin^2\theta(V_\varphi + \Omega)^2]$ stands for centrifugal force, and $(\frac{2}{3}V_r + 2\cot\theta V_\theta)\Omega$ the Coriolis force.

Considering an un-warped galaxy with two stationary spiral aims, the total mass-energy density and potential can be generally expanded by spherical harmonics $Y_{lm}(\theta, \varphi)$ with even m. To second order terms, we equivalently have the following approximation

$$\rho = \rho_0 + (\rho_1 + \rho_2\cos 2\varphi + \rho_3\sin 2\varphi)\sin^2\theta, \tag{3.12}$$

$$\Phi = \Phi_0 + (\Phi_1 + \Phi_2\cos 2\varphi + \Phi_3\sin 2\varphi)\sin^2\theta \tag{3.13}$$

where all (ρ_n, Φ_n) are functions of r, and $(\rho_0 \geq 0, \rho_1 \geq 0)$. Accordingly, the velocity of the stars in the disc, also to second order terms, should be

$$V_r = W_1\cos 2\varphi + W_2\sin 2\varphi, V_\theta = 0, \tag{3.14}$$

$$V_\varphi = \omega_0 + \omega_1 \cos 2\varphi + \omega_2 \sin 2\varphi \tag{3.15}$$

where all (W_n, ω_n) are functions of r.

Substituting (3.12) and (3.13) into (3.8), we get the relations between and Φ_n and ρ_n as follows

$$\partial_r^2 \Phi_0 + \frac{2}{r}\partial_r \Phi_0 + \frac{4}{r^2}\Phi_1 = 4\pi G \rho_0, \tag{3.16}$$

$$\partial_r^2 \Phi_1 + \frac{2}{r}\partial_r \Phi_1 - \frac{6}{r^2}\Phi_1 = 4\pi G \rho_1, \tag{3.17}$$

$$4c^{-2}\Omega^2 \Phi_k + \partial_r^2 \Phi_k + \frac{2}{r}\partial_r \Phi_k - \frac{6}{r^2}\Phi_k = 4\pi G \rho_k, (k = 2, 3). \tag{3.18}$$

Substituting (3.13)-(3.15) into (3.9) and (3.11), and constraining $\theta = \frac{1}{2}\pi$, we get respectively

$$\begin{aligned} 0 = {} & \frac{1}{4}\partial_r (W_1^2 + W_2^2) + \partial_r (\Phi_0 + \Phi_1) - \frac{1}{2} r (2\omega_0^2 + \omega_1^2 + \omega_2^2) + \omega_1 W_2 \\ & - \omega_2 W_1 + 2\left[(\omega_0 - \Omega) W_2 - r\omega_0\omega_1 + \frac{1}{2}\partial_r \Phi_2\right]\cos 2\varphi \\ & +2\left[(\Omega - \omega_0) W_1 - r\omega_0\omega_1 + \frac{1}{2}\partial_r \Phi_3\right]\sin 2\varphi + \cdots \end{aligned} \tag{3.19}$$

and

$$\begin{aligned} 0 = {} & \frac{1}{2}(r\partial_r \omega_1 + 2\omega_1) W_1 + \frac{1}{2}(r\partial_r \omega_2 + 2\omega_2) W_2 \\ & + \left[(r\partial_r \omega_0 + 2\omega_0) W_1 + 2(\omega_0 - \Omega) r\omega_2 + \frac{2}{r}\Phi_3\right]\cos 2\varphi \\ & + \left[(r\partial_r \omega_0 + 2\omega_0) W_2 + 2(\Omega - \omega_0) r\omega_1 + \frac{2}{r}\Phi_2\right]\sin 2\varphi + \cdots. \end{aligned} \tag{3.20}$$

The equation (3.10) automatically holds. Theoretically, we can solve (Ω, W_j, ω_j) and some relations among (ρ_k, Φ_k) by condition that the coefficients of (3.19) and (3.20) vanish.

According to the high-accurate observational data [20, 21, 22], we find the rotation curves of most spiral galaxies are approximately flat, then we can assume

$$\omega_0 = \frac{v}{r} - \Omega, (r \in [R_0, R_1]), \tag{3.21}$$

where v is a constant speed with typical value $|v| = 200 \sim 400$ km/s, $[R_0, R_1]$ is the effective region of (3.21), with their typical values as $R_0 = 100 \sim 500$ pc, and $R_1 = 10 \sim 60$ kpc to be approximately the visible radius of the galaxy. Equivalently, we can assume ($\omega_0 > 0$, $v > 0$, $\Omega > 0$) in calculation, because the dynamical equations (3.3)-(3.6) have reversal invariance under transformation $\phi \to -\phi$. We use the empirical condition (3.21) to replace the dynamical equations for the background, then the problem is greatly simplified. In addition, for a given galaxy, instead of constant v, we can use fitting function $v = v(r)$ in (3.21) to get more accurate results and larger effective domain [23, 24, 25]. However in this case, we can usually get numerical results only.

Substituting (3.21) into (3.19) and (3.20), by setting the coefficients of $(\sin 2\varphi, \cos 2\varphi)$ terms to zero, we can solve

$$\omega_1 = -K\left[\frac{1}{2}v\partial_r\Phi_2 + \frac{2}{r}(v - r\Omega)\Phi_2\right], \tag{3.22}$$

$$\omega_2 = -K\left[\frac{1}{2}v\partial_r\Phi_3 + \frac{2}{r}(v - r\Omega)\Phi_3\right], \tag{3.23}$$

$$W_1 = K\left[(v - r\Omega)r\partial_r\Phi_3 + 2v\Phi_3\right], \tag{3.24}$$

$$W_2 = K\left[(v - r\Omega)r\partial_r\Phi_2 + 2v\Phi_2\right], \tag{3.25}$$

$$K \equiv (2r^2\Omega^2 - 4rv\Omega + v^2)^{-1}. \tag{3.26}$$

By the zeroth order terms in (3.19) and (3.20), we have

$$\partial_r(\Phi_0 + \Phi_1) + \frac{1}{4}\partial_r(W_1^2 + W_2^2) = \frac{1}{2}r(2\omega_0^2 + \omega_1^2 + \omega_2^2) - \omega_1 W_2 - \omega_2 W_1 \tag{3.27}$$

and

$$(r\partial_r\omega_1 + 2\omega_1)W_1 + (r\partial_r\omega_2 + 2\omega_2)W_2 = 0. \tag{3.28}$$

Substituting (3.22)-(3.26) into (3.27) and (3.28), we get two constraints for $\Phi_k(r)$ of the background potentials.

Obviously, the disc satisfies the mass conservation law independent of the dark halo, so we have the 2-dimensional continuity equation for all stars and baryonic particles moving in the disc as follows [3, 11]

$$\begin{aligned} 0 &= \partial_t\Sigma + \nabla\cdot(\vec{V}\Sigma), \\ &= (\partial_t + V_r\partial_r + V_\phi\partial_\phi)\Sigma + (\partial_r V_r + \partial_\phi V_\phi + \frac{1}{r}V_r)\Sigma \end{aligned} \tag{3.29}$$

where Σ stands for the surface mass density of the stars and baryons in the disc. It should be mentioned that, (3.8)-(3.11) are expressed in spherical coordinate system due to the background gravity, but (3.29) is expressed in polar coordinate system. In the stationary case, (3.29) becomes

$$\left(V_r\partial_r + V_\varphi\partial_\varphi\right)\Sigma + \left(\partial_r V_r + \partial_\varphi V_\varphi + \tfrac{1}{r}V_r\right)\Sigma = \nabla\cdot\left(\vec{V}\Sigma\right) = 0. \tag{3.30}$$

(3.29) or (3.30) is the equation to describe the mass density of the stars.

4. Some Solutions to the Equations

4.1. Solution to a Barred Spiral Galaxy

The general solutions to the above underdetermined equation system are quite complicated and unnecessary. The most important case to understand the nature of spiral structure is the stable and terminal state of a galaxy, which is similar to the eigenstate of a micro particle. At first, we consider the case that all stars move in the orbits near circle, for which the analytic solutions can be solved. In this case the radial speed is a high order little term, then we have

$$W_1 = 0, W_2 = 0. \tag{4.1}$$

By (3.24) and (3.25), noticing the symmetry between (Φ_2, Φ_3), we get the equivalent solution except for an initial phase of φ as follows

$$\Phi_2 = -\tfrac{q}{r^2}(v - r\Omega)^2, \Phi_3 = 0 \tag{4.2}$$

where $q \geq 0$ is a constant. Substituting (4.1) and (4.2) into the above equations (3.16)-(3.18) and (3.22)-(3.28), we finally get

$$\omega_1 = \tfrac{q}{r^3}(v - r\Omega), \omega_2 = 0, \rho_3 = 0, \tag{4.3}$$

$$\rho_1 = -\rho_0 + \tfrac{1}{4\pi G}\left(\tfrac{q^2 v}{12r^2}(16r\Omega - 15v) - \tfrac{2v^2}{r^2}\ln\tfrac{r}{r_0} + \tfrac{2\Phi_0}{r^2} + \tfrac{v^2}{r^2}\right), \tag{4.4}$$

$$\rho_2 = \tfrac{q}{2\pi G r^4}\left((3r^2\Omega^2 - 6rv\Omega + 2v^2) - \tfrac{2r^2\Omega^2}{c^2}(v - r\Omega)^2\right), \tag{4.5}$$

$$0 = \Phi_0 + \Phi_1 + \frac{q^2}{24r^4}(6r^2\Omega^2 - 8vr\Omega + 3v^2) - v^2 \ln\frac{r}{r_0} \tag{4.6}$$

where $r_0 > 0$ is a constant with length dimension.

Substituting the solutions into (3.9) and (3.11), we can check the truncation error of the equations reads

$$\Delta(3.9) = -\frac{q^2}{2r^5}(v - r\Omega)^2\cos 4\varphi, \Delta(3.11) = -\frac{q^2}{2r^6}(v - r\Omega)^2\sin 4\varphi, \tag{4.7}$$

which are higher order terms. So the above solution is a good approximation. If $q = 0$, the solutions are exact, which correspond to the galaxy without spirals in the disc. To get more accurate results, one should introduce higher order terms such as $(\cos 4\varphi, \sin 4\varphi)$ terms in (3.12)-(3.15).

By (3.21)-(4.2), we get the speed of the stellar flow

$$V_r = 0, V_\theta = 0, V_\varphi = \frac{v}{r} + \frac{q}{r^3}(v - r\Omega)\cos 2\varphi. \tag{4.8}$$

Substituting it into (3.30), we get the continuity equation as $\partial_\varphi(V_\varphi\Sigma) = 0$, which yields the mass density distribution of the stars as

$$\Sigma(r, \varphi) = \frac{vr^2\varrho(r)}{vr^2 + q(v - r\Omega)\cos 2\varphi} \tag{4.9}$$

where $\varrho(r)$ is a density function determined by boundary conditions. Apparently, the density distribution of the stars display a profile of barred spiral galaxy.

To determine (ρ_0, ρ_1), we need another condition related to the dynamics of the black halo. In the case of $(\rho_0 \propto \rho_1)$, we can solve

$$\rho_0 \sim \frac{n^2 - n - 4}{4\pi G} q^2 \left(\frac{9v^2}{2(n-4)(n+3)r^6} - \frac{\Omega^2}{(n-2)(n+1)r^4}\right) + \frac{n^2 - n - 4}{4\pi G}\left(\frac{a}{r^{2+n}} + \frac{3v^2}{n(n-1)r^4}\right), \tag{4.10}$$

$$\rho_1 \sim \frac{(3-n)(n-2)}{n^2 - n - 4}\rho_0 \tag{4.11}$$

in which a and $\frac{1}{2}(\sqrt{17} + 1) \leq n \leq 3$ are constants. (4.10) and (4.11) provide a heuristic mass density for the dark halo in barred spiral galaxy within the range $[R_0, R_1]$.

4.2. Solution to a Spiral Galaxy

In an ordinary spiral galaxy, the potentials Φ_2 and Φ_3 should take the following form

$$\Phi_2 = P^{-1}\cos(\xi r + \varphi_0), \Phi_3 = P^{-1}\sin(\xi r + \varphi_0) \tag{4.12}$$

where ξ is a constant, φ_0 the initial phase of φ and $P(r)$ a function of r. Obviously we can set $\varphi_0 = 0$ by a translation of φ similar to the barred spiral case. Substituting (4.12) into (3.28), we get a linear equation for P

$$P'' - \frac{4v}{r(v-r\Omega)}P' + \left(\xi^2 + \frac{4\Omega^2}{v(v-r\Omega)} + \frac{4\Omega}{rv} + \frac{8}{r^2}\right)P = 0. \tag{4.13}$$

The equation (4.13) is independent of $\cos(\xi r + \varphi_0)$ and $\sin(\xi r + \varphi_0)$, which reflects (4.12) touches the nature of the spiral structure.

The solution to (4.13) can be expressed by the Hankel function with complex parameters, which is much complicated. Here we examine the simplest case of (4.13), the static case with $\Omega = 0$. The analytic solution to this case can be expressed in a clear form, and it is heuristic to reveal the nature of a spiral galaxy. In the general cases, the numerical simulation is more convenient and efficient.

Setting $\Omega = 0$, (4.13) becomes a Bessel-like equation

$$P'' - \frac{4}{r}P' + \left(\xi^2 + \frac{8}{r^2}\right)P = 0. \tag{4.14}$$

The solution to (4.14) is given by

$$P = \sqrt{r^5}[C_1 J_\alpha(|\xi|r) + [C_2 J_{-\alpha}(|\xi|r)], \alpha = \frac{1}{2}\sqrt{7} \tag{4.15}$$

where (C_1, C_2) are constants, $(J_\alpha, J_{-\alpha})$ are Bessel functions with complex parameters, which are defined by

$$J_\nu(x) = \left(\frac{x}{2}\right)^\nu \sum_{k=0}^{\infty} \frac{1}{k!}\frac{1}{\Gamma(\nu+k+1)}\left(\frac{x}{2}\right)^{2k}, (\nu = \pm\alpha). \tag{4.16}$$

Substituting (4.15) into (3.18) and (3.22)-(3.26), we can check (ρ_2, ρ_3) and speed all have spiral structure like (4.12). Again by (3.30), we find Σ also has spiral structure. Substituting the results into (3.27), we get a constraint for (Φ_0, Φ_1) similar to (4.6). Since the concrete expressions are long and complicated, and can be derived by straightforward calculation, we do not display them here.

DISCUSSION AND CONCLUSION

The above procedure provides a method to research the structure of a galaxy via dynamical approach, which naturally connects the hydrodynamics with empirical data. The above solutions provide manifest analytic relations and functions for a spiral galaxy, which are much helpful to understand the structure and property of the galaxy. These solution verifies that the spiral arms are stationary or even static stellar density wave distribution.

In the case of barred spiral galaxy with $\Omega \neq 0$, by (4.5), we find $\rho_2 = 0$ has a root $r = R \leq v/\Omega$, which means the bar vanishes at radius $r = R \sim R_1$. By (4.3), we get $\omega_1(R) \doteq 0$, which implies the density wave vanishes near $r = R$, and the naring of stars will form. The above solution is derived under the assumption (3.21), so the effective domain of the solution is also $[R_0, R_1]$, the effective domain of (3.21). These conclusions can be used to estimate the global angular speed

$$\Omega \approx \frac{v}{R_1} \sim 300 \text{ km/s/ (30kpc)} = 10 \text{ km/s/kpc} \sim 0.002''\text{/year},$$

which is less than the previously estimated pattern speed 30 ~ 60 km/s/kpc [26].

By (4.7) and (4.8), we learn all stars in the disc move in the almost circular orbits. This conclusion is reasonable and coincident with facts. Through some mechanisms and long-term evolution, the baryons were shifted and concentrated into the disc, and their orbits of motion became harmonious in a regular galaxy. Otherwise violent collisions will occur frequently among the moving stars, and then the galaxy becomes a hell.

By (4.3) and (4.5), we find $\omega_1 > 0$ and $\rho_2 > 0$ hold simultaneously. This means where the mass density of the background is larger, where the potential is lower and the stellar speed is higher. On the other hand, by (4.9) we find that the higher the stellar speed, the lower the number density. This means the mass density of the disc should be compensatory for that of the dark halo. Namely, the density of dark halo ρ should take minimum in the spiral arms where Σ take maximum or vice versa. This is a strange but interesting phenomenon, which reflects the complicated stream lines of the dark halo. This phenomenon might also be one reason why the dark matter and dark energy can be hardly detected in the vicinity of the solar system.

The relations (4.1) and (4.4)-(4.6) as well as (4.12) provide some information for the distribution and property of the total mass density and potential. How to combine the results with the dynamics of the background is an interesting problem, which might be a shortcut to study the properties of the dark matter and dark energy.

ACKNOWLEDGMENTS

The author is grateful to his supervisors Prof. Ta-Tsien Li for his encouragement.

REFERENCES

[1] B. Binney, B., and Tremaine, S. (1987). *Galactic Dynamics* (Ch. 3, Ch. 4, Ch. 6), Princeton University Press.

[2] Orlova, N. V., Korchagin, V. I., Kikuchi, N., Miyama, S. M., and Moiseev, A. V. (2002). Application of the global modal approach to the spiral galaxies. *Astrophys. Space Sci. 284*, 739-742; arXiv:astro-ph/0211538.

[3] Meidt, S. E., Rand, R. J., Merrifield, M. R., Debattista, V. P., and Shen, J. (2007). *Tests of the radial Tremaine-Weinberg method*, arXiv:0711.4104.

[4] Lin, C. C., and Shu, F. H. (1964). On the spiral structure of disk galaxies. *Astrophys. J. 140 (2)*, 646-655.

[5] Lin, C. C., and Shu, F. H. (1966). On the spiral structure of disk galaxies, II. Outline of a theory of density waves. *Proc. Natl. Acad. Sci. 55*, 229-234.

[6] Goldreich, P., and Lynden-Bell, D. (1965). II. Spiral arms as sheared gravitational instabilities. *Mon. Not. Roy. Astron. Soc., 130*, 125-158.

[7] Julian, W. H. and A. Toomre, A. (1966). Non-axisymmetric responses of differentially rotating disks of stars. *Astrophys. J. 146*, 810- 830.

[8] Somerville, R. S., Barden, M., Rix, H.-W., Bell, E. F., Beckwith, S. V. W., Borch, A., Caldwell, J. A. R., Haeussler, B., Heymans, C., Jahnke, K., Jogee, S., McIntosh, D. H., Meisenheimer, K., Peng, C. Y., Sanchez, S. F., Wisotzki, L., and Wolf, C. (2008). An explanation for the observed weak size evolution of disk galaxies. *Astrophys. J., 672(2),* 776-786; arXiv:astro-ph/0612428.

[9] Kendall, S., Kennicutt, R. C., Clarke, C., and Thornley, M. D. (2008). *Tracing spiral density waves in M81*. arXiv:0804.2380.

[10] Dehnen, W., and Binney, J. (1998). Mass models of the Milky Way. *Mon. Not. Roy. Astron. Soc. 294*, 429-438, arXiv:astro-ph/9612059.

[11] Vorobyov, E. I., and Theis, Ch. (2006). Boltzmann moment equation approach for the numerical study of anisotropic stellar disks. *Mon. Not. Roy. Astron. Soc. 373*, 197-208; arXiv:astro-ph/0609250.

[12] Vorobyov, E. I., and Theis, Ch. (2008). Shape and orientation of stellar velocity ellipsoids in spiral galaxies. *Mon. Not. Roy. Astron. Soc. 373 (3)*, 817–830; arXiv: 0709.2768.

[13] Romero-Gómez, M. Athanassoula, E., Masdemont, J. J., and García-Gómez, C. (2007). *The formation of spiral arms and rings in barred galaxies*. arXiv:0705.2958 [astro-ph].

[14] Romero-Gómez, M. (2008). *The Formation of spiral arms and rings in barred galaxies,* arXiv:0801.0366 [astro-ph].

[15] Weinberg, S. L. (1972). *Gravitation and cosmology* (Ch.9.8, Ch.10.1), John Wiley & Sons, New York.

[16] Gu, Y. Q. (2017). Functions of state for spinor gas in general relativity, *OALib. J.* 4, e3953; arXiv:0711.1243.

[17] Gu, Y. Q. (2006). *The Vierbein formalism and energy-momentum tensor of spinors,* arXiv:gr-qc/0612106.

[18] Gu, Y. Q. (2006). *The Simplification of spinor connection and classical approximation*, arXiv:gr-qc/0610001.

[19] Mori, M., and Umemura, M. (2006). Galactic evolutionary path from primeval irregulars to present-day ellipticals. *Nature* 440(7084), 644-647; arXiv:astro-ph/0512424.

[20] Sofue, Y., Tutui, Y. Honma, Y., Tomita, A. Takamiya, T., Koda, J., and Takeda, J. (1999). Central rotation curves of spiral galaxies. *Astrophys. J. 523*, 136-146; arXiv:astro-ph/9905056.

[21] Sofue, Y. (1996). Central rotation curves of galaxies. *Astrophys. J. 458*, 120-131; astro-ph/0010595.

[22] Bownstein, J. R., and Moffat, J. M. (2006). Galaxy rotation curves without non-baryonic dark matter. *Astrophys. J. 636*, 721-741; astro-ph/0506370.

[23] Salucci, P. (2007). The mass distribution in spiral galaxies. *Proc. IAU Symposium No. 244*, arXiv:0707.4370.

[24] Persic, M., Salucci, P. (1995). Rotation curves of 967 spiral galaxies. *Astrophys. J., Suppl. 99*, 501-541, arXiv:astro-ph/9502091.

[25] Persic, M., Salucci, P., and Stel, F. (1996). The universal rotation curve of spiral galaxies: I. The dark matter connection, *Not. Roy. Astron. Soc. 281(1)*, 27-47; arXiv:astro-ph/9506004.

[26] Widrow, L. M., Pym, B., and Dubinski, J. (2008). Dynamical blueprints for galaxies. *Astrophys. J., 679*(2), 1239-1259; arXiv:0801.3414.

In: Old Problems and New Horizons in World Physics
Editors: Volodymyr Krasnoholovets et al.
ISBN: 978-1-53615-430-6

Chapter 13

A REVIEW OF SELF-ORGANIZED CRITICALITY AS A MODEL OF SCIENTIFIC DEVELOPMENT

***Victor Christianto*[1,*] *and Florentin Smarandache*[2,∞]**
[1]Satyabhakti Advanced School of Theology – Jakarta Chapter, Indonesia,
[2]Department of Mathematics and Sciences,
University of New Mexico, Gallup, US

ABSTRACT

According to a conventional belief held by most science communities, sciences can advance by two ways: (a) incremental approach by virtue of scientific methods, (b) paradigm shift. The latter term has been advocated by a notorious science historian, Thomas Kuhn. Despite such widespread acceptance, such a term of paradigm shift is not often tested empirically.

Therefore, in this paper we review a recent work which uses citation analysis of journals published in recent years. This analysis reveals that scientific progress seems to follow Self-Organized Criticality. In the last section, we compare this citation analysis result with our model of the origin of the Universe.

Keywords: paradigm shift, realism interpretation, scientific progress, Self-Organized Criticality

PACS: 2010: 02, 03, 41, 98

---

* Author's Email: victorchristianto@gmail.com.
∞ Author's Email: florentin.smarandache@laposte.net.

INTRODUCTION

In its simplest form and according to a conventional belief held by most science communities, sciences can advance by two ways: (a) incremental approach by virtue of scientific methods, (b) paradigm shift. The latter term has been advocated by a notorious science historian, Thomas Kuhn. According to his proponents, one of the most interesting features of Thomas Kuhn's work in The Structure of Scientific Revolutions is its naturalism. But naturalism is just another philosophical strand which may or may not agree with empirical data itself. Despite such widespread acceptance, the fact is that such a term of paradigm shift is not often tested empirically.

There are other ways to describe innovation changes, namely: (c) Christensen's disruptive change, (d) Self-organized criticality. But they are rarely employed for describing scientific progress. Therefore, in this paper we review a recent work which uses citation analysis of journals published in recent years. This analysis reveals that scientific progress seems to follow Self-Organized Criticality.

A REVIEW OF 4 METHODS OF HUMAN KNOWLEDGE PROGRESS

There are some papers in literature which indicate those 4 methods, we will review briefly as follows:

a. *Incremental Approach:*
   by virtue of scientific methods, science advances by small steps. Until the 1950s, the hegemony of logical empiricism reached to its highest level - by the representatives of the logistic approach such as R. B. Braithwaite, Rudolf Carnap, Herbert Feigl, Carl G. Hempel and Hans Reichenbach. Prior to Kuhn's SSR, historians and philosophers of science considered the scientific enterprise to be a rational endeavor in which progress and knowledge are achieved through the steady, daily, rigorous accumulation of experimental data accredited facts and new discoveries [4].
b. *Paradigm Shift:*
   Thomas Kuhn's *Structure of Scientific Revolutions* (SSR) is believed to be one of the most important books in the 20th century. The book conceived a whole industry of commentary, interpretation, and exegesis. The growth of a new academic discipline – the sociology of science- came into existence around a shared paradigm following Kuhn's emphasis on the importance of communities of scientists. After the book was published researchers began to examine scientific disciplines much as sociologists studied social/cultural groups, and in which

science was regarded not as the most esteemed, untouchable product of the Enlightenment but as just another subculture. Yet, as Kuhn claimed "the philosophy and sociology of science cannot be practiced independently of each other". However, Kuhn saw the communities (not individuals) as the basic agents of science and he thought that communities must be characterized by the specific cognitive values to which they are committed [4].

After the 1960s and 70s, following Kuhn's historiography, and philosophers such as Paul Feyerabend, Imre Lakatos, Larry Laudan and Michael Polanyi have greatly contributed to the creation of an anti-positivistic philosophy of science as a new tradition. History of science after Kuhn has frequently taken a more consciously externalist line, in looking outside science for the causes of the content of science [4].

c. *Disruptive Change:*

In his article in Harvard Business Review, Clayton Christensen, differentiate between: Sustaining innovations and disruptive innovations [6]. This seems to follow Schumpeterian view of creative destruction. But, this paper will not focus on disruptive innovation. See also his more recent article in HBR 2015 [6a].

d. *Self-Organized Criticality:*

Self-organized criticality is a rich phenomenon as it combines self-organization and criticality to describe complexity. This concept was first introduced by P. Bak and the collaborators in the seminal paper in 1987, and also in his book [5]. SOC is a property of dynamical systems to organize its microscopic behavior to be spatial (and/or temporal) scale independent. That resembles of the critical behavior of the critical point of phase transitions. However, in contrast to the usual phase transitions, the systems displaying SOC do not require external tuning of the control parameters, i.e., the system evolutions, i.e., organizes itself into the critical behavior. Thus, the dynamical system organizes itself into a state with complex, but rather general structure. Complexity arises in the sense that no single characteristic event size exists, i.e., no scale present to guide the system's evolution. Despite the complexity, system exhibits simple statistical properties governed by power laws [7].

RESULTS OF CITATION ANALYSIS

Bolijka Tadic et al. have reported self-organized criticality pattern in online social behavior especially in knowledge creation process [2]. But the first convincing citation analysis to prove this pattern has been made by Loet Leydesdorff, Caroline S. Wagner, and Lutz Bornman [1]. As we know, journals play a crucial and institutionalized role in the validation of knowledge claims and in the incorporation of new knowledge into the archive of science. Given their role in the codification of knowledge, journals can be considered as an organizing layer of the scientific literature. Not incidentally, the Science Citation Index (SCI) and its derivates (the Social Sciences Citation Index (SSCI) and the Arts & Humanities Citation Index (AHCI)) were defined in terms of specific journal selections (Garfield 1972; 1979b), as is Scopus, the main competitor of the SCI since 2004 [1].

Method

In their study, Leydesdorff et al. use aggregated journal-journal citation relations as units of analysis instead of journals. Each relation specifically combines two journals. The journal-journal citation relation is a link in the networks of which journals are the nodes. The citation relations among the 10,000+ journals contained in the Journal Citation Reports (JCR) of the Science Citation Index and the Social Sciences Citation Index can be organized as a matrix of (10,000+) cells, each representing a unique relation between a citing and a cited journal. These (valued) relations can change over time to the extent that they can disappear or emerge; that is, turn from a zero into a positive value larger than or equal to one [1].

Result

According to Leydesdorff et al. [1]:

> "The fit with a power-law in Figure 7 suggests self-organized criticality (SoC) in the system of journal-journal citation relations. New knowledge claims in manuscripts continuously generate journal-journal citation relations potentially leading to an equivalent of "avalanches" of reconstructed inter-journal relations. These avalanches can occur anywhere; their effects may be very different; but the consequences are local, that is, within the discipline or specialty. The self organized criticality remains globally available for new critical transitions in the full range from minor, but more frequently occurring changes, to rare but disruptive ones. The comparison with earthquakes provides another metaphor.
>
> … This model of self-organized criticality differs from the Kuhnian model of normal science versus revolutionary science as phases in paradigm transitions (Kuhn 1962; Marx & Bornmann 2013; van den Daele & Weingart 1975). The flux of manuscripts with knowledge claims contain references to other journals which can be compared with the grains of sand that hit the sand pile or, in this case, the knowledge base as a construct. The effect can be an avalanche of any size depending on the state of the system at that specific place and time".

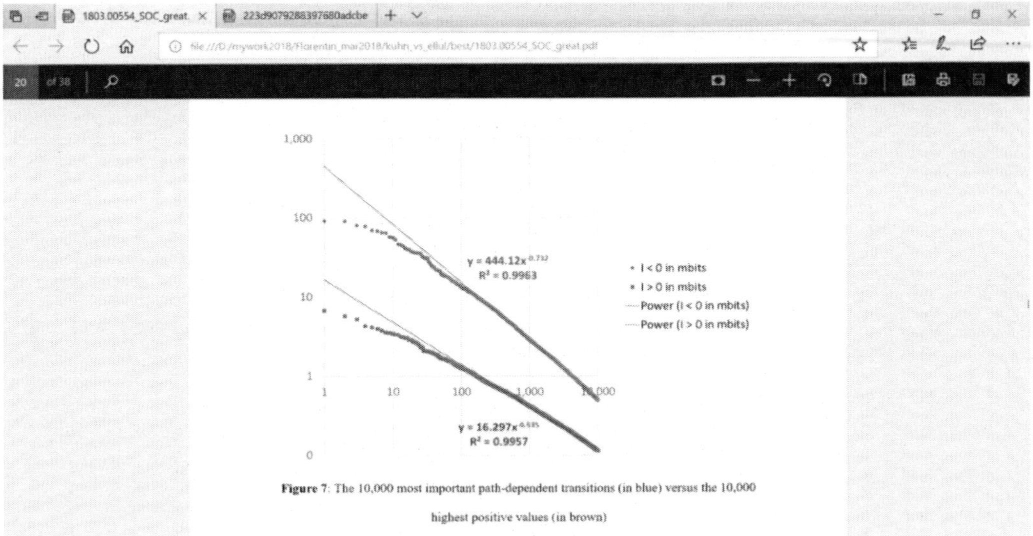

Diagram 1. Source: Leydesdorff et al. [1].

COMPARISON WITH OUR MODEL OF ABRUPT ORIGIN OF THE UNIVERSE

Now we will compare this citation analysis result with our proposed model of the origin of the Universe.

It has been known for long time that most of the existing cosmology models have a singularity problem. Cosmological singularity has been a consequence of excessive symmetry of flow, such as "Hubble's law." A more realistic one is suggested, based on Newtonian cosmology model, but here we include the vortical-rotational effect of the whole Universe.

In another paper, we obtained an Ermakov-type equation following Nurgaliev [8]. Then we solve it numerically using Mathematica 11. An interesting result from that simple computational simulation is shown in the following diagram [9].

From the above computational experiment, we conclude that the evolution of the Universe depends on the constants involved, especially on the rotational-vortex structure of the Universe. This certainly needs to be investigated in more detail.

One conclusion that we may derive especially from Diagram 2, is that our computational simulation suggests that it is possible to consider that the Universe has existed for long time in a prolonged stagnation period, then suddenly it burst out from being *empty and formless* (Gen. 1:2), to take its current shape with observed "accelerated expansion". Comparing our model of the abrupt origin of the Universe with the above citation analysis seems to reveal similarities. But whether such abrupt origin of the Universe also indicates SOC feature, remains open for further study.

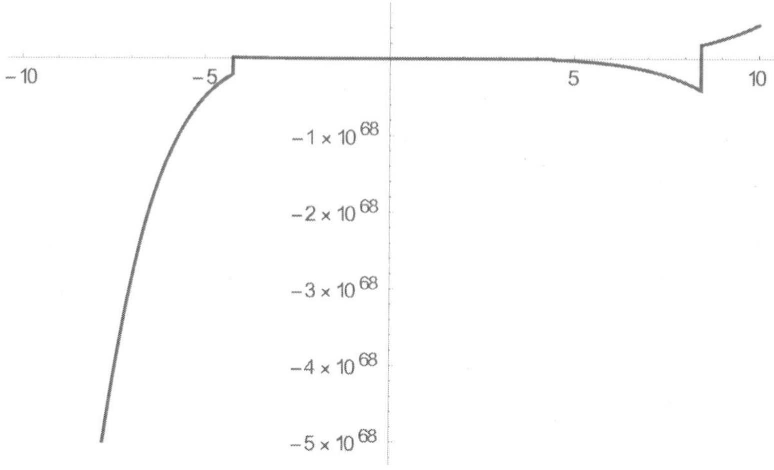

Diagram 2. Plot of Ermakov-type solution for A = 1, B = -10 (from [9]).

CONCLUDING REMARKS

Despite its enormous popularity in the past 5-6 decades, a paradigm shift view of scientific progress has not been tested quite often. Therefore, in this paper we reviewed recent work which uses journal-journal citation analysis over recent years. This analysis reveals that scientific progress seems to follow a Self-Organized Criticality pattern.

Comparing our model of abrupt origin of the Universe with the above citation analysis seems to reveal similarities. But whether such abrupt origins of the Universe also indicates an SOC feature, remains open for further study.

It can be expected that the above discussions will shed some light on such an old problem, especially in the context of modelling scientific progress based on empirical data (evidence based). This is reserved for further investigations.

ACKNOWLEDGMENT

Special thanks to Prof. Thee Houw Liong for bringing up future science and technology in a recent RG forum. Nonetheless, ideas presented here is our sole responsibility.

DOCUMENT HISTORY

version 1.0: March 27th, 2018, pk. 21:16
VC & FS

REFERENCES

[1] Loet Leydesdorff, Caroline S. Wagner and Lutz Bornman. *Discontinuities in Citation Relations among Journals: Self-organized Criticality as a Model of Scientific Revolutions and Change.* arXiv: 1803.00554 (2018).

[2] Bosiljka Tadic, Marija Mitrovic Dankulov and Roderick Melnik. *The mechanisms of self-organised criticality in social processes of knowledge creation.* arXiv: 1705.0982 (2017).

[3] Andrei P. Kirilyuk. *The last Scientific Revolution.*

[4] Turkan Firinci Orman. "Paradigm" as a Central Concept in Thomas Kuhn's Thought. *International Journal of Humanities and Social Science*, vol. 6, no. 10 (2016).

[5] Per Bak. *How Nature Works.* New York: Springer, 1996.

[6] Clayton M. Christensen and Michael Overdorf. Meeting the Challenge of Disruptive Change. *HBR On Point,* March-April 2000. www.hbr.org; [6a] Clayton M. Christensen *et al*. https://hbr.org/2015/12/what-is-disruptive-innovation.

[7] Vladyslav A. Golyk. *Self-organized criticality.* Url: https:// pdfs.semanticscholar.org/7084/223d9079288397680adcbe129c2ad323b45e.pdf.

[8] Ildus S. Nurgaliev. *E pur si muove!* arXiv: 1210.4091. [8a] see also Singularities are averted by vortices. *13th Gravitational Conference – International Conference on Gravitation, Cosmology and Astrophysics*, June 23-28, 2008. Peoples Friendship University of Russia (PFUR), Moscow, Russia. Abstracts. Published by PFUR, ISBN 978-5-209-03274-8, pp. 117 - 118; [8b] Also in *Gravitation and Cosmology,* vol. 16 no. 4 (2010) pp. 313 - 315.

[9] Victor Christianto, Florentin Smarandache and Yunita Umniyati. Solving Numerically Ermakov-type Equation for Newtonian Cosmology Model with Vortex. *Prespacetime Journal*, Oct. 2017. www.prespacetime.com.

[10] Victor Christianto and Florentin Smarandache. How Neutrosophic Logic may Resolve Dispute on the Origin of the Universe through re-reading Gen. 1:1-2. *Jurnal Teologi Amreta,* Vol. 1 no. 2, March 2018. url: http://jurnal.sttsati.org.

In: Old Problems and New Horizons in World Physics ISBN: 978-1-53615-430-6
Editors: Volodymyr Krasnoholovets et al.

Chapter 14

INERTON ASTRONOMY

Volodymyr Krasnoholovets *
Department of Theoretical Physics, Institute of Physics,
National Academy of Sciences of Ukraine, Kyiv, Ukraine

ABSTRACT

A new branch of astronomy / astrophysics is proposed, namely, inerton astronomy. The first observatory – Inerton Observatory, – which will measure inerton signals from outer space, can start imminently.

Keywords: inerton, inerton sigals, inerton astronomy

We propose to start the first observatory – *Inerton Observatory* – that will measure inerton signals from outer space. It will continuously produce mass flux (inerton rays) data related to the behavior of massive objects (planets, stars, galaxies, etc.) because any moving mass creates a measurable deformation of space. We have called these deformations inertons and the associated radiation is referred to as inerton radiation, or inerton rays. Inertons are carriers of mass and provide real quantum mechanical and gravitational interactions between any massive objects.

The appropriate submicroscopic concept has been presented in the recent book of the author [1] with the description of a number of experiments demonstrating the existence of inertons in condensed matter, plasma, nuclear physics and astrophysics. Inerton astronomy will be able to provide us with an inertonic map of the universe, with rich information about massive events occurring in the universe.

* E-mail: krasnoh@iop.kiev.ua.

Small spectroscopic devices, which are prototypes of a future inerton observatory, were designed by our team for measuring inertons at laboratory and field condition (Figure 1 and 2). It was found that at natural field conditions on the surface of the Earth, the number of signals recorded by the device depends on the orientation of the antenna and the current position of the sun. By using the devices shown in Figures 1 and 2 we measured the Earth's inerton field (in a range of frequencies 1 to 100 kHz) and the influence on it from a flux of solar inertons.

Figure 1. Spectroscopic device Rudra-1 that measures inertons (designed in 2011).

Figure 2. Spectroscopic device Rudra-2 with more modern design and electronics, which measures inerton signals (designed in 2018).

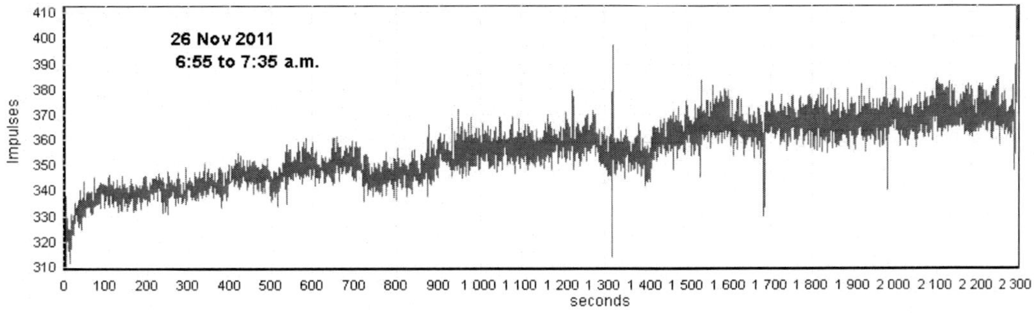

Figure 3. Trend analysis obtained with the use of the inerton spectroscopic device (measured in the outskirts of Kyiv, Ukraine).

A trend analysis conducted during about 40 minutes from the moment of dawn, 6:55 a.m. on 26 Nov. 2011 is presented in Figure 3. It is obvious that the intensity of an inerton flow increases along the east-west axis with the approaching sunrise.

Inertons from outer space can be recorded in a laboratory located on Earth. We may study the intensity of inerton signals, the direction from which they arrive, their frequency, the degree of homogeneity of signals and we may also measure the speed of free inertons, which by preliminary estimate is about two orders of magnitude of the speed of light. The appropriate equipment can be designed in such a way that it will be possible to measure amplitude and spectral and time characteristics of non-stationary inerton signals in a range of frequencies up to hundreds of MHz in real time. The measuring equipment may consist of several set-ups that will be launched in a synchronous operation.

REFERENCES

[1] V. Krasnoholovets, *Structure of Space and the Submicroscopic Deterministic Concept of Physics*, Apple Academic Press, Oakville, Canada; Waretown, USA (2017).

In: Old Problems and New Horizons in World Physics
Editors: Volodymyr Krasnoholovets et al.
ISBN: 978-1-53615-430-6

Epilogue

WHAT A WONDERFUL WORLD

Louis Daniel Armstrong

I see trees of green, red roses too
I see them bloom for me and you
And I think to myself what a wonderful world
I see skies of blue and clouds of white
The bright blessed day, the dark sacred night
And I think to myself what a wonderful world
The colors of the rainbow so pretty in the sky
Are also on the faces of people going by
I see friends shaking hands saying how do you do
They're really saying I love you
I hear babies crying, I watch them grow
They'll learn much more than I'll never know
And I think to myself what a wonderful world
Yes I think to myself what a wonderful world

Songwriters: George Douglas / George David Weiss / Bob Thiele

Leonardo da Vinci: *Salvator Mundi* (1500)

Leonardo da Vinci's *Salvator Mundi* is one of the greatest and most unexpected artistic rediscoveries of the 21st century. Its illustrious 500-year history, and the story of its re-emergence, restoration and authentication, is as fascinating as any of the bestselling thrillers about Leonardo's life and times.[1]

[1] https://www.christies.com/features/Salvator-Mundi-timeline-8644-3.aspx

EDITORS' CONTACT INFORMATION

Volodymyr Krasnoholovets*, *PhD

Senior Research Scientist, Department of Theoretical Physics,
Institute of Physics, National Academy of Sciences
of Ukraine, Kyiv, Ukraine
E-mail: krasnoh@iop.kiev.ua

Victor Christianto, DDiv

Satyabhakti Advanced School of Theology,
Jakarta Chapter, Jakarta, Indonesia
http://www.sci4God.com
Email: victorchristianto@gmail.com

Florentin Smarandache, PhD

Professor, Department of Mathematics and Sciences,
University of New Mexico, Gallup, New Mexico, US
E-mail: smarand@unm.edu

LIST OF CONTRIBUTORS

George Shpenkov, PhD, DrSci, Retired Professor, Independent Researcher, Bielsko-Biala, Poland. Email: g.shpenkov@gmail.com

Leonid Kreidik (1931-2002), MPhys, Minsk, Belarus.

Robert Neil Boyd, PhD, consulting physicist for Princeton Biotechnology Corporation, Dept. Information Physics Research. Email: rnboydphd@comcast.net

Adrian Klein, MDD, PhD, cognitive neuropsychology, PhD metaphysical sciences, Parapsychological Association, ECAO, ISPE, IQN, AAPS, AAAS. Affiliation: ECAO Aff., Israel. Email: adrian10@bezeqint.net

Akira Kanda, PhD, Professor, Omega Mathematical Institute.
Email: kanda@cs.toronto.edu

Mihai Prunescu, PhD, University of Bucharest. E-mail: mihai.prunescu@gmail.com

Renata Wong, Department of Computer Science and Technology, Nanjing University, China. Email: renata.wong@protonmail.com

Arnold Gorgels, Mathematical Physics Institute in Potsdam, Member of DPG, Germany.
Email: arnogorgels@gmail.com

Ying-Qiu Gu, PhD, School of Mathematical Science, Fudan University, China. Email: yqgu@fudan.edu.cn

INDEX

A

B

C

D

E

L

M

N

O

P

T

U

V

W

Y

Z

Related Nova Publications

Horizons in World Physics. Volume 299

Editor: Albert Reimer

Series: Horizons in World Physics

Book Description: *Horizons in World Physics. Volume 299* opens with a study on the analysis of photoluminescence spectra of II-VI-based thin films (pure II-VI semiconductor compounds – CdTe, ZnTe, CdSe, Cd1-xZnxTe ternary compound, CdS:Dy rare-earth element doped II-VI compound) at different substrate temperatures.

Hardcover ISBN: 978-1-53615-361-3
Retail Price: $250

Proceedings of the 2017 International Conference on "Physics, Mechanics of New Materials and Their Applications"

Editors: Ivan A. Parinov; Shun-Hsyung Chang, Ph.D.; and Vijay K. Gupta

Series: Physics Research and Technology

Book Description: This book presents new results of theoretical and experimental analysis of advanced materials and devices with previously given and improved characteristics, developed on the basis of methods of electric elasticity and physics of condensed matter.

Softcover ISBN: 978-1-53614-083-5
Retail Price: $310

To see complete list of Nova publications, please visit our website at www.novapublishers.com

Related Nova Publications

ADVANCES IN APPLIED PHYSICS RESEARCH

EDITOR: Gabriel Waters

SERIES: Physics Research and Technology

BOOK DESCRIPTION: This new book presents and discusses current research in the field of applied physics. Topics discussed include radiation effects in piezoelectric AlGaN/GaN heterostructures; fluorescence resonance energy transfer between CdTe quantum dots and organic dye; and band offsets in dielectric/InGaZnO4 thin film transistors.

HARDCOVER ISBN: 978-1-63483-226-7
RETAIL PRICE: $130

PHYSICS AND LOGIC OF LIFE

AUTHOR: Abir Igamberdiev

SERIES: Physics Research and Technology

BOOK DESCRIPTION: This book discusses the basic foundations of theoretical biology. Contrary to the objects of theoretical physics, the biological object contains a kind of ontological duality and refers to a fundamental wholeness of a living system.

HARDCOVER ISBN: 978-1-61942-664-1
RETAIL PRICE: $170

To see complete list of Nova publications, please visit our website at www.novapublishers.com